Edited by
Herbert E. Allen
Chin P. Huang
George W. Bailey
Alan R. Bowers

METAL SPECIATION *and* CONTAMINATION *of* SOIL

Library of Congress Cataloging-in-Publication Data

Metal speciation and contamination of soil / editors, Herbert E. Allen ... [et al.].

p. cm.

Includes bibliographical references and index.

1. Metals--Environmental aspects. 2. Metals--Speciation.
3. Soil remediation. I. Allen, Herbert E. (Herbert Ellis), 1939- .
TD879.M47M48 1994
628.5'5--dc20 94-29388
ISBN 0-87371-697-3

Lewis Publishers is an imprint of CRC Press

International Standard Book Number 0-87371-697-3

Library of Congress Card Number 94-29388
Printed in the United States of America
1 2 3 4 5 6 7 8 9 0

Printed on acid-free paper

PREFACE

"Soil is the thin skin at the earth's surface responsible for life being possible on this planet -- it is the life sustaining pedosphere. This biologically active porous and structured medium is an effective integrator and dissipater of mass flux and energy, the medium through which biomass productivity is sustained, the foundation upon which structures are built, the fabric in which organisms are anchored and housed (habitat), the repository of solid and liquid wastes and the living filter for bioremediation of waste products and water supplies" (Wilding, 1994). It is a basic component of ecosystems and is one of the most vulnerable to contamination and degradation through accidental or deliberate mismanagement.

Soil contamination and involvement of multimedia transport processes can adversely impact the quality of water and air, thereby reducing overall environmental quality. Use of soil as a synthetic waste repository for environmental cleanup purposes may reduce its quality. Agricultural sustainability will be adversely impacted.

This book resulted from a Workshop on Metal Speciation and Contamination of Soils held at Jekyll Island, Georgia, on May 22 - 24, 1991. The workshop assembled a cadre of experts in soil physical chemistry, environmental chemistry, microbiology, and engineering, oceanography, forestry, and geochemistry. This book contains most of the invited contributions, which together provide a comprehensive overview of metal speciation and contamination of soils from theoretical, experimental and pragmatic perspectives.

Authors were chosen who had comprehensive insight and expertise in metal speciation, redox chemistry, kinetics of metal reactions, spectroscopic characterization of metal ion reactions at surfaces, modeling hydrologic transport phenomena and colloid-associated transport of metals through the soil profile to groundwater, and remediation alternatives.

In the initial chapter Ulrich Förstner presents a global perspective and defines the magnitude of the overall problem. Pathways of metal uptake by plants and soils are enumerated, environmental factors that influence these processes are discussed, and the ecological and human health implications of these reactions are explored.

In the second chapter, Don Sparks presents an overview of metal sorption kinetics on soils. The author discusses soil constituents important in metal retention, provides time scales for kinetic reactions on soils, offers a critique of current kinetic models and data interpretation, presents methods that can be used to study metal sorption kinetics, and describes applications of the pressure jump relaxation technique to gain mechanistic information about metal sorption kinetics on soil components. Caution is suggested in distinguishing between chemical kinetics and transport kinetics when studying metal behavior in soils.

Janet Hering, in the third chapter, outlines the mechanisms and kinetics of chemical processes that influence metal transport -- particularly, sorption, complexation and dissolution reactions, -- in porous media and provides experimental data from a combination of laboratory and field studies to substantiate her observations.

Liyuan Liang and John McCarthy, in chapter 4, discuss the mechanisms of metal sorption onto surfaces of oxide colloids, factors influencing these reactions, and mechanisms governing the formation and transport of colloids. The occurrence and importance of subsurface colloidal transport of both bound-metal contaminants, and the transport of organic and inorganic colloids are evaluated.

In chapter 5, Joe Stucki and co-workers review principles of redox reactions in soils and sediments, provide a comprehensive summary of basic equilibrium thermodynamics, outline concepts of solid state electron transfer, and discuss acid-base phenomena. Insight is also provided on the known effects of changes in redox conditions on the behavior and physico-chemical properties of phyllosilicate minerals. Several different mechanisms are proposed and evaluated for the reduction or oxidation of structural iron in phyllosilicates. An assessment is made of the potential importance of redox reactions in phyllosilicate in predicting the transformation and transport of metal species in the vadose zone.

In chapter, 6, Terry Beveridge and co-workers review the literature on the structure of the bacterial cell wall and capsule; examine metal binding by these surfaces; discuss metal immobilization, mineralization and remobilization at bacterial surfaces and at clay mineral-bacterial cell wall composite surfaces; and examine factors influencing metal transport in natural systems.

In chapter 7, Thomas Theis and Ramesh Iyer treat the complex problem of parameterizing coupled chemistry-transport models for understanding and predicting trace metal chemical reactions in groundwater. The authors emphasize and demonstrate the importance of establishing meaningful error bounds on the parameter estimations obtained for solute sorption regardless of the experimental configuration used.

In chapter 8, Walter Wenzel and Winfried Blum deal with methodological problems involved in assessing metal mobility in soils. They identify several critical sources of error caused by soil sampling procedures, sample preparation steps and extraction techniques.

In chapter 9, Cliff Johnston and colleagues point out that the key to predicting subsurface transformation and transport of both metals and organic contaminants is a fundamental understanding at the nanoscale of the physico-chemical mechanisms of metal interactions at surfaces. The authors examine the use of metal cations as molecular probes to study the surface chemistry of smectite-water interactions. ^{23}Na NMR spectroscopy was used to evaluate the mobility and thus the hydration character of the cation and the extent to which the hydrated cation interacts with the clay surface.

Robert Peters and Linda Shem, in chapter 10, evaluate the metal removal efficiencies of several chelating agents as influenced by both hydrogen ion activity and the chelating agent concentration.

In chapter 11, Michael Amacher and co-workers evaluate the effect of mine waste on water quality in headwater streams through the use of field studies. Manganese and sulfur concentration changes were primarily the result of mine waste inflows mixing with stream waters whereas iron concentrations were controlled by precipitation of hydrous oxides.

In chapter 12, Jingyi Liu and co-workers summarize a decade of research on the assessment of metal contaminants dispersed in the aquatic environment of certain major rivers of the People's Republic of China. Identification and speciation work on metal contaminants and process level research defining metal speciation and transport are reported along with simulation experiments and model development work.

In the final chapter, Chris L. Bergren and colleagues evaluate a purged-water-management disposal strategy to remove such constituents as radionuclides, organic chemicals and inorganic chemicals from contaminated groundwater by applying this contaminated groundwater to the ground surface near the purged well. They also determine whether any of these constituents that initially are below regulatory levels will, after a long-term application period, accumulate in surface soils in excess of regulatory levels thereby becoming a hazardous waste.

We hope that the information in this book will provide the reader a greater insight and a more comprehensive understanding of the biogeochemical processes governing the behavior, transport, and bioavailability of heavy metals in contaminated soils and an improved appreciation of available alternative approaches to effectively remediate these metal-contaminated soils.

We wish to thank all authors for their contributions and for their consistent cooperation during the time required to produce this book. The expert technical assistant of Ms. Dana M. Crumety and Ms. Beverly A. Henderson, who provided expert secretarial support, is gratefully acknowledged. Finally, we wish to thank the U. S. Environmental Protection Agency for their financial support.

References:

Wilding, L.P. Changing Vision of Soil Science, *Geotimes*, 39:13-14, 1994.

Herbert E. Allen	Newark, Delaware
Chin P. Huang	Newark, Delaware
George W. Bailey	Athens, Georgia
Alan R. Bowers	Nashville, Tennessee

Herbert E. Allen is Professor of Civil Engineering at the University of Delaware, Newark, Delaware, U.S.A. Dr. Allen received his Ph.D. in Environmental Health Chemistry from the University of Michigan in 1974, his M.S. in Analytical Chemistry from Wayne State University in 1967, and his B.S. in Chemistry from the University of Michigan in 1962. He served on the faculty of the Department of Environmental Engineering at the Illinois Institute of Technology from 1974 to 1983. From 1983 70 1989 he was Professor of Chemistry and Director of the Environmental Studies Institute at Drexel University.

Dr. Allen has published more than 100 papers and chapters in books. His research interests concern the chemistry of trace metals and organics in contaminated and natural environments. He conducted research directed toward the development of standards for metals in soil, sediment and water that take into account metal speciation and bioavailability.

Dr. Allen is past-chairman of the Division of Environmental Chemistry of the American Chemcal Society. He has been a frequent advisor to the World Health Organization, the Environmental Protection Agency and industry.

Chin P. Huang is the Distinguished Professor of Environmental Engineering at the University of Delaware. He received his Ph.D. and M.S. in Environmental Engineering from Harvard University and his B.S. in Civil Engineering from the National Taiwan University, Taipei, Taiwan.

Dr. Huang has authored and co-authored over 150 research papers, book chapters, and conference proceedings. His research expertise in environmental physical chemical process includes the removal of heavy metals from dilute aqueous solutions, surface acidity of hydrous solids, and industrial waste management.

Dr. Huang's recent research interests are advanced chemical oxidation, *in-situ* and *ex-situ* treatment of hazardous wastes. Currently, he serves on the editorial board of the Journal of Environmental Engineering, Chinese Institute of Environmental Engineering, and the Advisory Board, Industrial Park Communication, Ministry of Economics, Taiwan.

George W. Bailey is a Research Soil Physical Chemist at the Environmental Research Laboratory, U.S. Environmental Protection Agency, Athens, Georgia, U.S.A. Dr. Bailey received his Ph.D. in Soil Chemistry and Mineralogy from Purdue University in 1961, his M.S. in Soil Mineralogy from Purdue University in 1958, and his B.S. in Agronomy from Iowa State University in 1955.

He carried out a 3-year National Institutes of Health postdoctoral study from 1961 to 1964 at Purdue University on soil-pesticide interactions. He held research positions in the U.S. Public Health Service, Federal Water Pollution Control Administration, Federal Water Quality Administration and research and administrative positions in the U.S. Environmental Protection Agency from the Agency's inception.

Dr. Bailey has conducted research and published extensively in the areas of modeling the transport and transformation of pesticides in agricultural watersheds, of investigating the surface chemistry of organic and inorganic pollutants at environmental surfaces, and of applying scanning probe microscopy and scanning tunneling spectroscopy to defining the structure, morphology, and reactivity of environmental surfaces as they influence pollutant speciation, bioavailability, and mobility in terrestrial and aquatic ecosystems.

Dr. Bailey is past chairman of the Division of Environmental Quality, American Society of Agronomy, and the Division of Soil Chemistry, Soil Science of America and Fellow, American Society of Agronomy. He is also past chairman of the Northeast Georgia Section of the American Chemical Society.

Alan R. Bowers is Associate Professor of Civil and Environmental Engineering at Vanderbilt University, Nashville, Tennessee, U.S.A. Dr. Bowers received his Ph.D. in Environmental Engineering from the University of Delaware in 1982, his M.C.E. in Environmental Engineering from the University of Delaware in 1978, and his B.C.E. in Civil Engineering from the University of Delaware in 1976. He has served on the faculty at Vanderbilt since 1982.

Dr. Bowers has published more than 35 papers, reports, and chapters in books. His research interests are in fate and transport of metals and organics in the natural environment, remediation of contaminated soils, toxicity reduction, and minimization of hazardous wastes.

Dr. Bowers is a registered professional engineer and has been consultant to the World Health Organzation, the U.S. Department of Defense, and the U.S. Department of Energy, as well as a variety of major industries. In addition, he is a member of the Board of Directors and is the Secretary-Treasurer of the International Chemical Oxidation Association.

CONTENTS

1 LAND CONTAMINATION BY METALS - GLOBAL SCOPE AND MAGNITUDE OF PROBLEM

Ulrich Förstner

1. INTRODUCTION 1
2. ECOLOGICAL AND HUMAN HEALTH IMPLICATIONS 2
3. SOURCES OF METAL POLLUTANTS IN SOIL 3
   - 3.1 Agricultural soil 3
   - 3.2 Industrial and municipal sites - "contaminated land" 6
4. METAL TRANSFER TO PLANTS - SCIENTIFIC BASES FOR REGULATIONS 8
   - 4.1 Metal transfer from soil to plants 10
   - 4.2 Soil parameters affecting pollutant transfer to plants 10
   - 4.3 Evaluation of threshold values for soil pollutants 13
   - 4.4 Criteria for regulations 14
   - 4.5 Remedial measures for metal-contaminated agricultural soils 17
5. REMEDIAL OPTIONS FOR METAL-RICH DREDGED MATERIALS, LANDFILL RELEASES AND CONTAMINATED LAND 17
   - 5.1 Stabilization of metal-rich dredged materials 17
   - 5.2 Demobilization of metals in solid waste materials 19
   - 5.3 Restoration of contaminated land 21
6. OUTLOOK 24

REFERENCES 24

2 KINETICS OF METAL SORPTION REACTIONS

Donald L. Sparks

1. INTRODUCTION 35
2. SOIL MINERALS IMPORTANT IN METAL SORPTION DYNAMICS 36
   - 2.1 Time scales for metal sorption reactions 37
3. RATE LAWS AND DATA INTERPRETATION 39

4. ADVANCES IN KINETIC METHODOLOGIES 45
5. CONCLUSIONS 56
REFERENCES 56

3 IMPLICATIONS OF COMPLEXATION, SORPTION AND DISSOLUTION KINETICS FOR METAL TRANSPORT IN SOILS

Janet G. Hering

1. INTRODUCTION 59
2. TRACE METAL CHEMISTRY AND MOBILITY 59
3. STUDIES OF REACTION MECHANISMS 61
3.1 The surface-controlled dissolution model 61
3.2 Ligand-promoted dissolution: role of surface complexes 63
3.3 Comparison of dissolution and desorption 66
3.4 Dissolution of iron oxides by a widely occurring anthropogenic ligand 70
3.5 Reactions of metal complexes at the mineral-water interface 72
4. FIELD OBSERVATIONS 75
4.1 Chemical composition of groundwater along a river-groundwater infiltration flow path: Glattfelden, Switzerland 75
4.2 Transport of radionuclides in groundwater 78
5. COMPARISON OF FIELD AND LABORATORY OBSERVATIONS 78
6. CONCLUDING REMARKS 79
APPENDIX
EXPERIMENTAL SECTION 80
1. Materials 80
2. Dissolution and desorption experiments 80
REFERENCES 81

4 COLLOIDAL TRANSPORT OF METAL CONTAMINANTS IN GROUNDWATER

Liyuan Liang and John F. McCarthy

1. INTRODUCTION 87
2. TWO PHASE SYSTEM: METAL ADSORPTION ON SOLID SURFACES 88

3. THREE PHASE SYSTEM: ROLE OF A MOBILE COLLOIDAL PHASE IN SUBSURFACE 93
4. PREDICTING TRANSPORT OF COLLOIDS AND COLLOID-ASSOCIATED METALS 97
5. A CASE STUDY 101
6. CONCLUDING REMARKS 105
REFERENCES 105

5 REDOX REACTIONS IN PHYLLOSILICATES AND THEIR EFFECTS ON METAL TRANSPORT

Joseph W. Stucki, George W. Bailey and Huamin Gan
1. INTRODUCTION 113
2. PHYLLOSILICATE STRUCTURES AND COMPOSITION 114
3. THERMODYNAMIC PRINCIPLES OF REDOX REACTIONS 121
3.1 Definitions 125
3.2 Equations of state 125
3.3 Partial molar Gibbs free energy and total potential 127
3.4 Activity 133
3.5 Chemical equilibrium 134
3.6 Applications to redox reactions 136
3.7 Different scales for reporting redox potentials 139
4. HETEROGENEOUS ELECTRON-TRANSFER REACTIONS 140
4.1 Electron transfer in the solid phase 141
5. PROCESSES AFFECTING METAL OXIDATION STATES 146
5.1 Complexation and solubility 146
5.2 Effect of organic ligand and pH on valence-state stabilization 147
5.3 Acid-base complex formation 149
6. REDOX PHENOMENA IN PHYLLOSILICATES 155
7. SUMMARY 168
REFERENCES 168

6 DETECTION OF ANIONIC SITES ON BACTERIAL WALLS, THEIR ABILITY TO BIND TOXIC HEAVY METALS AND FORM SEDIMENTABLE FLOCS AND THEIR CONTRIBUTION TO MINERALIZATION IN NATURAL FRESHWATER ENVIRONMENTS

Terry J. Beveridge, Susanne Schultze-Lam and Joel B. Thompson

1. INTRODUCTION 183
2. METAL BINDING AND MINERAL FORMATION 184
   - 2.1 Cell walls 185
   - 2.2 Capsules 187
3. MINERAL FORMATION 188
   - 3.1 In situ observations 189
   - 3.2 Fayetteville Green Lake 192
4. METAL TRANSPORT AND THE IMMOBILIZATION OF TOXIC HEAVY METALS 196
5. CONCLUSION 200

REFERENCES 200

7 TRACE METAL CHEMICAL REACTIONS IN GROUNDWATER: PARAMETERIZING COUPLED CHEMISTRY TRANSPORT MODELS

Thomas L. Theis and Ramesh Iyer

1. INTRODUCTION 207
2. SOLUTE PARTITIONING 209
3. EXPERIMENTAL APPROACHES 210
4. RESULTS AND DISCUSSION 211
   - 4.1 Batch experiments 211
   - 4.2 Column experiments 214
5. CONCLUSIONS 222

REFERENCES 223

8 ASSESSMENT OF METAL MOBILITY IN SOIL - METHODOLOGICAL PROBLEMS

Walter W. Wenzel and Winfried E.H. Blum

1. INTRODUCTION 227
2. MATERIALS AND METHODS 227
   - 2.1 Experiments 227
3. RESULTS 230
4. CONCLUSIONS 235

REFERENCES 235

9 VIBRATIONAL AND NMR PROBE STUDIES OF SAz-1 MONTMORILLONITE

Cliff T. Johnston, William L. Earl and C. Erickson

1.	INTRODUCTION	237
2.	EXPERIMENTAL	240
	2.1 Vibrational spectroscopy	240
	2.2 NMR spectroscopy	242
3.	RESULTS	243
	3.1 Desorption isotherms	243
	3.2 FTIR spectra of sorbed water	244
	3.3 NMR spectra of sodium ions	247
4.	CONCLUSIONS	249
	REFERENCES	251

10 TREATMENT OF SOILS CONTAMINATED WITH HEAVY METALS

Robert W. Peters and Linda Shem

1.	INTRODUCTION	255
2.	BACKGROUND	256
	2.1 Previous studies involving extraction of heavy metals from contaminated soils	259
3.	GOALS AND OBJECTIVES	263
4.	EXPERIMENTAL PROCEDURE	263
	4.1 Batch shaker test	263
5.	RESULTS AND DISCUSSION	265
	5.1 Preliminary experiments	265
	5.2 Batch extraction experiments	266
6.	SUMMARY AND CONCLUSIONS	271
	REFERENCES	272

11 EFFECT OF MINE WASTE ON ELEMENT SPECIATION IN HEADWATER STREAMS

Michael C. Amacher, Ray W. Brown, Roy C. Sidle and Janice Kotuby-Amacher

1.	INTRODUCTION	275
2.	MATERIALS AND METHODS	277
	2.1 Study site descriptions	277
	2.2 Water and sediment sampling	278
	2.3 Field measurements	279
	2.4 Sample preparation and analysis	279
	2.5 Data analysis	281
3.	RESULTS AND DISCUSSION	283
	3.1 Maybe Canyon	283

3.2 Daisy and Fisher Creeks 294
REFERENCES 307

12 ASSESSMENT OF METAL CONTAMINANTS DISPERSED IN THE AQUATIC ENVIRONMENT

Jingyi Liu, Hongxiao Tang, Yuhuan Lin and Meizhou Mao
1. INTRODUCTION 311
2. MERCURY DISPERSED IN RIVER-SEDIMENTS 313
2.1 Speciation and distribution of Hg 313
2.2 Organic associated Hg and stability of HgS 314
2.3 Complexation, adsorption, transport and fate of Hg 315
3. CHEMICAL STABILITY OF HEAVY METALS IN THE XIANG RIVER 316
3.1 Field work and direct measurements 316
3.2 Chemical equilibrium modeling 318
3.3 Release of metals from sediments 319
3.4 Chemical stability of metal contaminants in the Xiang River 319
4. ASSESSMENT OF METAL CONTAMINATION AT MINING AREA 320
4.1 Field survey and chemical analysis 320
4.2 Distribution of metals along the aquatic system 321
4.3 Simulation experiments 321
4.4 Preliminary study on water quality modeling 324
4.5 Site-specific sediment quality assessment 326
REFERENCES 328

13 APPLICATION OF INORGANIC-CONTAMINATED GROUND WATER TO SURFACE SOILS AND COMPLIANCE WITH TOXICITY CHARACTERISTIC (TCLP) REGULATIONS

Chris L. Bergren, Mary A. Flora, Jeffrey L. Jackson and Eric M. Hicks
1. ABSTRACT 331
2. BACKGROUND 332
3. METHODS 334
3.1 Selection and sampling of study sites 334

3.2 Selection and sampling of the purged water monitoring well 336
3.3 Laboratory study 340
3.4 Field study 342
4 RESULTS AND DISCUSSION 343
4.1 Comparison of spiked water to targeted trigger levels 343
4.2 Comparison of total TCLP inorganics with TCLP data 344
4.3 Comparison of total cation exchange capacities (CECs) 346
4.4 Comparison of initial, 4-liter, and 8-liter data 346
4.5 Comparison of total inorganics between the surface soils 347
4.6 Comparison of total inorganics between the subsurface soils 347
CONCLUSIONS 347
REFERENCES 348
SUBJECT INDEX 351

METAL SPECIATION *and* CONTAMINATION *of* SOIL

1

LAND CONTAMINATION BY METALS: GLOBAL SCOPE AND MAGNITUDE OF PROBLEM

Ulrich Förstner
Department of Environmental Science and Engineering
Technical University Hamburg-Harburg
D-2100 Hamburg 90
Germany

1. INTRODUCTION

In the past twenty years, the emphasis in soil metal chemistry switched from problems related to scarcity of plant nutrients to those arising from pollution because of excessive inputs from various anthropogenic sources. More recent concerns relate to contaminated soil, comprising heavily contaminated sites from abandoned waste deposits. Solution of such problems require a multi-disciplinary approach during investigation, assessment, and considerations of remedial action. This includes considerable experience on metal "speciation", i.e., on reactions such as precipitation-dissolution, adsorption-desorption, and complex formation in relation to pH, redox conditions and the content of soluble chelating agents.

Treatment of contaminated soil represents complex and challenging problems. In addition to the common predictive techniques for estimating contaminant losses - *a priori* techniques and pathway-specific tests - the interactive nature of various parameters, which may affect long-term mobility of metals, has to be studied. Examples involving considerations on metal speciation, such as "stabilization" or "containment" techniques, will be described from dredged material and municipal solid waste deposits as well as from the restoration of contaminated industrial sites.

0-87371-697-3/95/$0.00+$.50

2. ECOLOGICAL AND HUMAN HEALTH IMPLICATIONS

The shift of environmental problems at the expense of soil has become evident from the large-scale deterioration of forests, from regional difficulties with groundwater reclamation, from the problems with "abandoned landfills", and from various forms of soil degradation - e.g., erosion and urbanization, salinization and desertification - among which soil pollution is largely due to the activities of man and particularly to industrialization.

Metals are natural constituents in soils. However, over the 200 years following the beginning of industrialization, huge changes in the global budget of critical chemicals at the earth's surface have occurred, "challenging those regulatory systems which took millions of years to evolve" [1]. For example, the ratio of the annual mining output of a given element to its natural concentration in unpolluted soils, which can be used as an "Index of Relative Pollution Potential" [2] is particularly high for Pb, Hg, Cu, Cd and Zn, namely 10 to 30 times higher than for Fe or Mn.

Soil contamination may disrupt the delicate balance of physical, chemical and biological processes upon which the maintainance of soil fertility depends. Pollution of soils by heavy metals compounds may inhibit microbial enzyme activity and reduce the diversity of populations of soil flora and fauna. Transfer of metals to man may result from consuming contaminated plants or indirectly from consuming milk or meat from grazing animals that have consumed contaminated plants or soils. Specific concerns for land disposal of contaminated materials include the transport of heavy metals to both surface and ground waters, transport of pathogens to man through such pathways as crops grown in waste-amended soils and contamination of groundwater and surface water systems, as well as the export of nutrients to non-target ecological systems.

As for the mechanisms of toxicity, the most relevant is certainly the chemical inactivation of enzymes. All divalent transition metals readily react with the amino, imino and sulfhydryl groups of proteins; some of them (Cd, Hg) may compete with essential elements such as zinc and displace it in metalloenzymes. Some metals may also damage cells by acting as antimetabolites, or by forming precipitates or chelates with essential metabolites [3]. Soil biochemical processes considered especially sensitive to heavy metals are mineralization of N and P, cellulose degradation and possibly N_2-fixation [4].

3. SOURCES OF METAL POLLUTANTS IN SOIL

The increasing burden of chemical compounds induced a new area of interest in their behavior in soil systems. Whereas this interest was originally focused on elements naturally occurring in soils, the term “soil contaminant” was now introduced, referring to compounds which were originally absent in the system [5]. Land contamination has two different meanings. One is the slow but steady degradation of soil quality by the inputs of chemicals from various sources. The other - in the term "contaminated soil or land" - is the massive pollution of smaller areas, mainly by the dumping or leakage of industrial waste materials. Contaminated soil from an agricultural view mostly relates to adverse effects on foodstuff; waste dumping is usually connected with groundwater problems.

3.1 Agricultural soil

In addition to the natural constituents, trace elements enter the soil via beneficial agricultural additives such as lime, fertilizer, manure, herbicides, fungicides and irrigation waters as well as via potentially deleterious materials such as sewage sludge, municipal composts, mine wastes, dredged materials, fly ash and atmospheric deposits. Typical contents of trace elements in uncontaminated soil and in sources of potential contamination are listed in Table 1.

Soil composition varies widely and invariably reflects the nature of the parent material. Trace elements often range over two or three orders of magnitude; the principal factors determining these variations are the selective incorporation of particular elements in specific minerals during igneous-rock crystallization, relative rates of weathering, and the modes of formation of sedimentary rocks. In the absence of pollution, total metal contents are related to soil parent material, organic matter content, soil texture and soil depth [6].

The normal ranges for a number of metals in the 1986 CEC Council Directive [7] are much narrower than the normal ranges reported in Table 1. This applies to Cr, Co, Cu, Pb, and Ni. The derived mean values or typical contents for Cr and Ni are close or greater than the upper end of the normal range reported in the directive. There appears therefore to be a need to modify some of the values for upper permissible concentrations in soils, particularly for Cr which is known to be immobile in soils and to be relatively harmless to animals when ingested [9].

Table 1. Contents of Trace Elements, as mg/kg Dry Material, in Uncontaminated Soil and Selected Sources of Potential Soil Contamination (after Ure and Berrow [6]; Berrow and Burridge [8]; Berrow and Reaves [9]). CEC Council Directive of June 12, 1986 [7]. Critical Values in Additives are Printed in Bold Face.

	Typical Soil		CEC-1986 Directive	Fertilizer*	Sewage Sludge	Municipal Compost	Fuel Ash	Atmos. Fallout[a]
B	10	(0.9-1000)	-	P(30)	50	-	200	5.5
Be	6	(0.5-30)	-	All Low	3	15	23	-
Cd	0.4	(<0.1-8)	1-3	P(50)	12	10	10	0.25
Co	8	(0.3-200)	-	All Low	12	30	-	1.6
Cr	50	(0.9-1500)	-	P(200)	250	120	280	1.4
Cu	12	(<1-390)	50-140	FYM(20)	800	800	320	8.8
Hg	0.06	(>0.01-5)	1-1.5	All Low	4.4	-	-	0.05
Mn	450	(<1-18300)	-	L(500)	400	500	640	4.9
Mo	1.5	(0.1-28)	-	P(4)	5	8	40	0.14
Ni	25	(0.1-1520)	30-75	All Low	80	120	270	7.3
Pb	15	(1-890)	50-300	P(100)	700	1200	330	11.0
V	90	(<1-890)	-	P(50)	60	100	360	2.3
Zn	40	(1.5-2000)	150-300	P(150)	3000	2000	360	29

*Fertilizers typically with highest content; P = phosphate fertilizers, FYM = farmyard manure, L = limestone or dolomite commonly encountered values shown in brackets. [a] Atmospheric fallout expressed as mg/kg in topsoil to 20 cm depth, estimated 100 year accumulation.

Most of the materials added to improve cultivated soils, for example, lime, inorganic nitrogenous or potassic fertilizers, and farmland manure, have low trace element contents and when used at normal rates are unlikely to affect trace element levels in soils and crops [8]. Phosphates, however, can often contain significant amounts of several trace elements. High levels of cadmium in phosphate fertilizers, sometimes around 100 mg/kg, can produce increases both in soil contents and in plant uptake. In many countries, such as in the Federal Republic of Germany, phosphate rocks from different source areas were used for production of phosphate fertilizers. Low-Cd source materials came from USA and USSR phosphate mines. Increasing imports from other countries - rock phosphates from Senegal and Togo exhibit 10- to 100-fold higher Cd-contents - suggest problems that could become reality which seem to be overestimated at present [10]. Technological measures that might be taken to eliminate cadmium are very limited [11] or at least expensive [12], although production of phosphoric acid is reducing cadmium concentrations (the cadmium being removed with the gypsum produced). However, high elimination rates only occur with phosphates that have low cadmium contents [13].

Irrigation of arid and semi-arid soils can pose critical situations by increased salinization, i.e. by accumulation of salts of metals such as sodium, calcium and magnesium. This type of pollution induces reduction in plant growth and yield through three different mechanisms [14]: (1) reduced water availability to plants through osmotic effects, (2) uptake and accumulation of toxic levels of certain metals and non-metals in plants, and (3) nutrient imbalances. Ameliorative methods to reduce salinity problems are based mainly on improved drainage and leaching of excess salts.

The catastrophic event of Itai-itai disease in the Jintsu River catchment area of Japan was caused by waste materials from a zinc mine situated some 50 km upstream from the afflicted villages [15]. It was found that the Cd drainage from the mine (together with other metal pollutants) had discharged into the Jintsu River and had accumulated in paddy field soil in the basin. The Cd was later absorbed by crops, such as paddy rice and soybeans, grown on the polluted soil [16]. Other cases where rivers draining mine wastes have been used to irrigate paddy fields were studied by Asami [17] on the Ichi and Maruyama River basin situated between the Japan Sea (in the north) and Seto Inland Sea (in the south).

A Working Group of the World Health Organization on the health risks of chemicals in sewage sludge applied to land [18] suggested that cadmium is the most important contaminant because it can be accumulated from the soil by certain food plants. Similar effects have been found for cadmium in waste compost [19] and a study for Switzerland has prognosed that if today's practice of using compost would be continued, a critical level of 3 ppm Cd in soil would be surpassed in 20-30 years [20]. Addition of coal fly ash to soil could produce problems of salinity and through elevated concentrations of boron, molybdenum and selenium [21,22]. Dredged materials affects groundwater quality and agricultural products, as exemplified from the polder area of the Rhine River estuary [23]; calculations by Kerdijk [24] of dispersion processes of pollutants in groundwater suggest that chloride, showing conservative behavior, will appear in the adjacent polders in the year 2100 approximately, the heavy metals one to three centuries, and pesticides several thousand years later. Long-term, continuous high-rate artificial groundwater recharge of wastewater could result in substantial trace metal enrichment of the affected soil and, in turn, may render the land in question unsuitable for subsequent uses [25]; this is valid for cadmium even at relative low concentrations in wastewater.

In Table 1, last column, the average amount of metals that would accumulate over 100 years from atmospheric deposition at 8 widely distributed U.K. sites during 1972-1975 has been calculated (after [8]). An elemental deposition of 2.5 kg/ha is assumed to be equivalent to an increase of 1 mg/kg in the top 20 cm of soil. It is suggested that contribution from atmospheric sources to soil contamination is particularly important for Cd, Cu, Hg, Pb, and Zn. Deposition rates reported from the U.S.A., The Netherlands and West Germany give similar values, indicating that this source of trace elements in soil is of considerable importance in industrialized countries.

The contribution of long-range atmospheric transport to trace metal pollution of surface soils has largely been ignored, although it has been known for many years that a very significant part of the aerosols released to the atmosphere from high temperature processes are dispersed over large areas. In particular, elements such as Pb, Cd, and As that form volatile compounds tend to be preferentially concentrated on small particles having long residence times in the atmosphere [26]. The significance of long-range atmospheric transport to the supply of trace metals to terrestrial ecosystems has been clearly evidenced for lead in soils from Norway [27].

3.2 Industrial and municipal sites - "contaminated land"

Metal-containing waste materials which may affect groundwater pollution include municipal solid wastes, sewage sludge, dredged material, industrial by-products, wastes from mining and smelting operations, filter residues from wastewater treatment and atmospheric emission control, ashes and slags from burning of coal and oil, and from incineration of municipal refuse and sewage sludge. The problem of soil pollution was introduced with the accidential detection of large scale contamination from industrial waste deposits which had been handled inproperly, such as Love Canal in USA, Lekkerkerk in The Netherlands, and Hamburg/Georgswerder.

With regard to loss of groundwater resources ascribable to contamination, a United States Library of Congress Report from 1980 lists 1360 well closings in a 30 year span, which can be broken down as follows [28]: metal contamination, 619 wells; organic chemical contamination, 242, including 170 from trichloroethylene used to emulsify septic tank grease; pesticide contamination, about 200 wells; industrial (not defined) sites, 185 wells; leachate from municipal solid waste landfills, 64 wells; 26 wells affected by high concentrations of chlorides; and 23 wells polluted by nitrates. Altogether, there is a significant effect of heavy metal pollution - about 40 percent in this study - on the loss of groundwater resources.

In an early inventory of 700 cases of soil pollution in The Netherlands [29], the group of solvents - tri- and perchloroethylene - were the major contaminants in 120 sites; oil and oil-like product are responsible for pollution of 150 sites; other typical forms of industrial contamination originate from gas works and production of biocides. Pollution involving high concentration of heavy metals forms the largest group of soil contamination pollution. Generally, there are many sites where mixtures of organic and inorganic pollution have been found, and these sites pose particular problems for restoration.

More recent surveys undertaken in five EC countries indicate that about 8,900 contaminated industrial sites exist, which require immediate treatment either because they present an environmental health problem or cannot be re-used without being decontaminated. The area of contaminated land is about 0.2% of the land areas in these countries. Other 14,000 potentially contaminated sites in Denmark, Germany and The Netherlands have been designated as requiring further investigation [30].

In the United States, the National Priority List of 1986 [31], developed by the EPA, exhibits a total of approximately 1000 sites that pose significant environmental or health risks. About 40% of these sites reported metal problems. The majority of these reported metals are combined with organics, but a significant number reported only metals or metals with inorganics. Most of the metal problems are connected with two or more metals (70%), whereas 30% were associated with only one metal (Table 2).

Although some of the National Priority (Superfund) Site descriptions simply identified heavy metals as a problem, many of them specified the metals [32]. The metals most often cited as a problem are lead, chromium, arsenic, and cadmium, each of which is cited as a problem at more than 50 sites. Copper, zinc, mercury and nickel are cited as problems at over 20 sites each.

Only a few industries or activities account for most of the NPL sites with metal problems; these are in particular metal plating, chemical, mining and smelting, battery recycling, wood treating, oil and solvent recycle, and nuclear processing industries. Not surprisingly landfills account for a high percentage (40%) as chemical wastes were often dumped in municipal landfills.

Typically, elevated concentrations of trace elements in groundwaters have been found in mining areas and in the vicinity of industrial waste deposits. A spectacular case was reported by Balke *et al.* [33] from the area of Nievenheim in the lower reaches of the Rhine River, where a zinc processing plant has infiltrated waste water onto the substratum. In the groundwater, the concentrations of arsenic surpassed 50 mg/L; maximum concentrations have been measured for cadmium of 600 μg/L, for thallium 800 μg/L, mercury 50 μg/L, and zinc 40 mg/L. It was shown that the concentration of the salts and trace elements in the contaminated groundwater had only very insignificantly decreased, even 18 months after the percolation was stopped.

Apart from the direct dumping of wastes, typical effects of local accumulations of air-borne pollutants have been found. For example, the close correlation between Pb, Cu, Zn, and As concentrations in surface samples of soils in some parks of Hamburg can be related to waste emissions from industrial plants, coal power stations, and refuse incineration plants [34].

4. METAL TRANSFER TO PLANTS - SCIENTIFIC BASES FOR REGULATIONS

With regard to intoxication of humans by excessive metal concentrations, ingestion intake in most cases is more significant than inhalation intake. In general, higher fluxes of metals to man occur from solid foodstuff than from water. For solid food, the soil-plant-man pathway generally is a much more critical route for metals than is the soil-plant-animal-man route. Data in Table 3, which is compiled from different sources [35], indicate that are relatively small differences between the average intake and the intake considered "tolerable", particularly for cadmium, arsenic, lead, and mercury. The average biological half-live of cadmium has been estimated to be about 18 years, and this is extremely long. Epidemiological surveys suggest that a continous oral intake of 200 μg Cd per day could cause an increased prevalence of kidney damage in persons over 50 years of age. According to market basket studies in United States and Canada, the daily intake of cadmium is in the range of 50 μg to 80 μg from food.

Table 2. Summary of Metals in National Priority (Superfund) Sites in the United States of America ([32] after Data from Ref. [31] and Bates, Private Communication 1988).

Summary of Metal Problems at NPL Sites (1986)

952 total NPL sites analyzed (703 listed, 249 proposed)

41% (389)	report metal problems
26% (244)	report metals with organcs
14% (133)	report metals, but not organics
1% (12)	unclear if organics are present
29% (113)	of sites reporting metals, report only one metal
71% (276)	of sites reporting metals, report multiple metals

Metal/	# of sites
Pb	133
Cr	118
As	77
Cd	65
Cu	49
Zn	40
Hg	32
Ni	24
Ba	10
Ag	10
Fe, Ra, U, Th, Mn, Se	48

# of sites	Industry	Metals most often reported
154	landfill/chemical waste dump	As Pb Cr Cd Ba Zn Mn Ni
43	metal finishing/plating/electronics	Cr Pb Ni Zn Cu Cd Fe As
35	chemical/pharmaceutical	Pb Cr Cd Hg As Cu
28	mining/ore processing/smelting	Pb As Cr Cd Cu Zn Fe Ag
21	federal (DOD, DOE)	Pb Cd Cr Ni Zn Hg As
19	battery recycle	Pb Cd Ni Cu Zn
18	wood treating	Cr Cu As
16	oil and solvent recycle	Pb Zn Cr As
13	nuclear processing/equipment	Ra Th U
5	pesticide	As
5	vehicle and drum cleaning	As
3	paint	Pb Cr Cd Hg
29	other	As Pb Hg Cr
389	total	

Table 3. Metal Intake by Human Nutrition (Ref. [35], after Data from Various Sources)

Element	Intake with nutrition (mg, average per week)	Provisional tolerable weekly intake (mg)
Arsenic	0.2-0.3	1
Cadmium	0.284	0.525
Copper	up to 700 mg	15
Mercury	0.063	0.35
Nickel	2-4	-
Thallium	n.d.*	0.1

* not determined

4.1 Metal transfer from soil to plants

The limit values for concentrations of heavy metals in soils, such as those given in the 1986 CEC Council Directive (Table 1), are an important indication of the potential burden of plants, but in some cases they need correcting as well as complementing with analysis of plants [36]. In addition, the further pathways to the consumer have to be considered when establishing critical values for pollutants in soils (Table 4).

From data in Table 4 it may be concluded [36] that the "passage poisons" Zn, Cu and Ni in the foods chain are more toxic to the plant themselves than to humans and animals which may eventually eat these plants. The "accumulation poisons" Pb, Cd, and Tl, on the other hand, are usually tolerated by plants in greater amounts than what is recommended for their utilization as food and feed. Cr and Hg are translocated from soils and roots into the plant shoots to such a small extent that toxicologically critical contents are hardly to be expected.

4.2 Soil parameters affecting pollutant transfer to plants

A global investigation on the interactions of micronutrients such as Mo, B, Cu, Fe, Mn and Zn between soils and food plants by Sillanpää [43] indicates - after statistical evaluation - the influence of characteristic parameters such as soil texture, organic carbon content, cation exchange

Table 4. Transfer and Critical Concentrations of Trace Metals in the Soil-Plant-Animal Chain (after Sauerbeck, [37]).

Element	Transfer Coeff. Plant/Soil[1]	Critical Concentrations (μg/g dry weight) for Plant Growth		in Animal Feed[2,3]
Cd	1 - 10	5 - 10	>	0.5 - 1
Tl	1 - 10	20 - 30	>	1 - 5
Pb	0.01 - 0.1	10 - 20	=	10 - 30
Zn	1 - 10	150 - 200	<	300 - 1000
Cu	0.1 - 1	15 - 20	<	30 - 100
Ni	0.1 - 1	20 - 30	<	50 - 60
Cr	0.01 - 0.1	1 - 2	<<	50 - 3000

[1] Only order of magnitude; for individual cases dependent on soil properties and plant species [38,39]

[2] Threshold values, where growth inhibition of particularly sensitive plant species begins [40]

[3] Various domestic animals [41,42]

capacity, calcium carbonate equivalent, and pH-values - with particularly strong effects of the latter factor. According to Haque and Subramanian [44] these factors interact in the following ways, which affect transfer rates of copper, lead and zinc from soil to plants:

(1) Soil pH has a direct relationship to the availability of metals as it affects their solubility and their capacity to form chelates in the soil. It has been noted by Stevenson and Ardakani [45] that the stability of metal humic complexes increases from pH 3.5 to pH 5 (log K: Cu-8.7 > Pb-6.1 > Fe-5.8 > Ni-4.1 > Mn-3.8 > Zn-2.3); this might be attributed to the dissocation of functional groups in the fulvic and humic acid molecule ([46]; see discussion by Cottenie [47]).

(2) Soil organic matter has an essential function in the accumulation and transport of metals as well as in delaying their circulation in the soil. Therefore, the toxicity of metals may be both increased as

well as decreased by organic substances under specific environmental conditions.

(3) Phosphate content of soils also exhibits ambiguous effects on metal availabilities; examples of both increase and decrease of the metal uptake at elevated phosphate contents have been reported [48].

(4) Plant physiology and exposure time has a direct influence on the accumulation of metals by plants. Soil pH, species of chelators, and quantity of metal in soil affects the passage of metal between different parts of plants, e.g. from root to shoot. Competition and synergisms of metals has been evidenced, for example, by Miller *et al.* [49] indicating an increased cadmium uptake by corn shoots at elevated soil lead concentration. Haque and Subramanian [44] suggested that antagonism is due to competition for the same site of element; synergistic effects are then thought to be due to the damage caused by one of the various elements.

One major adverse effect of dredged materials after land disposal relates to contamination of crops [50]. Factors controlling plant availability are the nature and properties of the contaminant, of the soil/dredged material, and of the plant species. Predominant substrate factors are texture, concentration and nature of organic matter and pH. High pH, high clay and organic matter contents reduce the plant availability of most metals [51]. Dredged materials often have higher silt and organic matter contents than the corresponding soils, and this may result in lower initial bioavailabilities of the contaminants. In the long run, however, the organic matter will be partly decomposed with a corresponding increase in bioavailability. Another major difference between dredged materials and soils usually is the higher content of oxidizable sulfide compounds, such as iron sulfide in the former substrate. This results in problems with less buffered dredged sediment.

For example, in the Hamburg harbor area where contaminated mud is still pumped into large polders for sedimentation, agriculture is no longer permitted at these sites. Due to the low carbonate content, metals are easily transferred to crops during lowering of pH and permissible limits of cadmium have been surpassed in as much as 50% of wheat crops grown on these materials [52]. This effect can be explained by the ability of certain bacteria (*Thiobacillus thiooxidans* and *T. ferrooxidans*) to oxidize sulfur and ferrous iron; while decreasing the pH from 4-5 to about 2, and the process of metal dissolution from dredged sludge is enhanced. High concentrations of metals have been measured in oxidized pore waters from sedimentation polders in the Hamburg harbor area [53].

Plants differ in their ability to accumulate heavy metals from dredged material (see review by Van Driel and Nijssen [51]). Monocotyledonous plant species accumulate less metals than dicotyledonous species [47], but difference between species, varieties and subspecies may mask these differences. Moreover, growth conditions, nutrient status, temperature and transpiration rate determine growth rate and element uptake. The organs of the plant also show different abilities to accumulate metals: for most plant species seeds and fruits accumulate less metal than leaves and roots. Plant bioassays have been developed at the U.S. Army Waterways Experiment Station, Vicksburg, MS [54]. For fresh water conditions mostly the marsh plant Cyperus esculentus is usually used, which may be cultivated in an aerobic or anaerobic rooting environment under upland and flooded growing conditions. For salt water marsh conditions Spartina alterniflora and some other salt water marsh plant species are applied.

Particularly for less buffered soil/dredged material systems, liming may be effective in the depression of metal transfer into plants. In calcareous river sediments, e.g. dredged material from the Rotterdam harbor basins, liming apparently has no effect [51]. The use of a soil cover layer system generally seems to be more effective. Model calculations show that in the case of a soil cover of a least 1.0 m there is no net transfer of dissolved contaminants from the contaminated soil to the clean soil cover. A prerequisite is that the groundwater level is fixed at 1 m below the surface. Experiments with plants show a decrease in the accumulation of metals by the plants, but roots of deep-rooting plants can still reach the contaminated soil layer and contribute to the metal status of the above-ground plant organs.

4.3 Evaluation of threshold values for soil pollutants

Soil properties such as texture, pH-value, ion exchange capacity and organic matter contents strongly influence the availability of pollutants to plants. For metals, considerable differences in the uptake exist depending on the plant families, plant species, and plant organs concerned. Guideline values, therefore, as stated before, need complementing with analyses of plants [36].

Among the various factors affecting transfer of metal pollutants to plants, pH-conditions in many instances are most important. Generally the lowering of pH by one unit will increase metal solubility by factor 10. Figure 1 demonstrates for the example of zinc, how the permissible limit of total metal concentration in soil is affected by pH as the dominant factor with respect to metal solubility. At pH 7 the limit of 1 mg Zn/L in soil equilibrium solution (which already may lead to slight depressions in yield for cabbage [56]) would be attained at approximately 1200 mg Zn/kg in soil. However, at pH 6, maximum permissible Zn-concentrations in solution would be reached with 100 mg Zn/kg soil, at pH 5 it would be reached even at 40 mg/kg. It seems that under the latter conditions, pH 5, adverse effects can be found even in unpolluted soils.

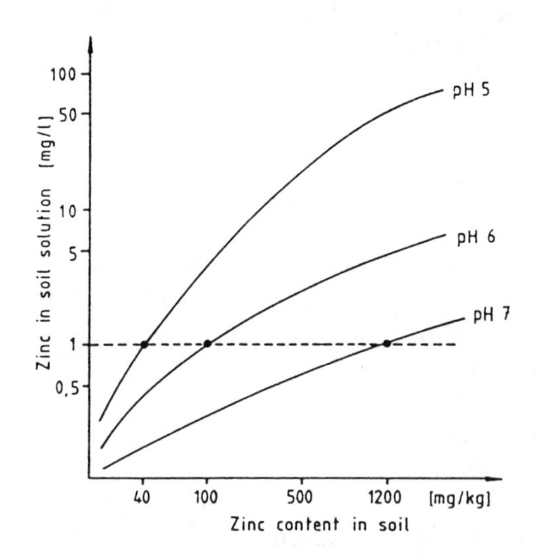

FIGURE 1. ***Solubility of Zinc in Relation to pH and Total Zn-Content of Soils Not or to a Different Degree Polluted with Zinc [55].***

4.4 Criteria for regulations

In the discussion of environmental problems with metals it is increasingly recognized that the ecotoxicological significance of metal inputs is determined rather by the specific form and reactivity of a metal compound than by its accumulation rate. This experience typically refers to the biological effects of contaminated solid materials on agricultural land and to surface-associated pollutants that may become partly mobilized in the terrestrial milieu by changes of the pH and redox conditions, by increased salinities or concentrations of organic chelators.

In regulatory work on soil quality there is a shift from fixed levels (for example, CEC regulations 1986, Table 1) to a certain range of limitations, dependent from the composition of the soil, i.e., soil pH, cation exchange capacity, and/or content of organic matter (that is light or heavy soil). An initial effort has been undertaken by U.S. EPA, which already in 1978 proposed a three-class differentiation based on the cation exchange capacity [57].

In some traditional industrial areas and near hazardous waste disposal sites, the toxic contamination of soil and groundwater may represent a health risk for the residents of these areas. In order to provide a basis for administrative decisions on redevelopment measures the Dutch authorities have developed guideline values for contaminant levels in soil and groundwater. Soil guidelines for metals are related to the percentage of clay and organic matter, respectively (Table 5); value "A" is the reference level, at level "B" more detailed examination should be made, and at level "C" detailed examination should be undertaken with regard to redevelopment measures.

Another approach relates to the more soluble fractions, which provides some indications about the future behaviour of metal pollutants in soil. An initial example is given in Table 6 by the Swiss Ordinance for Pollutants in Soil [60], where limitations were proposed on the basis of sodium nitrate soluble proportions.

Subsequent to the catastrophic cases of metal pollution from irrigation water in Japan, maximum allowable limits of heavy metals in agricultural soil have been set for Cd, Cu, and As. The requirements for the designation of agricultural land soil pollution policy areas in Japan are based on the maximum allowable limits (MALs) of Cd in rice (1 mg/kg in unpolished rice) and on the concentration of Cu and As in paddy soil (125 mg Cu/kg by 0.1 N HCl extraction; 15 mg As/kg by 1 N HCl extraction). Results of the basic survey in Japan proved that 9.5% of paddy soil is polluted by Cd. Rehabilitation of the polluted paddy soils has been made by covering the polluted surface soil with a 25-30 cm deep layer of unpolluted soil, thus reducing the Cd contents in rice below 0.1 mg/kg, which is the Cd concentration in the rice from unpolluted field under usual agricultural practice [61].

Table 5. Dutch Guideline Values for Metal Concentrations in Soil and Groundwater for the Judgement of Contaminant Levels in Soil ([58] after [59]).

Metal	Soil (mg/kg dry weight)			Groundwater (μg/L)		
	A	B	C	A	B	C
Arsenic	15 + 0.4 (L + H)[a]	30	50	10	30	100
Barium	200	400	2000	50	100	500
Cadmium	0.4 + 0.007 (L + 3H)	5	20	1.5	2.5	10
Cobalt	20	50	300	20	50	200
Copper	15 + 0.6 (L + H)	100	500	15	50	200
Chromium	50 + 2L	250	800	1	50	200
Lead	50 + L + H	150	600	15	50	200
Mercury	0.2 + 0.0017 (2L + H)	2	10	0.05	0.5	2
Molybdenum	10	40	200	5	20	100
Nickel	10 + L	100	500	15	50	200
Tin	20	50	300	10	30	150
Zinc	50 + 1.5 (2L + H)	500	3000	150	200	800

[a]L = percentage of clay; H = percentage of organic matter in soil

Table 6. Guideline Values for Tolerable Metal Concentration in Agricultural Soil. Values According to the Swiss Ordinance on Soil Contaminants ([60]; cited from [Ref. [58]).

Metal	Total Extractable by HNO_3 mg/kg dry weight	Soluble in 0.1 mol/L $NaNO_3$ mg/kg dry weight
Cadmium	0.8	0.03
Copper	50	0.7
Lead	50	1.0
Nickel	50	0.2
Zinc	200	0.5

4.5 Remedial measures for metal-contaminated agricultural soils

In most cases, ameliorative means are not available at reasonable costs to reduce metal concentrations in soils that have been contaminated by inputs from fertilizers, sewage sludge, municipal waste compost, fly ash, dredged materials and atmospheric deposition. Agricultural-chemical measures such as pH-increase, addition of limestone, loam, clay and organic substances, as well as application of ion exchanger are expensive and in many cases of limited success [62]. Other remedial action on moderately contaminated soils include physical-mechanical measures such as deep-ploughing, covering with clean soil or soil exchange; more effective measures have been introduced by changing management (variation of cultivation periods, irrigation, optimization of nutrient supply) and utilization methods (selection of appropriate plant species); final steps include preparation of the contaminated crops by mixing, cleaning or boiling (see compilation by Sauerbeck [62]). However, these measures can only be tolerated for a limited period of time, and action has to be undertaken at the source, particularly with respect to the inputs from air emissions and from excessively contaminated sewage sludge.

5. REMEDIAL OPTIONS FOR METAL-RICH DREDGED MATERIALS, LANDFILL RELEASES AND CONTAMINATED LAND

5.1 Stabilization of metal-rich dredged materials

With respect to the transfer of pollutants from highly contaminated solid material into the ecosphere it has been shown that one of the major mechanisms of pollutant release is oxidation of organic and sulfidic particulate matter. In the case of polluted sediments, it has been suggested that following oxidation of the surface material, the ecosystem rapidly recovers and that an oxidized, bioturbated surface layer constitutes an effective barrier against the transfer of most trace metals from below into the overlying water. Nonetheless, it is indicated that dispersing contaminated sediments generally is much more problematic than "containing" them [63].

Incorporation in naturally formed minerals, which remain stable over geological times, constitutes favourable conditions for the immobilization of potentially toxic metals in large-volume waste materials both under environmental safety and economic considerations. There is a particularly low solubility of metal sulfides, compared to the respective carbonate, phosphate, and oxide compounds. One major prerequisite is the microbial reduction of sulfate; thus, this process is particularly important in the marine environment, whereas in anoxic freshwaters milieu there is a tendency for enhancing metal mobility due to the formation of stable complexes with ligands from decomposing organic matter. Marine sulfidic conditions, in addition, seem to repress the formation of mono-methyl mercury, one of the most toxic substances in the aquatic environment, by a process of disproportionation into volatile dimethyl mercury and highly insoluble mercury sulfide [64]. A summary of the positive and negative effects of anoxic conditions on the mobility of heavy metals, arsenic, methyl mercury and organochlorine compounds in dredged sludges is given in Table 7.

Table 7. Summary of Positive and Negative Effects of Anoxic Sulfidic Conditions in Sludges (after Kersten [65]).

Element or Compound	Advantageous Effects	Disadvantageous Effects
Heavy Metals (e.g., cadmium)	Sulfide precipitation	Formation of mobile polysulfide and organic complexes under certain conditions with low $Fe(OH)_3$ concentrations; strong increase of mobility under acidic conditions
Metalloids (e.g., arsenic)	Capture by sulfides	Highly mobile under post-oxic and neutral/ slightly alkaline conditions
Methyl mercury	Degradation/inhibition of CH_3Hg^+ formation by precipitation of HgS	Formation of mobile polysulfide complexes, especially at low Fe concentrations

Most stabilization techniques aimed for the immobilization of metal-containing dredged materials are based on additions of cement, water glass (alkali silicate), coal fly ash, lime or gypsum [66]. Generally, maintenance of neutral or alkaline pH-values favors adsorption or precipitation of soluble metals [67].

5.2 Demobilization of metals in solid waste materials

"Geochemical and biological engineering" emphasize the increasing efforts of using natural resources available at the disposal site for reducing negative environmental effects of all types of waste material, in particular of acid mine wastes. Practical examples for improvement of storage quality of metal-containing waste, including measures for recultivation of old and recent mining waste disposal sites and physicochemical methods of water processing, have been reviewed [68,69].

In municipal solid waste landfills initial conditions are characterized by the presence of oxygen and pH-values between 7 and 8. During the subsequent "acetic phase", pH-values up to 5 were measured due to the formation of organic acids in a more and more reducing milieu; concentrations of organic substances in the leachate are high. In a transition time of 1 to 2 years, chemistry of the landfill changes from acetic to methanogenic conditions; the methanogenic phase is characterized by higher pH-values and a significant drop of BOD_5 (biochemical oxygen demand)-values from more than 5,000-40,000 mg/L in the acetic phase to 20-500 mg/L. Long-term evolution of a "reactor landfill", subsequent to the methanogenic phase, is still an open question. Oxidation of sulfidic minerals by intruding rainwater may mobilize trace metals, and the impact on the underlying groundwater could be even higher if a chromatography-like process, involving continuous dissolution and reprecipitation during passage of oxidized water through the deposit, would preconcentrate critical elements prior to final release with the leachate [70].

Experimental investigations performed by Peiffer [71] on long-term development of sewage sludge materials provide detailed insight into the sequence of processes taking place in the post-methanogenic stage of such deposits. Transition from anoxic to oxic conditions involves a pH-decrease from 6.7 to 6.4, an accumulation of sulfate on the expense of sulfide, and a release of Mn- and Ca-ions, either by cation exchange (by protons) or by an indirect redox effect via oxidation of Fe^{2+} to Fe^{3+}; organic substances act as an acid buffer. Time-dependent release of zinc and cadmium is similar to calcium and manganese, whereas lead and copper are not remobilized under these conditions. It has been inferred by Peiffer [71] that due to slow oxidation kinetics of the sparingly soluble metal sulfides ZnS, CdS, PbS and CuS, ion exchange will become the rate-determining mechanism in this system. From the current pH-decrease it can be expected that zinc and cadmium are being exchanged for protons, whereas lead and copper do not, because of their stronger bonding to the solid substrate. Because of their eminent practical significance, these initial findings need further confirmation. The same is valid for the effect of residual organic carbon in municipal solid waste incinerator slag as a potential proton producer due to microbial degradation to CO_2 [72].

Chloride is very mobile. Sodium, ammonium, potassium and magnesium are moderately mobile, and zinc, cadmium, iron and in most cases also manganese are only partly mobile. In particular, the heavy metals zinc and cadmium show very restricted mobility in the anaerobic zone of even very coarse aquifer materials. Comparison of inorganic groundwater constituents upstream and downstream of 33 waste disposal sites in West Germany (in unconsolidated Tertiary/Pleistocene sediments; Figure 2, from Arneth *et al.* [73]) indicates characteristic differences in pollutant mobilities. High contamination factors (CF-values: contaminated mean/uncontaminated mean) have been found for boron, ammonium, nitrate and arsenic; the latter element may pose problems during initial phases of landfill operations [74].

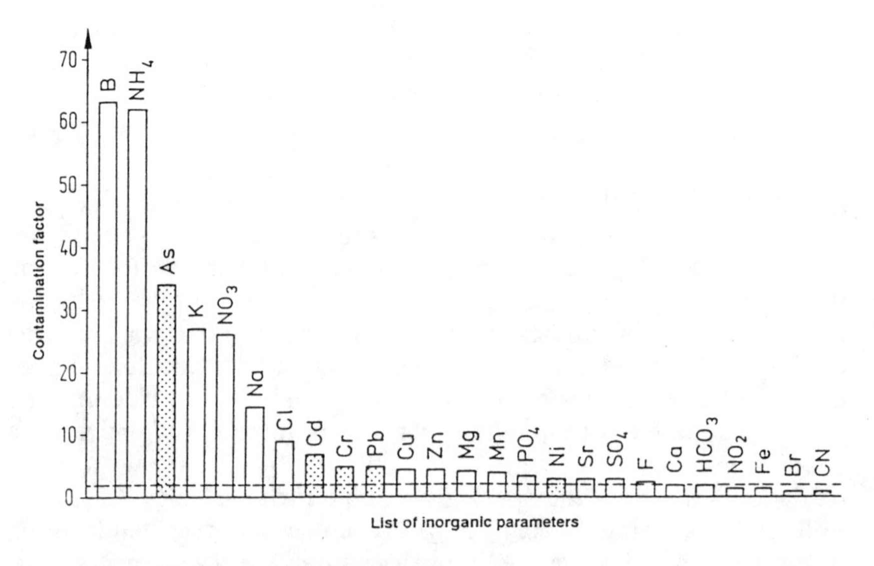

FIGURE 2. Influence of Waste Disposal on Ground Water Quality from 33 Sites in Germany [73].

Results by McLellon *et al.* [75], Atwell [76], Kelly [77], and Shuster [78], among others, indicated that the trace element contents of surface drainage water from landfills are mainly influenced by the adsorption capacities of the specific soil. Fuller and Korte [79], Griffin *et al.* [80] and Farquhar and Constable [81] from their studies on the attenuation of metal pollutants in municipal landfill leachate by clay minerals demonstrated that the potential usefulness of clay materials as liners for waste disposal sites depends to a large extent on the pH of the leachate solution; removal of heavy metal cations takes place through precipitation as hydroxides and carbonates or by coprecipitation with Fe and Mn hydrous oxides. Exchange of resident ions is also active in removing cations with alkaline earth metals being released in most cases and possibly being discharged from the liner (see review by Farquhar and Parker [82]). For metal-finishing sludges involving acid solutions, disposal in limestone-lined, segregated landfills may be a safe and economical alternative [83].

With regard to the immobilization of contaminants in municipal and industrial waste materials the term "final storage quality" has been brought into discussion [84]. Solid residues with final storage quality should have properties very similar to the earth crust (natural sediments, rocks, soil; Table 8).

Treatment includes several possible steps, e.g. assortment, thermal, chemical and biological treatments. In most cases this standard is not attained by simply incinerating municipal waste, i.e., by reduction of organic fractions only. There is, in particular, the problem of easily soluble minerals, such as sodium chloride. Measures before incineration include the separate collection of (organic) kitchen and garden wastes (containing some chlorine and sulfur), which can be transferred into compost; a major decrease of chlorine content, however, would require a significant reduction of PVC in municipal solid waste. After incineration, washing of the residues can be performed either with neutral or acidified water. Another possibility is to put the electrostatic precipitator dusts into a thermal process to remove metals like zinc, cadmium and lead as chlorides at high temperatures (>1200°C). These salts should then be used as educts for metal production [85].

5.3 Restoration of contaminated land

Various techniques and methods of remedial action have been applied at waste disposal sites, which can partly be used for the restoration of contaminated soil [86-88], and such contaminated sites can present complex and challenging problems. Solutions to such problems require a multi-disciplinary approach during investigation, assessment, and any remedial action. The problems are not only technical in nature, attention must also be given to the social and financial aspects [89].

A compilation of methods and estimated costs for both on-site and *in-situ* treatments is given in Table 9.

Liquid effluents from waste materials include direct and indirect discharges of landfill leachates and process waters from cleaning-up of contaminated soil. In principle, the techniques that are used for the purificartion of domestic and industrial wastewater can also be used for the treatment of polluted groundwater.

Table 8. Comparison of Inventories of Chemical Components in the Two Landfill Alternatives and in the Earth's Crust.

Reactor Landfill	Final Storage	Earth's Crust
Major Solid Constituents		
Putrescible Waste	Silicates, Oxides	Quarz, Fe-Oxide
Grease Trap Waste	[Gypsum, NaCl][1] (Char)[2]	Clay, Carbonates (Gypsum, NaCl) Kerogenic Comp.
Minor Solid Constituents		
Organic Micropollutants	Organic Micropollutants	-
Metals in Reactive Chemical Forms	Metal-Bearing Minerals Mainly Oxides	Metals Mainly in Inert Forms
Dissolved Constituents		
Protons, Electrons Organic Compounds Dissolved Salts	(Protons) (Organic Residues) [Dissolved Salts][1]	(pH: Acid Rain) (Humic Acids) (Dissolved Salts)

[1][Partial Extraction during Pretreatment]
[2](Minor Constituent)

For metal-rich effluents reasonable results are generally achieved by chemical and physicochemical methods. Chemical oxidation is successful at sites where there is a rather high level of free cyanide pollution (such as with electroplating plants; the usual techniques for removing complex cyanides, e.g. from gaswork sites, is coagulation/flocculation). If the heavy metals are adsorbed on suspended matter, they can be removed to a significant extent by sedimentation. Problems may arise in precipitation-based systems when large fluctuations occur in the composition of the influent. Tests carried out at a groundwater treatment plant on the site of a former electroplating factory provided better removal efficiency with sulfide precipitation than alkaline precipitation or co-precipitation with Fe(III) [92].

Table 9. Costs of "On-site" and "*In-situ*" Techniques in Remedial Action on Contaminated Land (after data from Ref. [90,91]).

Method	Disadvantage of the Remedial Action Method	Cost Estimation (US$/m^3) "On-site"	"*In-situ*"
Extraction	Problems with Fine-grained Materials; Soil is "Cleaned"	25-100	?
Thermal Treatment	Consumption of Energy + Time Effect for Volatile Cpds.	40-230	150-400
Chemical Treatment	Heterogeneity of Cpds.; Post-Treatment may be Needed Fast and Specific Reactions	10-100	130-150
Microbial Treatment	Toxicity of Some Pesticides to Microbes; Natural Process	25-50	50-100
Stabilization/ Encapsulation	Process Well Under Control Transportable ("On-Site"); *In-situ* Dispersion Difficult	25-500	80-150

Some treatments of contaminated land can result in an ultimate solution to the problems; the contamination is removed or rendered harmless. Other treatments leave the contaminants in the ground. It has been stressed by Smith [89] that these "stabilization" or "containment" options rely on engineering solutions and on engineers for their execution. They should be treated like any other engineering operation. Thus, critical points in the design should be identified and proper design and safety factors calculated from an assessment of the risks. As in civil and structural engineering, major difficulties arise from the simulation of field conditions in the laboratory. Long-term research is needed in order to develop more technically sound and cost effective remedial measures [89].

Future efforts will not only be aimed for chemically stabilizing critical compounds in their deposits but in particular for recycling valuable components in waste materials. For example, concentrations of metals such as lead, zinc, and silver in certain fractions of metal bearing wastes - including metal sludge from electroplating, heat treating, inorganic pigment manufacture, lime treatment of spent pickle liquor and emission control sludge from waste combustion could well compete with natural resources of some elements [93]. In the view of possible restrictions on land disposal of metal-bearing wastes it can well be predicted that in the not too distant future, large-scale waste processing for metal recovery will become economically feasible.

6. OUTLOOK

The enormous problems arising from the historic thoughtless dissipation of chemicals in the environment has become obvious in the last ten years, and especially with the opening of Eastern Europe. After solving a few spectacular cases with intensive efforts, it now seems that financial restrictions may inhibit the use of many newly developed remediation techniques (such as chemical extraction, high temperature incineration, and some biological procedures) to sanitize contaminated land on a large scale (i.e., for areas of hundreds of square kilometers). Although this relates primarily to parts of Eastern Europe, the tens of thousands of old landfill sites which must be excavated, treated and recultivated will soon strain even more prosperous economies. It may well be that economic considerations will strengthen the popularity of geochemically engineered solutions in these situations, as well as for large waste masses such as mine residues and dredged materials.

REFERENCES

1. Wood, J.M. and H.K. Wang, "Microbial Resistance to Heavy Metals," *Environ. Sci. Technol.* 17:582A-590A (1983).

2. Förstner, U. and G. Müller, "Heavy Metal Accumulation in River Sediments: A Response to Environmental Pollution," *Geoforum* 14:53-61 (1973).

3. Nriagu, J.O., Ed. "Changing Metal Cycles and Human Health," p. 445 Dahlem Koferenzen, Springer-Verlag, (Berlin, 1984).

4. Domsch, K.H., G. Jagnow and T.H. Anderson, "An Ecological Concept for the Assessment of Side-effects of Agrochemicals on Soil Microorganisms," *Residue Review* 86:65-105 (1983).

5. Haan, F.A.M. and W.M. van Riemsdijk, "Behaviour of Inorganic Contaminants in soil," in *Contaminated Soil,* J.W. Assink and W.J. Van Den Brink, Eds. pp. 19-32. Martinus Nijhoff Publishers, (Dordrecht, 1986).

6. Ure, A.M. and M.L. Berrow, "The Chemical Constituents in Soils," in *Environmental Chemistry* H.J.M. Bowen Ed. pp. 127-224. The Royal Chemical Society, (London, 1982).

7. Council Directive of June 12, 1986 “On the Protection of the Environment, and in Particular of the Soil, When Sewage Sludge is Used in Agriculture," Official Journal of the European Communities No. L 181/6-12, (Brussels, 1986).

8. Berrow, M.L. and J.C. Burridge, "Sources and Distribution of Trace Elements in Soils and Related Crops," in *Proc. Internat. Conf. Heavy Metals in the Environment, London, Sept. 1979* pp. 304-311. CEP Consultants, (Edinburgh, 1979).

9. Berrow, M.L. and G.A. Reaves, "Background Levels to Trace Elements in Soils," in *Proc. First Internat. Conf. Environmental Contamination, London, July 1984* pp. 333-340. CEP Consultants, (Edinburgh, 1884).

10. Sauerbeck, D., "The Environmental Significance of the Cadmium Content in Phosphorous Fertilizers," *Plant Research Dev.* 19:24-34 (1984).

11. Smith, J.L. and L.W. Bierman, "Possible Means for Controlling Cadmium Levels in Phosphate Fertilizers," *Cadmium Seminar Rosslyn, VA, November 20-21 1980*, pp. 153-173 The Fertilizer Institute, (Washington, D.C. 1980).

12. Kirner, A., "Phosphorsäure und Düngemittelindustrie, Technische Möglicheiten und Aufwand für Minderungsmaßnahmen," *Sachverständigenanhörung der Bundesregierung zu Cadmium 1981, Block E7*, pp. 8-14. Umweltbundesamt (Berlin, 1982).

13. Sauerbeck, D. and E. Reitz. “Zur Cadmiunbelastung von Mineraldüngern in Abhängigkeit von Rohstoff und Herstellungsverfahren," *Landwirtschaftl. Forsch.* 37:685-696 (1981).

14. Mattigod, S.V. and A.L. Page, "Assessment of Metal Pollution in Soils," in *Applied Environmental Geochemistry* I. Thornton, Ed. pp. 355-394. Academic Press, (London, 1983).

15. Kobayashi, J., "Relation Between the "Itai-Itai" Disease and the Pollution of River Water by Cadmium from a Mine," *Proc. 5th Internat. Conf. Advanced Water Pollution Research, San Francisco and Hawaii, Vol. I-25*, pp. 1-7. Pergamon Press, (Oxford, 1971).

16. Morishita, T., "The Jinzu River Basin: Contamination of Soil and Paddy Rice with Cadmium Discharges from Kamioka Mine," in *Heavy Metal Pollution in Soils of Japan,* K. Kitagishi and I. Yamane Eds. pp. 107-124. Japan Soil Sci. Soc., (Tokyo, 1981).

17. Asami, T., "The Iche and Maruyama River Basins: Soil Pollution by Cadmium, Zinc, Lead and Copper Discharged from Ikuno Mine," in *Heavy Metal Pollution in Soils of Japan,* K. Kitagishi and I. Yamane, Eds. pp. 227-236. Japan Soil Sci. Soc., (Tokyo, 1981).

18. Dean, R.B. and M.J. Suess, "The Risk to Health of Chemicals in Sewage Sludge Applied to Land," *Waste Management and Res.* 3:251-278 (1985).

19. Häni, H. and F. Klötzli, "Schwermetalle in Klärschlamm und Müllkompost," in *Metalle in der Umwelt - Verteilung, Analytik und Biologische Relevanz*, E. Merian, Ed. pp. 153-162. VCH Verlagsgesellschaft, (Weinheim, 1984).

20. Keller, L. and P.H. Brunner, "Waste-related Cadmium Cycle in Switzerland," *Ecotoxicology Environ. Safety* 102:1189-1193 (1976).

21. Page, A.L., A.A. Elseewi and I.R. Straughan, "Elemental Emissions and Chemical Properties of Fly Ash from Coal-fired Power Plants with Reference to Environmental Impact," *Residue Rev.* 71:83-120 (1979).

22. Petruzzelli, G., L. Lubrano and S. Cervelli, "Effects of Heavy Metals in Fly Ash Applied to Different Soils," in *Trans. XIII, Congress of the International Society of Soil Science August 1986, Vol. II,* pp. 427-428 (1986).

23. Salomons, W., W. Van Driel, H. Kerdijk and R. Boxma, "Help! Holland is Plated by the Rhine (Environmental Problems Associated with Contaminated Sediments)," in *Effects of Waste Disposal on Groundwater and Surface Water, Proc. Exeter Symp., July 1982*, No. 139, pp. 255-269. IAHS-Publishers, (1982).

24. Kerdijk, H.N., "Groundwater Pollution by Heavy Metals and Pesticides from a Dredge Spoil Dump," in *Quality of Groundwater*, P. Glasbergen and H. van Lelyveld, Eds. pp. 279-286. Elsevier , (Amsterdam, 1981).

25. Chang, A.C. and A. L. Page, "Soil Deposition of Trace Metals During Groundwater Recharge Using Surface Spreading," in *Artificial Recharge of Groundwater*, T. Asano, Ed. pp. 609-626. Butterworths, (Boston, 1985).

26. Steinnes, E., "Heavy Metal Pollution of Natural Surface Soils from Long-range Atmospheric Transport," in *Transactions XIII, Congress of the International Society of Soil Science, Hamburg, August 1986, Vol. II*, pp. 504-505. (1986).

27. Allen, R.O. and E. Steinnes, "Contribution from Long-range Atmospheric Transport to the Heavy Metal Pollution of Surface Soil," in *Proc. Internat. Conf. Heavy Metals in the Environment, London, September 1979*, pp. 271-274. CEP Consultants, (Edinburgh, 1979).

28. Groundwater Strategies, *Environ. Sci. Technol.,* 14:1030-1035 (1980).

29. De Kreuk, J.F., "Microbiological Decontamination of Excavated Soil," in *Contaminated Soil*, J.W. Assink and W.J. Van Den Brink, Eds. pp. 669-678. Martinus Nijhoff, (Dordrecht, The Netherlands 1986).

30. Gieseler, G., "Contaminated Land in the EC.," in *Contaminated Soil'88*, K. Wolf, W.J. Van Den Brink and F.J. Colon, Eds. pp. 1555-1562. Kluwer Academic Publishers, (Dordrecht, The Netherlands 1988).

31. National Proirities List Fact Book, HW 7.3, U.S. Environmental Protection Agency p. 94, (Washington, D.C. 1986).

32. Wilmoth, R.C., S.J. Hubbard, J.O. Burckle and J.F. Martin, "Production and Processing of Metals: Their Disposal and Future Risks," in *Metals and Their Compounds in the Environment Chapter L.2,* E. Merian, Ed. pp. 19-65. VCH Verlagsgsellschaft (Weinheim, 1991).

33. Balke, K.D., H. Kussmaul and G. Siebert, "Chemische und thermische Kontamination des Grundwassers durch Industriewasser," *Z. Dtsch. Geol. Ges.* 127:447-460 (1973).

34. Lux, W., B. Hintze and H. Piening, "Heavy Metals in the Soils of Hamburg," in *Transactions XIII Congress of the International Society of Soil Science, Hamburg, August 1986 Vol. II* pp. 376-377 (1986).

35. Förstner, U., "Umweltschutztechnik - Eine Einführung," p. 487. Springer-Verlag, (Berlin, 1990).

36. Kloke, A., D.R. Sauerbeck and H. Vetter, "The Contamination of Plant and Soils with Heavy Metals and the Transport of Metals in Terrestrial Food Chains," in *Changing Metal Cycles and Human Health*, J.O. Nriagu, Ed. pp. 113-141. Dahlem Konferenzen, Springer-Verlag, (Berlin, 1984).

37. Sauerbeck, D., "Funktionen, Güte und Belastbarkeit des Bodens aus agrikulturchemischer Sicht," *Kohlhammer Verlag* p. 259 (Stuttgart 1985).

38. Chaney, R.L., "Plant Accumulation of Heavy Metals and Phytooxicity Resulting from Utilization of Sewage Sludge and Sludge Compost on Cropland," in *Proc. 1977 Nat. Conf. Composting of Municipal Residues and Sludges*, pp. 86-97. Inform. Transfer Inc., (Rockville, MD 1978).

39. Lisk, D.J., "Trace Metals in Soils, Plants and Animals," *Adv. Agron.* 24:267-325 (1972).

40. Sauerbeck, D., "Auswirkung des sauren Regens auf landwirtschaftlich genutzte Böden," *Landbauforsch. Völkenrode* 33:201-207 (1983).

41. Mineral Tolerance of Domestic Animals, Subcommittee of Mineral Toxicity in Animals, National Academy of Science, (Washington, D.C. 1980).

42. Underwood, E.J., "Trace Elements in Human and Animal Nutrition," p. 288 Academic Press, (New York, 1977).

43. Sillanpää, M., "Micronutrients and the Nutrient Status of Soils: A Global Study," *FAO Soils Bull. 48* p. 442 (Rome, 1982).

44. Haque, M.A. and V. Subramanian, "Copper, Lead and Zinc Pollution of Soil Environment," *CRC Crit. Reviews Environ. Control March 1982*, pp. 13-67. (1982).

45. Stevenson, F. and M.S. Ardakani, "Organic Matter Reactions Involving Micronutrients in Soils," in *Micronutrients in Agriculture*, pp. 36-58. Soil Sci. Soc. Amer. Inc. (Madison, WI 1972).

46. Verloo, M., "Komplexvorming van Sporenelementen met Organische Bodemkomponenten," Doctoral Dissertation State University Ghent, (Belgium, 1974).

47. Cottenie, A., Ed. *Trace Elements in Agriculture and in the Environment*, Laboratory of Analytical and Agrochemistry, State University Ghent, (Belgium, 1981).

48. Saeed, M. "Phosphate Fertilization Reduces Zinc Adsorption by Calcareous Soil," *Plant and Soil* 48:641-647 (1977).

49. Miller, J.E., J.J. Hassett and D.E. Koeppe, "Interactions of Lead Cadmium on Metal Uptake and Growth of Corn Plants," *J. Environ. Qual.* 6:18-26 (1977).

50. Förstner, U., *Contaminated Sediments*, Lecture Notes in Earth Sciences 21, p. 157 Springer-Verlag, (Berlin, 1989).

51. Van Driel, W. and J.P.L. Nijssen, "Development of Dredged Material Disposal Site: Implications for Soil, Flora and Food Quality," in *Chemistry and Biology of Solid Waste - Dredged Material and Mine Tailings*, W. Salomons and U. Förstner, Eds. pp. 101-126. Springer-Verlag, (Berlin, 1988).

52. Herms, U. and L. Tent, "Schwermetallgehalte im Hafenschlick sowie in landwirtschaftlich genutzten Hafenschlick-Spülfeldern in Raum Hamburg," *Geol. Jb. F* 12:3-11 (1982).

53. Maaβ, B., G. Miehlich and A. Gröngröft, "Untersuchungen zur Grundwassergefährdung durch Hafenschlick-Spülfelder, II, Inhaltsstoffe in Spülfeldsedimenten und Porenwässern," *Mitt. Dtsch. Bodenkundl. Ges.* 43/I:253-258 (1985).

54. Folsom, B.L., C.R. Lees and D.J. Bates, *Influence of Disposal Environment on Availability and Plant Uptake of Heavy Metals in Dredged Materials*, Technical Report EL-81-12, U.S. Army Corps of Engineers Waterways Experiment Station p. 186 (Vicksburg, MS 1981).

55. Herms, U. and G. Brümmer, "Einfluß der Bodenreaktion auf Löslichkeit und tolerierbare Gesamtgehalte an Nickel, Kupfer, Zink, Cadmium und Blei in Böden und kompostierbaren Siedlungsabfällen," *Landwirtschaftl. Forschung* 33:408-423 (1980).

56. Hara, T. and J. Sonoda, "Comparison of the Toxicity of Heavy Metals to Cabbage Growth," *Plant and Soil* 50:127-133 (1979).

57. Proposed Classification Criteria - Solid Waste Disposal Facilities, U.S. Environmental Protection Agency, Federal Register 43, 4942 (1978).

58. Ewers, U. "Standards, Guidelines, and Legislative Regulations Concerning Metals and Their Compounds," in *Metals and Their Compounds in the Environment Chapter I.20b*, pp. 687-711. VCH Verlagsgesellschaft, (Weinheim, 1991).

59. Leidraad Bodemsanering (Dutch Guidelines for Soil Restoration), Aflevering 4, Noverber 1988, Staatsuitgeverij's, Gravehage, (The Netherlands, 1988).

60. Swiss Ordinance for Pollutants in Soil. Federal Office for Environmental Protection, (Berne, Switzerland 1986).

61. Asami, T., "Soil Pollution by Metals in Japan," in *Transition XIII, Congress of the International Society of Soil Science Hamburg, August 1986, Vol. II*, pp. 222-223. (1986).

62. Sauerbeck, D., "Möglichkeiten zum Schutz der Pflanzenproduktion auf belasteten Böden," in *Belastungen der Land- und Forstwirtschaft durch Äußere Einflüsse, Schriftenreihe Agrarspektrum 11*, W. Henrichsmeyer et al., Eds. pp. 205-229. DLG-Verlag, (Frankfurt, 1986).

63. Förstner, U., W. Ahlf, W. Calmano and M. Kersten, "Mobility of Pollutants in Dredged Materials - Implications for Selecting Disposal Options," in *Role of the Ocean as a Waste Disposal Option*, G. Kullenberg, Ed. pp. 597-615. D. Reidel Publishers, (Dordrecht, 1986).

64. Craig, P.J. and P.A. Moreton, "The Role of Sulphide in the Formation of Dimethyl Mercury in River and Estuary Sediments," *Mar. Pollut. Bull.* 15:406-408 (1984).

65. Kersten, M., "Geochemistry of Priority Pollutants in Anoxic Sludges: Cadmium, Arsenic, Methyl Mercury, and Chlorinated Organics," in *Chemistry and Biology of Solid Waste - Dredged Material and Mine Tailings*, W. Salomons and U. Förstner, Eds. pp. 170-213. Springer-Verlag, (Berlin, 1988).

66. Calmano, W., "Stabilization of Dredged Mud," in *Environmental Management of Solid Waste - Dredged Material and Mine Tailings*, W. Salomons and U. Förstner, Eds. pp. 80-98. Springer-Verlag, (Berlin, 1988).

67. Gambrell, R.P., C.N. Reddy and R.A. Khalid, "Characterization of Trace and Toxic Materials in Sediments of a Lake Being Restored," *J. Water Pollut. Control Fed.* 55:1201-1213 (1983).

68. Salomons, W. and U. Förstner, Eds. "Chemistry and Biology of Solid Waste - Dredged Materials and Mine Tailings," p. 305. Springer-Verlag, (Berlin, 1988).

69. Salomons, W. and U. Förstner, Eds. "Enviromental Management of Solid Waste: Dredged Materials and Mine Tailings," p. 396. Springer-Verlag, (Berlin, 1988).

70. Förstner, U., M. Kersten and Wienberg, "Geochemical Processes in Landfills," in *The Landfill - Reactor and Final Storage*, Lecture Notes in Earth Sciences 20, P. Baccini, Ed. pp. 39-81. Springer-Verlag, (Berlin, 1989).

71. Peiffer, S., *Biogeochemische Regulation der Spurenmetallöslichkeit während der anaeroben Zersetzung fester kommunaler Abfälle,* Dissertation: Universität Bayreuth, p. 197 (1989).

72. Krebs, J., H. Belevi and P. Baccini, "Long-term Behavior of Bottom Ash Landfills," Proc. 5th ISWA Internat. Solid Wastes Exhibition and Conference, September 11-16, 1988 pp. 371-376. (Copenhagen, 1988).

73. Arneth, J.-D., G. Milde, H. Kerndorff and R. Schleyer, "Waste Deposit Influences on Ground Water Quality as a Tool for Waste Type and Site Selection for Final Storage Quality," in *The Landfill - Reactor and Final Storage*, Lecture Notes in Earth Sciences 20, P. Baccini, Ed. pp. 399-415. Springer-Verlag, (Berlin, 1989).

74. Blakey, N.C., "Behavior of Arsenical Wastes Co-disposed with Domestic Solid Wastes," *J. Water Pollut. Control Fed.* 56:69-75 (1984).

75. McLellon, W.M., D.H. Vickers, J.F. Charba and G.I. Bengstrom, "Environmental Impact Assessment of a Sanitary Landfill in a Higher Water Table Area," *Proc. 29th Industrial Waste Conf., Purdue Univ. Ext. Ser.* 145:94-102 (1974).

76. Atwell, J.S., "Identifying and Correcting Groundwater Contamination at a Land Disposal Site," *Proc. 4th Natl. Congr. Waste Management* Res. Rec. U.S. Environmental Protection Agency SW-Sp. pp. 278-298, (Cincinnati, Ohio 1976).

77. Kelly, W.E., "Ground-water Pollution Near a Landfill," *J. Environ. Eng, Div. Amer. Soc. Civ. Eng.* 102:1189-1193 (1976).

78. Shuster, K.A. "Leachate Damage Assessment: Case Studies," U.S. Environmental Protection Agency, EPA/530 - SW-509. SW-514, SW-517, (Cincinnati, OH 1976).

79. Fuller, W.H. and N. Korte, "Attenuation Mechanisms Through Soil," in *Gas and Leachate from Landfills*, U.S. Environmental Protection Agency Report EPA-600/9-76-004, E.J. Genetelli and J. Cirello, Eds. pp. 125-156. (Cincinnati, OH 1977).

80. Griffin, R.A., R.R. Frost, A.K. Au, G.D. Robinson, G.D. Shimp and N.F. Shimp, "Attenuation of Pollutants in Municipal Landfill Leachate by Clay Minerals. II, Heavy Metal Adsorption" Illinois State Geological Survey, *Environ. Geol. Notes 79,* p. 47 (1977).

81. Farquhar, G.J. and T.W. Constable, *Leachate Contaminant Attenuation in Soil*, Waterloo Research Institute, Project No. 2123, p. 152 University of Waterloo, (Canada, 1978).

82. Farquhar, G.J. and W. Parker, "Interactions of Leachates with Natural and Synthetic Envelopes," in *The Landfill - Reactor and Final Storage*, Lecture Notes in Earth Sciences 20, P. Baccini, Ed. pp. 175-200. Springer-Verlag, (Berlin, 1989).

83. Regan, R.W. and C.E. Draper, "Segregated Landfilling of Metal Finishing Sludge: Concept Evaluation Studies," *Proc. Internat. Conf. Heavy Metals in the Environment, New Orleans*, pp. 245-247 CEP Consultants, (Edinburgh, 1987).

84. Baccini, P., Ed. p. 438 "The Landfill - Reactor and Final Storage," Lecture Notes in Earth Sciences 20, Springer-Verlag, (Berlin, 1989).

85. Baccini, P., "The Control of Heavy Metal Fluxes from the Anthroposphere to the Environment," in *Proc. Internat. Conf. Heavy Metals in the Environment Vol. 1* J.P. Vernet Ed. pp. 13-23. CEP Consultants, (Edinburgh, 1989).

86. Assink, J.W. and W.J. Van Den Brink, Eds. *Contaminated Soil Proc. First Intern. TNO Conf. On Contaminated Soil, November 11-15, 1985*, p. 923. Martinus Nijhoff, Publisher (Dordrecht, 1986).

87. Wolf, K., J. Van Den Brink and F.J. Colon, Eds. *Contaminated Soil '88*, Kluwer Academic Publishers, p. 1661. (Dordrecht, The Netherlands, 1988).

88. Arendt, F., M. Hinsenveld and W.J. Van Den Brink, Eds. *Contaminated Soil '90,* Kluwer Academic Publishers, p. 1602. (Dordrecht, 1990).

89. Smith, M.A., Ed. *Contaminated Land - Reclamation and Treatment*, Plenum Press, p. 385. (New York, 1985).

90. Rulkens, W.H., J.W. Assink and W.J.Th. Van Gemert, "On-site Processing of Contaminated Soil," in *Contaminated Land - Reclamation and Treatment*, M.A. Smith, Ed. pp. 37-90. Plenum Press, (New York, 1985).

91. Jessberger, H.L., "Techniques for Remedial Action at Waste Disposal Sites, in *Contaminated Soil*, J.W. Assink and W.J. Van Den Brink, Eds. pp. 587-599. Martinus Nijhoff Publishers, (Dordrecht, 1986).

92. Van Liun, A.B., "The Treatment of Polluted Groundwater from the Clean-up of Contaminated Soil," in *Contaminated Soil '88*, K. Wolf, W.J. Van Den Brink and F.J. Colon, Eds. pp. 1167-1174. Kluwer Academic Publishers, (Dordrecht, The Netherlands 1988).

93. Ball, R.O., G.P. Verret, P.L. Buckingham and S. Mahfood, "Economic Feasibility of a State-wide Hydrometallurgical Recovery Facility," in *Metal Speciation: Separation and Recovery* J.W. Patterson and R. Passino Eds. pp. 690-709. Lewis Publishers, (Chelsea, MI 1987).

2

KINETICS OF METAL SORPTION REACTIONS

Donald L. Sparks
Department of Plant and Soil Sciences
University of Delaware
Newark, DE 19717

1. INTRODUCTION

Macroscopic measurements of metal sorption reactions on soils are ubiquitous in the soil and environmental sciences literature. Metal sorption has been described using an array of empirical, semi-empirical and surface complexation models including Freundlich, Langmuir, Temkin, constant capacitance and triple layer. In some cases, these models equally well describe sorption data and are often useful for describing metal sorption over a range of pH and ionic strength values. However, many of these models have numerous adjustable parameters, and, thus, it is not surprising that sorption data can be well described using them. In effect, most sorption models that employ macroscopic data are often curve-fitting exercises. Despite this, many investigators have used them to make mechanistic interpretations about metal sorption on surfaces. However, some of these models can describe several different sorption mechanisms. Thus, conformity of material balance data to a particular model does not prove that a particular mechanism is operational.

I do not wish to imply that equilibrium-based modeling of metal sorption on clay minerals, oxides, sediments, and soils is useless. To the contrary, much useful information has been obtained from macroscopic studies. However, one cannot glean mechanistic information from such approaches. Moreover, equilibrium studies are often not appropriate to simulate field conditions since soils are seldom,

0-87371-697-3/95/$0.00+$.50

if ever, at equilibrium with respect to ion and molecular transformations and interactions.

To properly understand the fate of metals in soils and particularly to comprehend their mobility with time, kinetic investigations are necessary. Such time-dependent data can also be used to derive mechanisms for metal sorption. Of course, to definitively ascertain sorption mechanisms, one should employ surface spectroscopic or microscopic techniques.

Only within the last 30 years, and particularly in the past decade, have kinetic investigations of metal sorption on soils appeared in the literature. Soils are indeed complex, heterogeneous systems, and the application of kinetics to such solid surfaces is arduous and fraught with pitfalls. While many advances have been made in describing the kinetics of metal reactions on soils [1], I believe we are only in the infancy of this important area of soil and environmental chemistry. In this paper, I wish to describe some of the recent advances in the area of kinetics of metal sorption on soils and soil components. Unfortunately, space does not allow for a comprehensive review of the topic, but the reader can consult a number of other publications if he or she is interested in more comprehensive treatments [1-4]. The objectives of this review are to discuss several aspects of metal sorption kinetics. These include soil components that are important in metal retention; time scales for reactions on soils; a critique of kinetic models and data interpretations; methods that can be used to study metal sorption kinetics; and applications of pressure-jump relaxation to glean mechanistic information about metal sorption kinetics on soil constituents.

2. SOIL MINERALS IMPORTANT IN METAL SORPTION DYNAMICS

Secondary clay minerals, hydrous oxides, carbonates, and humic substances are primary sorbents in soils for metals, and their physical, chemical, and mineralogical properties largely dictate the kinetics of these reactions. Some of the relevant properties of these soil constituents are given in Table 1. The clay minerals are assemblages of silica tetrahedral sheets ($Si_2O_5^{2-}$ repeating unit) bound to octahedral sheets that are composed of ions such as Al, Mg, and Fe that are coordinated to O^{2-} and OH^-. Clay minerals can be classified as either 1:1 (e.g., kaolinite) or 2:1 (e.g., smectite) depending on the arrangement of the tetrahedral and octahedral sheets.

Other important soil minerals include oxides and hydroxides such as goethite, an Fe-oxide, gibbsite, an Al-oxide, and birnessite, a Mn-oxide. These materials arise from primary silicate weathering or by hydrolysis and desilication of secondary clay minerals such as kaolinite.

Carbonates and sulfates are also significant sorbents in soils, particularly those in arid regions. Humic substances including humic acid, fulvic acid and humin are also primary sorbents of metals and even in small quantities can drastically affect metal retention in soils.

These inorganic and organic soil constituents can exhibit both permanent and variable charge. Permanent charge arises primarily from ionic substitution when an ion substitutes for another of similar size in a crystal structure. For example, if Al^{3+} substitutes for Si^{4+} in the tetrahedral layer of a phyllosilicate or if Mg^{2+} substitutes for Al^{3+} in the octahedral layer, a net negative charge exists on the mineral. Permanent charge is invariant with pH and is created during the crystallization of aluminosilicates. Smectite and vermiculite are permanent charge minerals.

With variable charged surfaces, such as kaolinite, goethite, gibbsite, and humic substances, the net charge changes with pH and is positive at lower pH and negative at higher pH. The principal source of variable charge on soils is the protonation and deprotonation of functional groups on the colloid surfaces such as hydroxyl (-OH), carboxyl (-COOH), phenolic ($-C_6H_4OH$), and amine ($-NH_2$). In most soils, there is a combination of both permanent and variable charge surfaces.

2.1 Time scales for metal sorption reactions

The type of soil component can drastically affect the rate of metal sorption [3]. Sorption reactions can involve physical sorption, outer-sphere complexation (electrostatic attraction), inner-sphere complexation (ligand exchange), and surface precipitation and can occur on time scales of microseconds to months. Metal sorption reactions are often rapid on clay minerals such as kaolinite and smectite and much slower on mica and vermiculite, particularly if metals such as NH_4^+, Cs^+, and K^+ with small hydrated radii are studied. An example of this is shown for K-Ca exchange kinetics (Figure 1). Here, the kinetics of sorption are greatly affected by the structural properties of the clay minerals. With kaolinite, only easily available planar external sites are available for exchange.

Table 1. Important Characteristics of Secondary Clay Minerals.[a]

Mineral	Type	Chemical Formula	Layer Charge	Cation Exchange Capacity	Surface Area	Permanent Charge	Variable Charge
				cmol kg^{-1}	x10^{3}m^{2}kg^{-1}		
Kaolinite	1:1	$[Si_4]Al_4O_{10}(OH)_8\ nH_2O$(n=0 or 4)	<0.01	1-2	10-20	No	Yes
Montmorillonite	2:1	$M_x[Si_8]Al_{3.2}Fe_{0.2}Mg_{0.6}O_{20}(OH)_4$	0.5-1.2	80-120	600-800	Yes	No
Vermiculite	2:1	$M_x[Si_7Al]Al_3Fe_{0.5}Mg_{0.5}O_{20}(OH)_4$	1.2-1.8	120-150	600-800	Yes	No
Mica	2:1	$K_2Al_2O_5[Si_2O_5]_3Al_4(OH)_4$	1.0	20-40	70-120	Yes	No
Chlorite	2:1 [b]	$(Al(OH)_{2.55})_4[Si_{6.8}Al_{1.2}]$ $Al_{3.4}Mg_{0.6}O_{20}(OH)_4$	[c]	20-40	70-150	Yes	Yes
Allophane	-----	$Si_3Al_4O_{12}\bullet nH_2O$	-----	10-150	70-300	No	Yes

[a] Bohn et al. [17]

[b] with hydroxide interlayer

[c] variable

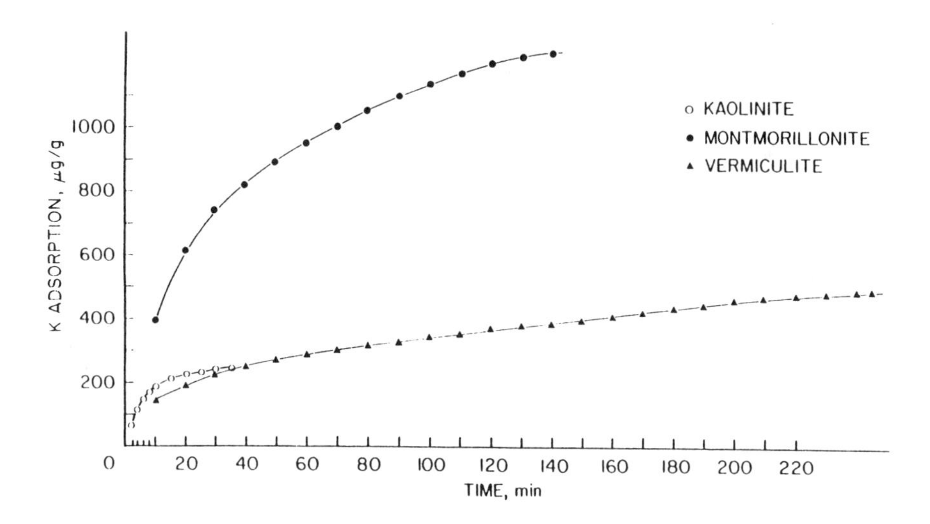

FIGURE 1. Rate of Potassium Adsorption on Three Clay Minerals.

With montmorillonite, the interlayer is expansible and enables rapid sorption. In vermiculitic and micaceous minerals, the interlayer space is restricted, and tortuous, slow, mass transfer reactions predominate. Thus, the kinetics are slow. Additionally, metal sorption on soil minerals like vermiculite involves two to three different reaction rates, high rates on external sites, intermediate rates on edge sites, and low rates on interlayer sites [3,5].

Metal sorption kinetics on oxides, hydroxides, and humic substances depend on the type of surface and metal being studied, but generally are rapid. For example, reactions of molybdate, sulfate, selenate, and selenite on goethite occurred on millisecond time scales [6-8]. Half-times for bivalent Pb, Cu, Cd, Zn sorption on peat ranged from 5 to 15 seconds [9].

3. RATE LAWS AND DATA INTERPRETATION

While one may wish to determine the chemical kinetics of metal sorption on soils, viz., reaction rates and the molecular processes by which these reactions occur where transport is not limiting, due to the heterogeneity of soils, one is usually determining a combination of chemical kinetics and kinetics events. Kinetics is a generic term that refers to time-dependent phenomena, e.g., film, particle, and surface diffusion, and chemical kinetics [2].

Consequently, the determination of mechanistic rate laws for soils, which assume that only chemical kinetics are operational and that transport-controlled kinetics that entail physical aspects of the soil are absent, is difficult. The heterogeneity of soils including varying particle sizes, sorption sites, and porosities promotes transport processes.

Therefore, in almost all instances, when one studies the kinetics of metal sorption reactions on soils, one is measuring apparent rate laws since mass transfer and transport phenomena predominate. A number of transport processes can occur depending on the type of solid. These include: transport in the soil solution, transport across a liquid film at the solid/liquid interface, transport in liquid filled macropores, diffusion of sorbate occluded in micropores, and diffusion processes in the bulk of the solid (Figure 2). Any one or more of these transport phenomena and/or chemical reaction could be rate-limiting. In most metal sorption reactions, transport phenomena are rate-limiting [10].

Data for metal sorption kinetics on soils have been described by an eclectic group of models including ordered expressions, e.g., zero-, first-, second-, fractional-, and Elovich, power function, and parabolic diffusion expressions. Several studies have shown that some or all of these models can equally well describe kinetic data. Also, the fit of experimental data to these models has unfortunately often been used as a basis for elucidating mechanisms for metal sorption.

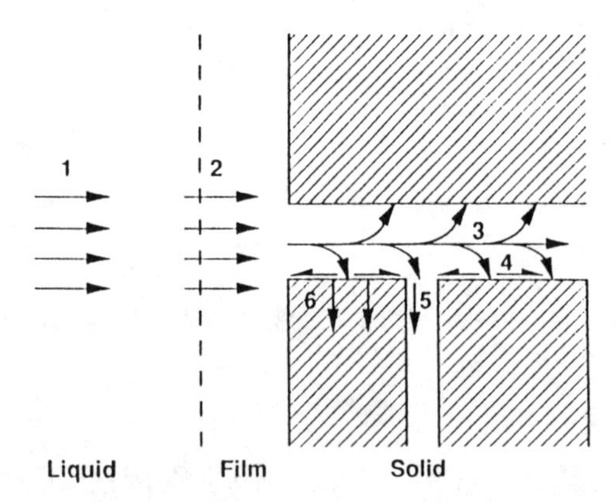

*FIGURE 2. **Transport Processes in Solid-liquid Soil Reactions. Non-activated processes: 1) Transport in the soil solution; 2) Transport across a liquid film at the solid liquid interface; 3) Transport in a liquid-filled micropore. Activated processes: 4) Diffusion of a sorbate at the surface of the solid; 5) Diffusion of a sorbate occluded in a micropore; 6) Diffusion in the bulk of the solid. Aharoni and Sparks [10].***

In fact, it is usually difficult to apply simple kinetic equations to describe soil chemical reactions since the reacting surfaces are heterogeneous and one cannot differentiate transport effects from chemical kinetics. Aharoni and Sparks [10] have shown that a generalized empirical equation can be derived by examining the applicability of power function, Elovich, and apparent first-order equations to experimental data. These equations are:

Power - function

$$q = kt^{\nu} \tag{1}$$

where q is the quantity adsorbed at time t, and k and ν are constants

Elovich

$$q = A + (1/b)\ln(t + t_{\circ}) \tag{2}$$

where A, b, and t_o are constants

Apparent first - order

$$q/q_{\infty} = 1 - \alpha \ \exp(-\beta t) \tag{3}$$

where q_{∞} is the maximum quantity adsorbed and α and β are constants.

Differentiating these equations and writing them as explicit functions of the reciprocal of the rate (Z) = $(dq/dt)^{-1}$, one can show that the plot of Z vs. t for an experimental isotherm should be convex if equation (1) is operational, linear if Equation (2) is applicable, and concave if Equation (3) is appropriate (Figure 3). However, Aharoni and Sparks [10] have shown that Z vs. t plots for soil chemical reactions are usually S-shaped: convex at small t, concave at large t, and linear at some intermediate t range. This finding suggests that the kinetics follow some abstruse function that can be approximated by the power equation at small t, by the Elovich equation at an intermediate t, and by an apparent first-order equation at large t. One often assumes S behavior even where one of these three empirical equations seems to apply over the entire reaction period. The fact that equations like power function, Elovich, and apparent first-order are approximations to which the general Z equation reduces at certain limited ranges of adsorption indicates why no consistent theoretical derivation can be found for these equations. It also helps explain why there is no correlation between the applicability of any of these equations and the nature of the retention processes, viz., dissimilar processes are fitted by the same equation, and similar processes are fitted by different equations. One must then critically question the meaning of conformity of rate data to various empirical and different ordered equations.

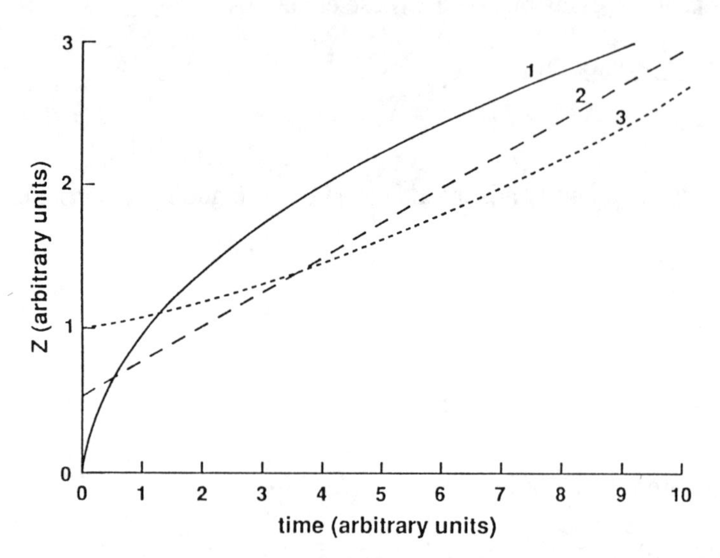

FIGURE 3. ***Plots of Z Against t Implied by the Simple Empirical Equations. 1) Plot according to $Z = (dq/dt)^{-1} = (1/\upsilon\ k)t^{1-\upsilon}$ implied by Equation 1; 2) Plot according to $Z = (dq/dt)^{-1} = b(t+t_o)$ implied by Equation 2; and 3) Plot according to $Z = (dq/dt)^{-1} = (1/q_\infty\beta\alpha)\exp(\alpha t)$ implied by Equation 3. Aharoni and Sparks [10].***

The meaning of S-shaped Z(t) plots such as that in Figure 4 can be explained by homogeneous and heterogeneous diffusion models. In homogeneous diffusion, equations for diffusion yield S-shaped Z(t) plots in which the final and initial curved parts predominate whereas equations for heterogeneous diffusion give S-shaped Z(t) plots in which the intermediate linear part is dominant.

Diffusion in a homogeneous medium can be expressed as:

$$q/q_\infty = 1 - \sum_{n=0}^{n=\infty} a_n \exp(-b_n\, t/\tau) \qquad (4)$$

where q is the amount of metal sorbed at some time, t, q_∞ is the maximum amount of metal sorbed, and a_n and b_n are parameters with an infinite number of discrete values determined by the integers n. The relations $a_n(n)$ and $b_n(n)$ depend on the geometry of the system.

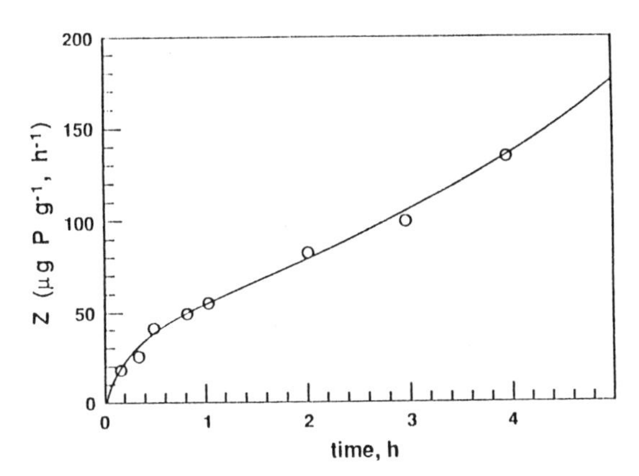

FIGURE 4. ***Sorption of Phosphate by a Soil Plotted as Z Against t. Aharoni and Sparks [10].***

In Equation (4), τ can be determined by integrating Fick's equation, and can be defined as:

$$\tau = r^2 / D \qquad (5)$$

where r is the maximum length of the diffusion path and D is the diffusion coefficient.

Equation (4), with any set of a_n and b_n, can be rewritten in a form that reduces at small t to

$$q / q_\infty = k_s (t / \tau)^{1/2} \qquad (6)$$

and that reduces at large t to

$$q / q_\infty = 1 - k_f \exp(-\alpha_f \, t / \tau) \qquad (7)$$

where k_s in Equation (6) and k_f and α_f in equation (7) are parameters determined by a_n and b_n.

Equation (6) is similar in form to Equation (1) and leads to a plot of Z vs. t that is convex. Equation (7) is similar to Equation (3) and leads to a plot of Z vs. t that is concave. This implies that the plot of Z vs. t corresponding to Equation (4) is S-shaped and must be convex at small t, concave at large t, and have an inflection point [10]. At an intermediate region, the plot approaches a straight line and can be represented by Equation (2), which is conveniently formulated in terms of t_p, the time at which a plot of $Zr([d(q/q_\infty)/d(t/\tau)]^{-1})$ vs. t/τ has an inflection point.

$$q/q_\infty = A_r + (q/b_r)\text{In}(t/t_p + t_r) \tag{8}$$

Equation (8) has only one adjustable parameter t_p, which depends on t; $A_r + b_r$ and t_r are fixed for each geometry. Plots of Zr vs. t/τ are shown for Equations (4), (6) and (7) in Figure 5.

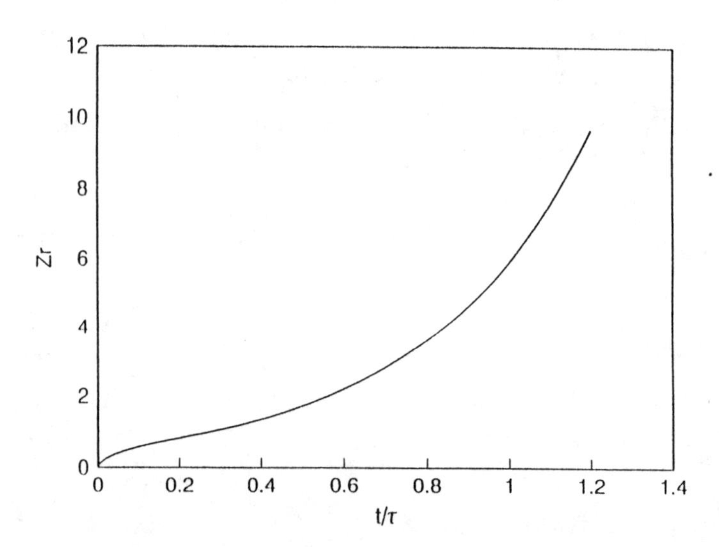

FIGURE 5. Plots of Zr Against t/τ According to Equation 4. Aharoni and Sparks [10].

In homogeneous diffusion it is assumed that a soil constituent is mainly responsible for sorption of the solute and that the rate-controlling process is diffusion of the sorbed solute in that soil constituent. It is further assumed that particle size and diffusion coefficients do not vary considerably and that the flux of the sorbate is parallel.

Heterogeneous diffusion considers a more frequently encountered case, one in which a metal diffuses in solid media such as soils that have different properties. The overall process can be viewed as an array of diffusion processes characterized by their t values, which includes the effect of the diffusion coefficient and the length of the diffusion path [Equation (5)]. Considering the overall uptake q at time t as the sum of the component uptakes q_{τ_t} one can write,

$$q = \int_{\tau_i}^{\tau_m} q_{\tau_\infty}\left(q_{\tau_t} / q_{\tau_\infty}\right) d\tau \tag{9}$$

where q_{τ}, is the uptake due to the process characterized by τ at $t \rightarrow \infty$, and τ_i and τ_m are the smallest and largest τ in the system. Equation (9) can be solved by introducing a function that relates $(q_{\tau_t}/q_{t_\infty})$ to the time such as Equation (4) and by introducing a distribution function relating q_{τ_∞} to τ (see further details in [10]).

4. ADVANCES IN KINETIC METHODOLOGIES

Without question, one of the most important aspects of studying metal sorption kinetics on soils is the type of method one utilizes. A plethora of batch and flow techniques have been employed to study metal reactions in soils. None of them is a panacea for kinetic analyses, each having advantages and disadvantages. These methods are extensively reviewed in several publications including Sparks [1] and Amacher [11].

Only a few salient features of these methods will be provided in this review. Batch methods involve agitation of the sorptive and sorbent by stirring or shaking and separation of the two by centrifugation or rapid filtration. Flow techniques include continuous and stirred-flow methods. Batch techniques are often hampered by inadequate mixing, alterations in the surface chemistry of the solid due to over-mixing and resultant increases in reaction rate, failure to remove desorbed metals which can affect further metal release that may lead to secondary precipitation, and reaction rates that are dependent on degree of mixing. Continuous flow methods, while employing small solid-to-solution ratios and effecting removal of desorbed species, often result in non-dispersed systems, pronounced transport phenomena, and dilution errors [1]. In both batch and flow methods, apparent rate laws are determined.

The stirred-flow method developed by Carski and Sparks [12] and refined by Seyfried *et al.* [13] is a good one for studying metal sorption on soils, provided reactions are slower than 0.6 minutes. Attributes of this method include the following: (1) reaction rates are independent of the physical effects of porous media, (2) the same apparatus can be used to measure both equilibrium and kinetic parameters, (3) continuous (or incremental) measurements allow monitoring of reaction progress, (4) experimental conditions such as flow rate and soil mass can be easily manipulated, (5) the technique is suitable over a wide range of textures and colloidal particle sizes, (6) the sorbent is dispersed, and dilution errors can be accounted for, (7) desorbed species are constantly removed, and (8) perfect mixing occurs [13]. Recently, tests involving flow rate, concentration, and stopped-flow experiments have been presented and verified using the stirred-flow method. Hence, one can differentiate between instantaneous equilibrium and kinetic phenomena and kinetic reactions that are concentration dependent and concentration independent [14,15].

As pointed out earlier, many metal sorption reactions occur on millisecond time scales and consequently cannot be measured using batch and flow methods. For such reactions, chemical relaxation methods must be employed. Examples of transient relaxation techniques include pressure-jump, temperature-jump, concentration-jump, and electric field pulse. Detailed descriptions of these are provided in Sparks [1] and Sparks and Zhang [16]. All of these relaxation techniques are based on the principle that the equilibrium of a reaction mixture is rapidly perturbed by some external factor such as pressure, temperature, or electric field strength. Rate information can then be obtained by following the approach to a new equilibrium by measuring the relaxation time via a particular detection system such as conductivity. The perturbation is small, and, therefore, the final equilibrium state is close to the initial equilibrium. Because of this, all rate expressions are reduced to first-order equations regardless of reaction order or molecularity (see [1]). Therefore, the rate equations are linearized, simplifying determination of complex reaction mechanisms. A general linearized rate equation can be derived as:

$$\tau^{-1} = \frac{1}{\tau} = k_1(C_A C_B) + k_{-1} \tag{10}$$

where k_1 and k_{-1} are formed and backward rate constants, respectively and C_A and C_B are the concentrations of reactants A and B at equilibrium. Further experimental and theoretical aspects of chemical relaxation can be found in Sparks [1] and Sparks and Zhang [16].

Recently, Zhang and Sparks [6-8] have successfully employed pressure-jump relaxation to study molybdate, sulfate, selenate and selenite adsorption/desorption on goethite, a frequently occurring oxide in soils. Some salient details of these experiments will be discussed.

Kinetics of molybdate adsorption on goethite yielded a double relaxation. The faster relaxation was completed by 0.04 seconds while the slower relaxation terminated by 0.20 seconds. Relaxations were not obtained for a goethite -$NaNO_3$ suspension, the supernatant solution of a goethite-molybdate suspension and only a goethite suspension. However, the concentration of molybdate in suspension significantly decreased after it was equilibrated with goethite. These results indicated that the relaxations were due to adsorption/desorption of molybdate at the goethite/water interface [6].

Combining information from p-jump relaxation studies and from the overall equilibrium partitioning and the reaction stoichiometry, a two-step reaction was proposed to describe the mechanism of molybdate adsorption/desorption on goethite [6]:

$$XOH_2^+ + MoO_4^{2-} \underset{\text{Step 1}}{\leftrightarrows} XOH_2^+ - MoO_4^{2-} \underset{\text{Step 2}}{\leftrightarrows} XMoO_4^- + H_2O \quad (11)$$

Step 1 involved an outer-sphere complexation reaction, whereas Step 2, which was rate-limiting, involved inner-sphere complexation. Using the kinetic model in Equation (11) and the stoichiometry of the overall reactions, i.e., one molybdate ion replaces one ligand from the protonated surface, relationships between the fast reciprocal relaxation time (τ_2^{-1}), the slow reciprocal relaxation time (τ_1^{-1}) and reactant concentrations were derived [6].

If the mechanism proposed in Equation (11) is consistent with the experimental relaxation data, then plots of the linearized rate equations for the two reactions in Equation (11) should generate straight lines, and the slopes and intercepts will give the forward and backward intrinsic rate constants (k_1^{int}, k_{-1}^{int}, k_2^{int} and k_{-2}^{int}), respectively for the two steps. This was indeed the case as shown in Figures 6 and 7. From these linear relationships, intrinsic rate constants were calculated, and these revealed that Step 1 had the highest forward rate constant, which was about 10 times higher than its backward rate constant. On the other hand, the backward rate constant for Step 2 was much higher than the forward rate constant. Step 2, the ligand exchange process, was rate-limiting.

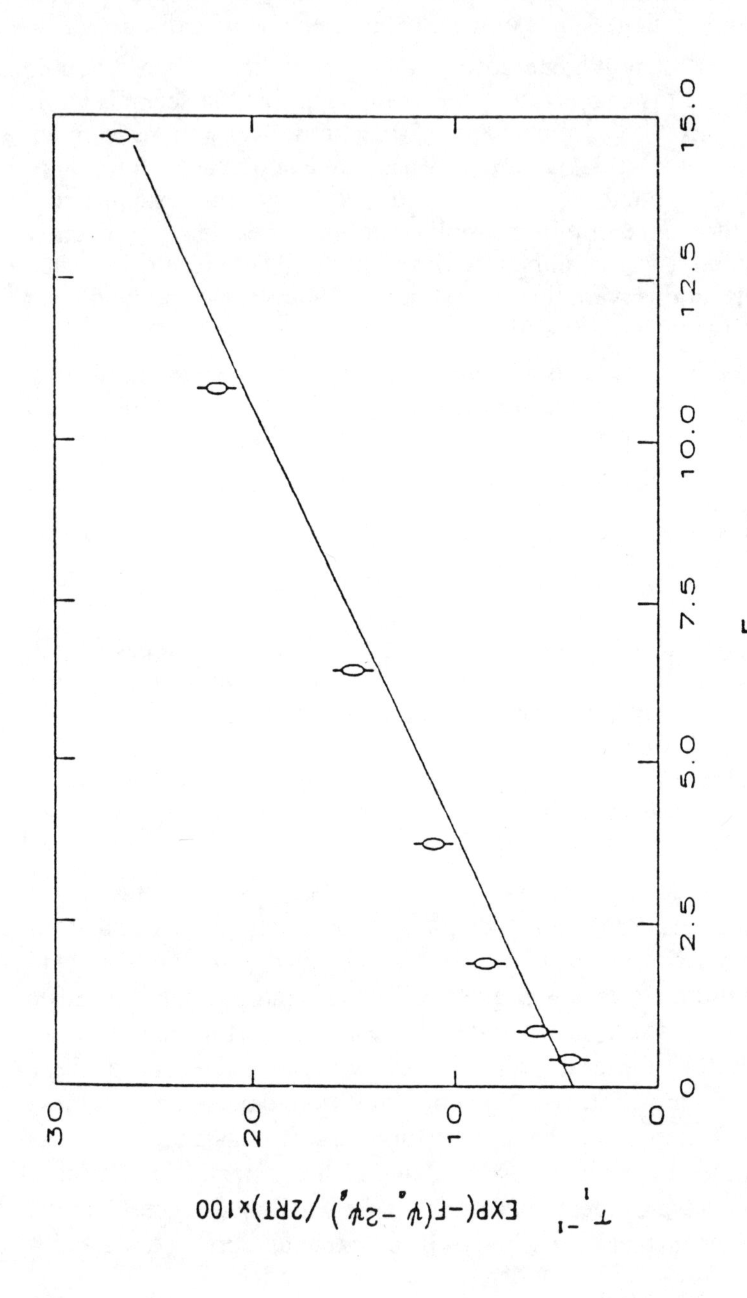

FIGURE 6. Plot of Linearized Rate Equation for Fast Reciprocal Relaxation Time (See Equation 32). Zhang and Sparks [8].

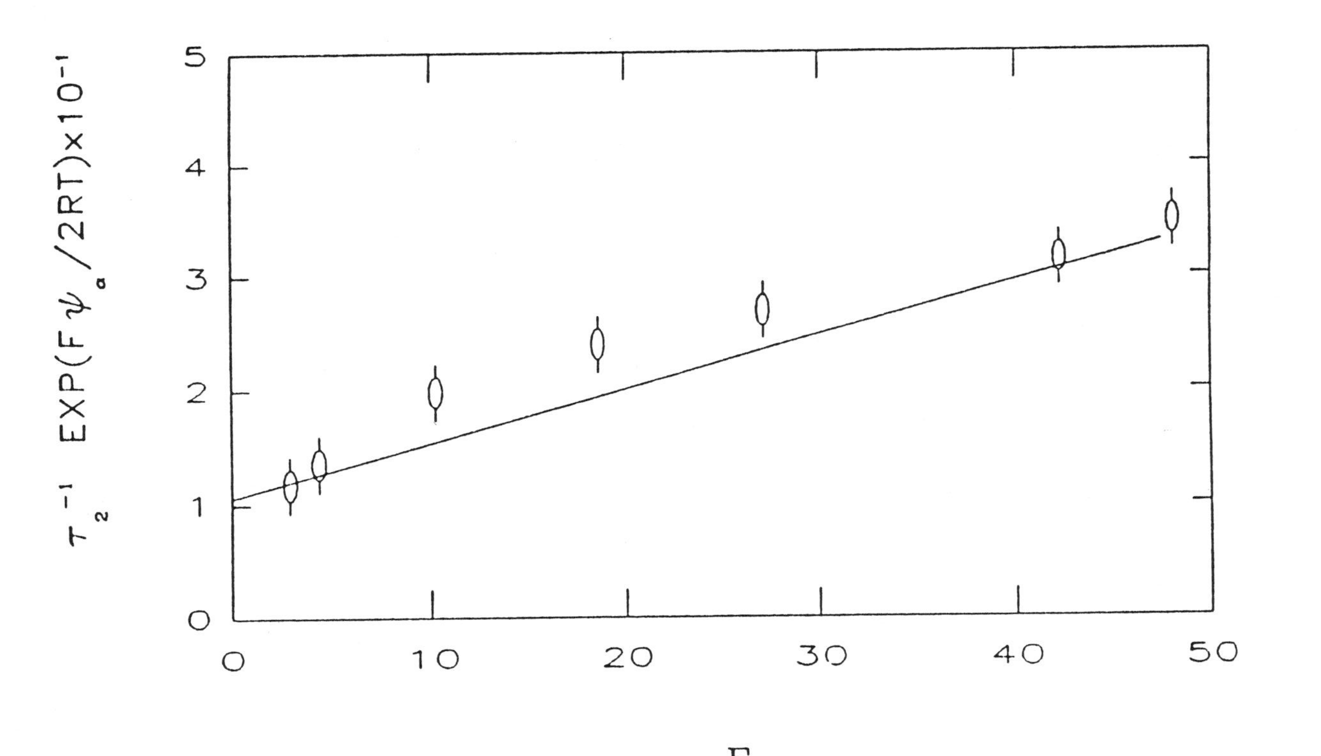

FIGURE 7. Plot of Linearized Rate Equation for Slow Reciprocal Relaxation Time (See Equation 33). Zhang and Sparks [8].

Zhang and Sparks [8] also used pressure-jump relaxation to study mechanisms of selenate and selenite adsorption/desorption on goethite. Selenate adsorption occurred mainly under acidic conditions (Figure 8).

In the pH range that was investigated, the dominant selenate species was SeO_4^{2-} since the pK_2 for selenious acid is 2. As pH increased, SeO_4^{2-} adsorption rapidly decreased. At pH 2.98, the total percent of adsorption was 93. At pH >7.2, no adsorption occurred. Selenate adsorption was described very well with the triple layer model. With this model it was assumed that SeO_4^{2-} occurred via outer-sphere surface complexation [8].

Zhang and Sparks [8] observed a single relaxation for selenate adsorption on goethite. Based on this and the equilibrium modeling, it was assumed that SeO_4^{2-} adsorption involved outer-sphere complexation and could be described as:

$$XOH + H^+ + SeO_4^{2-} \rightleftarrows XOH_2^{2+} - SeO_4^{2-} \quad (12)$$

where XOH represents 1 mol of reactive surface hydroxyl bound to a Fe ion in goethite.

A linearized relationship between reciprocal relaxation time and the concentration of species in suspension for the reaction in Equation (12) was derived as:

$$\tau^{-1} = k_1\left([XOH]\left[SeO_4^{2-}\right] + [XOH]\left[H^+\right] + \left[SeO_4^{2-}\right]\left[H^+\right]\right) + k_{-1} \quad (13)$$

where the terms in the brackets are the concentrations of species at equilibrium. Since the reaction is conducted at the solid/water interface, then the electrostatic effect has to be considered in calculating the intrinsic rate constants. Using the triple layer model to obtain electrostatic parameters, a first-order equation is derived,

$$\tau^{-1} = \exp\left(\frac{-F\left(\psi_\alpha - 2\psi_\beta\right)}{2RT}\right) = k_1^{int}\left[\exp\left(\frac{-F\left(\psi_\alpha - 2\psi_\beta\right)}{RT}\right)\right] \times$$

$$\left([XOH]\left[SeO_4^{2-}\right] + [XOH]\left[H^+\right] + \left[SeO_4^{2-}\right]\left[H^+\right]\right) + k_{-1}^{int} \quad (14)$$

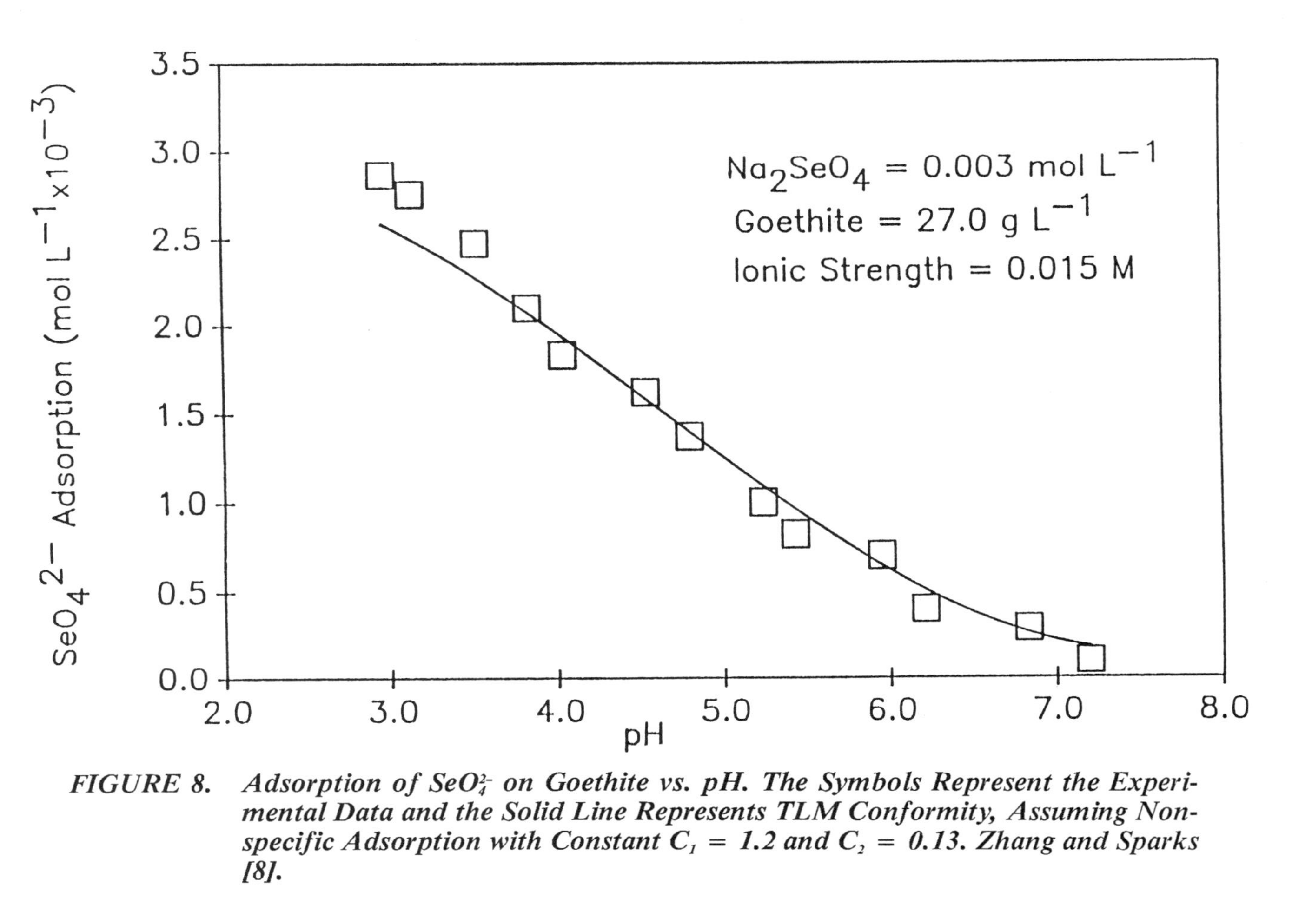

FIGURE 8. Adsorption of SeO_4^{2-} on Goethite vs. pH. The Symbols Represent the Experimental Data and the Solid Line Represents TLM Conformity, Assuming Nonspecific Adsorption with Constant C_1 = 1.2 and C_2 = 0.13. Zhang and Sparks [8].

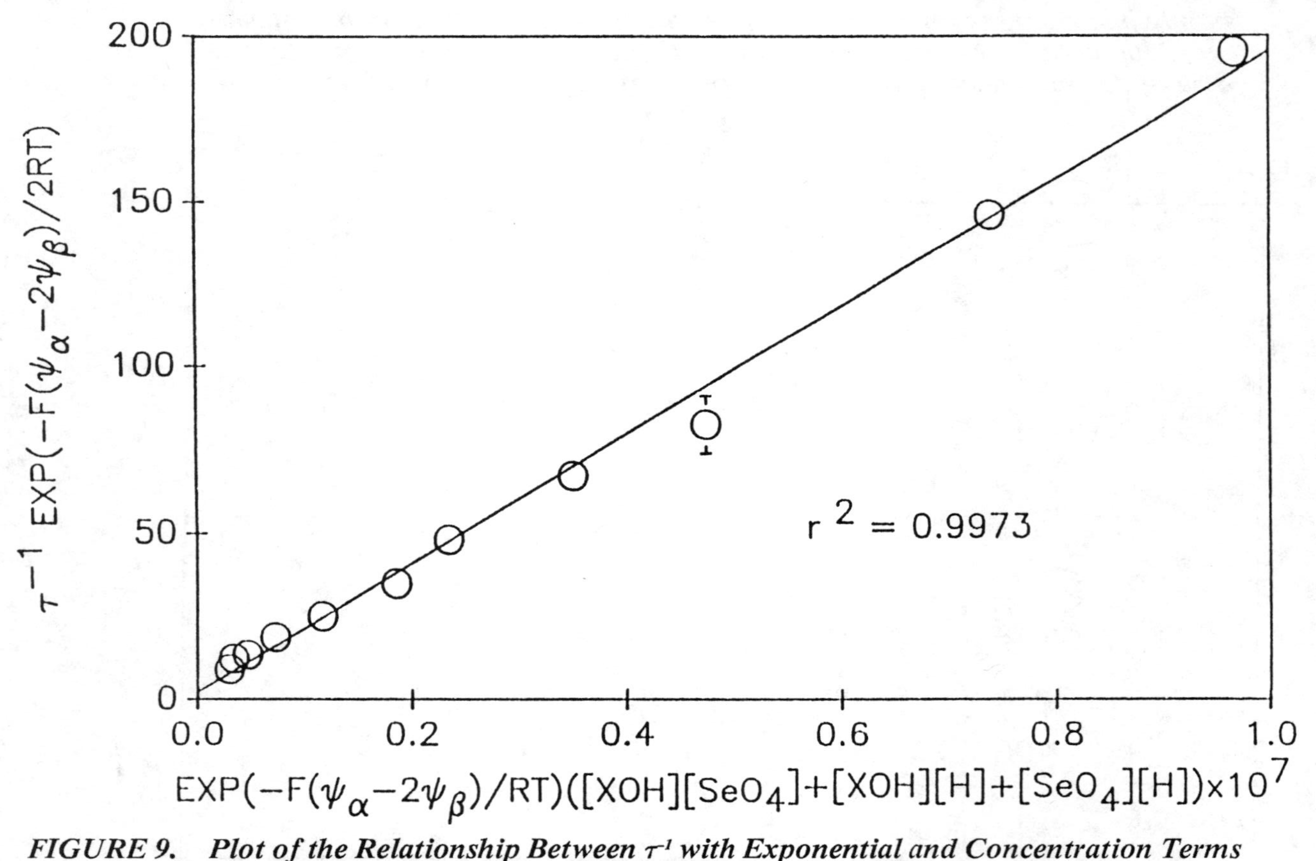

FIGURE 9. ***Plot of the Relationship Between τ^{-1} with Exponential and Concentration Terms as Proposed in Equation 14. Zhang and Sparks [8].***

Table 2. Intrinsic Rate and Equilibrium Constants for SeO_4^{2-} Adsorption and Desorption on Geothite[a].

k_1^{int} ($mol^{-2}L^2s^{-1}$)	3.52×10^8
k_1^{int} (s^{-1})	3.34
log $K_{kinetic}^{int}$	8.02
log K_{model}^{int}	8.64

[a] *From Zhang and Sparks [8].*

A plot of this relationship was linear (Figure 9), indicating the mechanism proposed in Equation (12) was correct. Additionally, calculation of the intrinsic equilibrium constant (K_{kin}^{int}) from the kinetic study where $K_{kin}^{int} = k_1^{int}/k_{-1}^{int}$, and comparing it to the intrinsic equilibrium constant obtained from triple layer modeling (K_{model}^{int}) showed excellent conformity (Table 2). This is further evidence that an outer-sphere complexation mechanism for SeO_4^{2-} adsorption on goethite is correct [8].

Total selenite adsorption on goethite decreased with an increase in pH (Figure 10). Using the modified triple layer model (TLM), that assumes metals could be adsorbed in either the alpha (inner-sphere complexes) or beta (outer-sphere complexes) layers, it was found that in the pH range studied, selenite adsorbed on goethite to form monovalent and bivalent selenite-Fe complexes, since selenite existed as both SeO_3^{2-} and $HSeO_3^-$ in suspension over the pH range studied. The amounts of both complexes, $XHSeO_3^o$ and $XSeO_3^-$ (Figure10) were significant in suspension. The amount of $XHSeO_3^-$ dropped precipitously at about pH 8.3 (pK_2 = 8.24 for $HSeO_3^-$); the amount of $XSeO_3^-$ increased with pH until pH 8.3 and then dropped as the total adsorption decreased. The solid line in Figure 10 represents the total adsorption as predicted by the TLM; it matched the experimental data very well. Adsorption of $HSeO_3^-$ and SeO_3^{2-}, as predicted from the TLM, are also shown in Figure 10.

Pressure-jump relaxation curves revealed a double relaxation. Zhang and Sparks [8] hypothesized a comprehensive two-step adsorption mechanism to describe selenite adsorption on goethite:

$$
\begin{array}{ccccc}
 & \overset{k_1}{\nearrow\!\!\swarrow}{}_{k_{-1}} & XOH_2^+ - HSeO_3^- & \underset{k_{-3}}{\overset{k_3}{\rightleftharpoons}} & XHSeO_3 + H_2O \\
XOH + 2H^+ + SeO_3^{2-} & & \updownarrow K_s & & \updownarrow K_6 \\
 & \overset{k_2}{\searrow\!\!\nwarrow}{}_{k_{-2}} & XOH_2^+ - SeO_3^{2-} + H^+ & \underset{k_{-4}}{\overset{k_4}{\rightleftharpoons}} & XSeO_3^- + H^+ + H_2O \\
 & & \text{Step 1} & & \text{Step 2}
\end{array}
\qquad (15)
$$

Step 1 involves the formation of outer-sphere surface complexes (XOH_2^+ -$HSeO_3^-$ and XOH_2^+ -SeO_3^{2-}) in the β layer. Step 2 is a ligand-exchange process, where the adsorbed selenite enters the α layer and replaces a ligand from the goethite surface to form the inner-sphere surface complexes ($XHSeO_3^0$) and $XSeO_3^-$). Two protolytic equilibria (K_5 and K_6) are rapidly established compared to the complexation reactions.

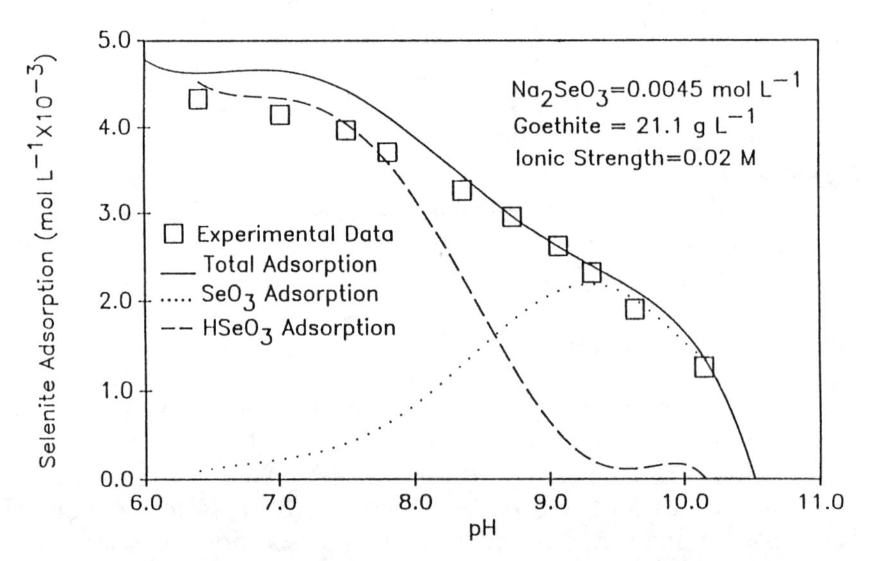

FIGURE 10. Selenite Adsorption vs. pH in the Selenite-Goethite System. The Lines Represent the Prediction of the TLM when Inner-sphere Surface Complex Formation is Assumed. From Zhang and Sparks [8].

To confirm the mechanisms in Equation (15), Zhang and Sparks [8]:(1) derived equations to show the relationships between the reciprocal relaxation times and the concentrations of species involved in the reaction; (2) solved the equations so that rate constants could be obtained for the expected mechanism; and (3) used the respective equation, τ_{obsd}^{-1} values, and other concentration terms in the equation to calculate the four unknown intrinsic rate constants (k^{int}) at four of the pH levels studied. These rate constants were then inserted into the rate equation for other pH levels that were studied to calculate τ^{-1}_{calcd} values. If the assumed mechanism is acceptable, the τ^{-1}_{calcd} values should match the τ^{-1}_{obsd} values. Also, the intrinsic equilibrium constants for formation of $XHSeO_3^{o}$ and $XSeO_3^{-}$ determined from the intrinsic rate constants ($K^{int}_{kinetics}$), which were calculated from the above derived equations should be consistent with those obtained from equilibrium studies (K^{int}_{model}). As shown in Table 3 the $\tau^{-1}_{1,obsd}$) and $\tau^{-1}_{2,obsd}$ values agreed well with the $\tau^{-1}_{1,calcd}$ and $\tau^{-1}_{2,calcd}$ values, confirming the mechanism in Equation (15). The $K^{int}_{kinetics}$ and K^{int}_{model} values, although not shown, also compared well, further validating the mechanism in Equation (15).

From the above cited studies, it would appear that the use of pressure-jump relaxation and other relaxation methods have much to offer in the study of metal sorption reactions on soil components. An especially attractive approach for ascertaining metal sorption mechanisms on soils would be to combine relaxation approaches with *in situ* surface spectroscopic techniques.

Table 3. Relaxation Data for Selenite Adsorption and Desorption on Goethite as a Function of pH at 298 K and an Ionic Strength of 0.02M[a].

pH	$\tau^{-1}_{1,obsd}$) s^{-1}	$\tau^{-1}_{1,calcd}$ s^{-1}	$\tau^{-1}_{2,obsd}$ s^{-1}	$\tau^{-1}_{2,calcd}$ s^{-1}
6.41	1.6	41.6	9.12	8.54
7.02	46.5	45.1	10.64	9.84
7.50	57.7	58.7	13.62	13.03
7.81	64.1	65.2	14.64	15.21
8.36	79.6	80.3	18.20	19.89
8.73	105.0	105.9	26.19	27.55
9.07	171.5	171.8	29.57	31.68
9.32	249.4	269.7	39.98	42.58
9.64	357.1	357.3	90.50	91.69

[a] *From Zhang and Sparks [8].*

5. CONCLUSIONS

In this review, an attempt has been made to provide a succinct overview of metal sorption kinetics on soils. Important soil minerals were described, background on rate laws and data interpretation *vis a' vis* metal sorption on soils was discussed, advances in kinetic methodologies were reviewed, and application of relaxation techniques to the study of metal sorption kinetics on goethite was provided.

While much progress has been made in understanding the rates of metal retention on soils, much is left to be done. There is particular need to obtain more experimental kinetic parameters that can be used in transport models to predict the fate of metals from contaminated soil, particularly if we are going to develop effective strategies for decontaminating soils; and there is a need to combine kinetic and *in situ* spectroscopic studies to definitively determine mechanisms for metal retention/release from soils.

REFERENCES

1. Sparks, D.L., *Kinetics of Soil Chemical Processes*, Academic Press, (New York, 1989).

2. Sparks, D.L., "Kinetics of Soil Chemical Processes: An Overview," *Trans. Int. Congress of Soil Sci.* Vol. 2 pp. 4-9 (1990).

3. Sparks, D.L., "Reaction Kinetics in Soils," in *Encyclopedia of Earth Systems Sci.*, W.A. Nirenberg, Ed. Academic Press, (New York, 1991).

4. Sparks, D.L. and D.L. Suarez, "Rates of Soil Chemical Processes," *Soil Sci. Soc. Am. Spec. Publ.*, Soil Sci. Soc. Am., (Madison, WI 1991).

5. Jardine, P.M. and D.L. Sparks, "Potassium-Calcium Exchange in a Multireactive Soil System: I. Kinetics," *Soil Sci. Soc. Am. J.* 48:39-45 (1984).

6. Zhang, P.C. and D.L. Sparks, "Kinetics And Mechanisms of Molybdate Adsorption/Desorption at the Goethite/Water Interface Using Pressure-Jump Relaxation," *Soil Sci. Soc. Am. J.* 53:1028-1034 (1989).

7. Zhang, P.C. and D.L. Sparks, "Kinetics and Mechanisms of Sulfate Adsorption/Desorption on Goethite Using Pressure-Jump Relaxation," *Soil Sci. Soc. Am. J.* 54:1266-1273 (1990).

8. Zhang, P.C. and D.L. Sparks, "Kinetics of Selenate and Selenite Adsorption/Desorption at the Goethite/Water Interface," *Environ. Sci. Technol.* 24:1848-1855 (1990).

9. Bunzl, K., W. Schmidt and B.Sansoni, "Kinetics of Ion Exchange in Soil Organic Matter. IV. Adsorption and Desorption of Pb^{2+}, Cu^{2+}, Cd^{2+}, Zn^{2+} and Ca^{2+} by Peat," *J. Soil Sci.* 27:32-41 (1976).

10. Aharoni, C. and D.L. Sparks, "Kinetics of Soil Chemical Reactions - A Theoretical Treatment," in *Rates of Soil Chemical Reactions. Soil Sci. Soc. Am. Special Publ.*, D.L. Sparks and D.L. Suarez, Eds. pp. 1-18. (Madison, WI 1991).

11. Amacher, M., "Methods of Obtaining and Analyzing Kinetic Data," in *Rates of Soil Chemical Reactions. Soil Sci. Soc. Am. Special Publ.*, Soil Sci. Soc. Am., D.L. Sparks and D.L. Suarez, Eds. pp. 19-59. (Madison, WI 1991).

12. Carski, T.H. and D.L. Sparks, "A Modified Miscible Displacement Technique for Investigating Adsorption-Desorption Kinetics in Soils," *Soil Sci. Soc. Am. J.* 49:1114-1116 (1985).

13. Seyfried, M.S., D.L. Sparks, A. Bar-Tal and S.Feigenbaum, "Kinetics of Calcium-Magnesium Exchange on Soil Using a Stirred-Flow Reaction Chamber," *Soil Sci. Soc. Am. J.* 53:406-410 (1989).

14. Bar-Tal, A., D.L. Sparks, J.D. Pesek and S. Feigenbaum, "Analysis of Adsorption Kinetics Using A Stirred-Flow Chamber: I. Theory and Critical Tests," *Soil Sci. Soc. Am. J.* 54:1273-1278 (1990).

15. Eick, M.J., A. Bar-Tal, D.L. Sparks and S. Feigenbaum, "Analyses Of Adsorption Kinetics Using A Stirred-Flow Chamber: II Potassium-Calcium Exchange on Clay Minerals," *Soil Sci. Soc. Am. J.* 54:1278-1282 (1990).

16. Sparks, D.L. and P.C. Zhang, "Relaxation Methods for Studying Kinetics of Soil Chemical Phenomena," in *Rates of Soil Chemical Processes. Soil Sci. Soc. Am. Spec. Publ.*, Soil Sci. Soc. Am., D.L. Sparks and D.L. Suarez, Eds. pp. 61-94. (Madison, WI 1991).

17. Bohn, H.L., B.L. McNeal and G.A. O'Connor, *Soil Chemistry,* Wiley & Sons, (New York, 1985).

3

IMPLICATIONS OF COMPLEXATION, SORPTION AND DISSOLUTION KINETICS FOR METAL TRANSPORT IN SOILS

Janet G. Hering
Civil and Environmental Engineering Department
University of California
Los Angeles, CA 90024

1. INTRODUCTION

Trace metal contaminants are introduced into soil and sediment environments through a variety of human activities, such as waste disposal, mining and smelting. Land disposal of radioactive wastes has resulted in subsurface contamination by metallic, as well as non-metallic, radionuclides near disposal sites [1-3]. The mobility of trace metals in soils and sediments is of particular concern because of the potential for bioaccumulation, food chain magnification, and human exposure. Although metals may be strongly sorbed to soils, clays, and oxides, such association does not guarantee immobilization of contaminant metals in the subsurface. Observed migration of radionuclides in groundwater has been attributed both to organic complexation [1-3] and to colloidal transport [4].

2. TRACE METAL CHEMISTRY AND MOBILITY

The transport of metals within aquifers (or model systems such as soil columns) is influenced by both chemical and physical processes [5] (shown schematically in Figure 1). Adsorption-desorption and precipitation-dissolution reactions directly affect the partitioning of metals between solid and aqueous phases, while complexation and oxidation-reduction reactions affect metal reactivity (e.g., solubility and

0-87371-697-3/95/$0.00+$.50

bioavailability). The effects of chemical processes on the mobility of metals are illustrated by the column experiments shown in Figure 2.

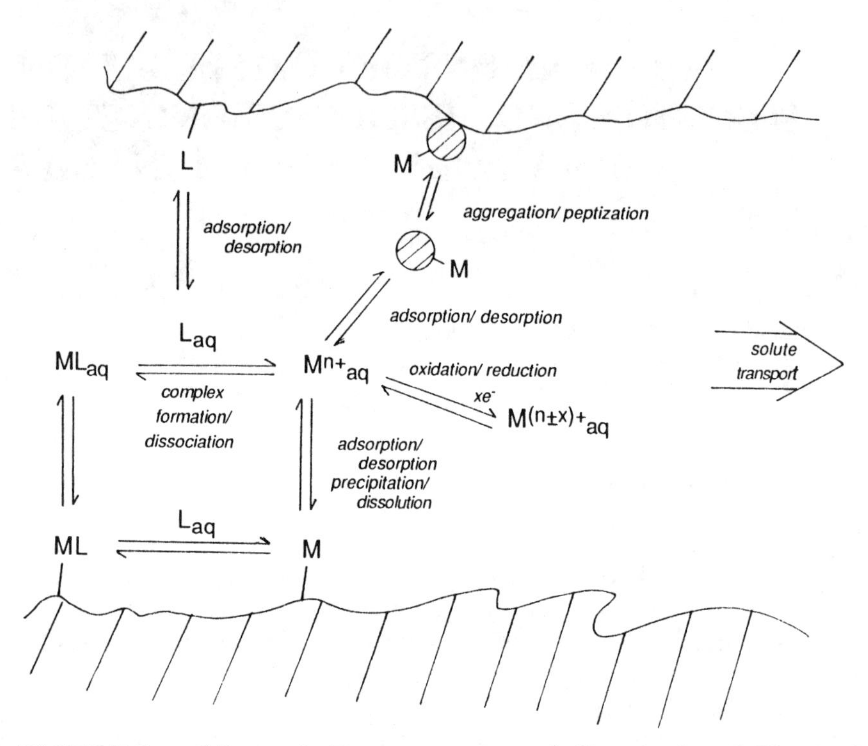

FIGURE 1. ***Schematic Representation of Chemical and Physical Processes Affecting Metal Transport in the Subsurface Environment.***

The retention of Hg^{2+} in the column is markedly decreased by increasing Cl^- or H^+ concentrations; the former effect may be explained by increased inorganic complexation of Hg^{2+} by Cl^- and the latter by competition between H^+ and Hg^{2+} for surface binding sites [6-8].

The effects of chemical processes on metal transport have most often been modeled by assuming the chemical reactions to be in (local) equilibrium [8-11] or at steady state [12,13]. If, however, the rates of chemical reactions are comparable to (or slower than) the rates of physical processes involved in metal transport, chemical kinetics must be considered explicitly. The incorporation of chemical kinetics into transport models has been discussed elsewhere [14,15] and will not be addressed here. This paper will concentrate on the mechanisms and kinetics of chemical processes that influence metal transport, particularly complexation, sorption and dissolution, with a comparison of field and laboratory observations.

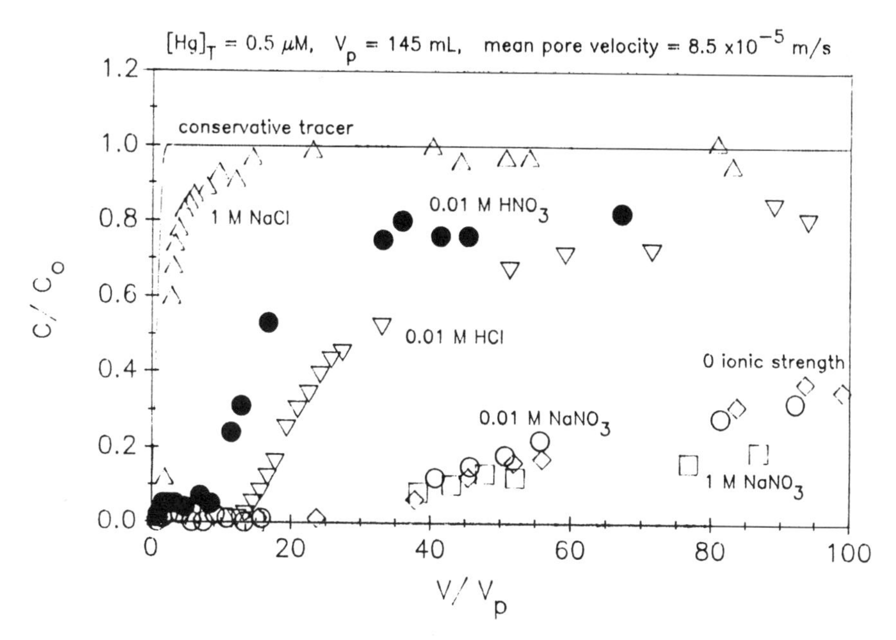

FIGURE 2. ***Breakthrough Curves for Mercury on a Sand Column: Effects of Ionic Strength, Inorganic Ligands, and pH. Mercury solutions prepared with chloride (Δ, ∇, ◇) or nitrate (o, •, □) salts (adapted from [7] with permission).***

3. STUDIES OF REACTION MECHANISMS

Both dissolution of minerals and desorption of metal ions from mineral surfaces can be facilitated by organic complexing agents. Such ligand-promoted dissolution and desorption reactions will be the focus of this paper. For discussion of reductive and oxidative processes, which involve a change in the oxidation state of the constituent metal of the reacting mineral, the reader is referred to other reviews [16,17].

3.1 The surface-controlled dissolution model

The model of surface-controlled dissolution, which holds that dissolution rates are limited by the rates of chemical reactions occurring at mineral surfaces [18-22], may be used as a framework for the discussion of both dissolution and desorption reactions. The process of ligand-promoted dissolution (outlined schematically in Figure 3) includes the following three general steps: sorption of the ligand and formation of a surface complex, detachment of a surface metal center (as a complex with the added ligand), and regeneration of the surface. Transport of reactants or products between the bulk solution and the surface is, in most cases, rapid compared to the rates of chemical reactions occurring at the mineral surface and may thus be neglected in describing the overall dissolution rate [23,24]. In the model of surface-controlled dissolution, detachment is taken to be the rate-limiting step.

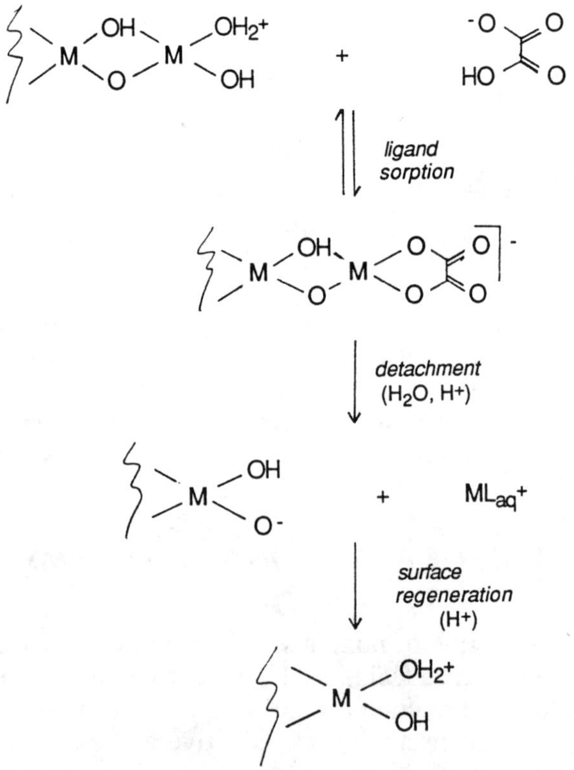

ligand sorption

detachment (H_2O, H^+)

+ ML_{aq}^+

surface regeneration (H^+)

FIGURE 3. Schematic Representation of the Process of Ligand-Promoted Dissolution of Oxide Minerals.

Thus, under conditions where neither the ligand nor the solid is significantly depleted during the reaction, steady state dissolution kinetics (i.e., a constant rate of dissolution over time) should be observed where the dissolution rate is proportional to the concentration of the surface complex.

The rate law for ligand-promoted dissolution would then be of the form

$$rate = \frac{d[M]}{dt} = k\{\equiv L\} \tag{1}$$

where $\{\equiv L\}$ is the surface concentration of the ligand and d[M]/dt is the increase in the dissolved concentration of M, the constituent metal of the dissolving mineral. Usually the dissolution rate is normalized for the surface area of the reacting mineral; then for $\{\equiv L\}$expressed in units of mol/m^2, the rate constant for dissolution, k, will have units of reciprocal time. Often, the surface concentration of the ligand can be related to its dissolved concentration, $[L]_{diss}$, by the Langmuir expression

$$\{\equiv \mathrm{L}\} = \frac{\mathrm{KS_T[L]_{diss}}}{1 + \mathrm{K[L]_{diss}}} \tag{2}$$

where S_T is the total concentration of sites on the mineral surface (in mol/m^2) and K is the equilibrium constant for sorption of L at the surface. In this case, the dissolution rate will also exhibit a Langmuir-type dependence on the dissolved ligand concentration; that is, the dissolution rate will reach a plateau at high dissolved ligand concentrations due to saturation of the surface.

3.2 Ligand-promoted dissolution: role of surface complexes

The results of dissolution experiments support the surface-controlled dissolution model. As shown in Figure 4a, steady-state dissolution kinetics is observed for the dissolution of δ-Al_2O_3 at pH 3.5 both in the presence and absence of the ligand oxalate. Enhanced dissolution is observed with increasing oxalate concentrations but not in proportion to the dissolved oxalate concentration. Rather, the ligand-promoted dissolution rate is directly proportional to the surface concentration of the ligand (as shown for oxalate and several other ligands in Figure 4b) [25,26].

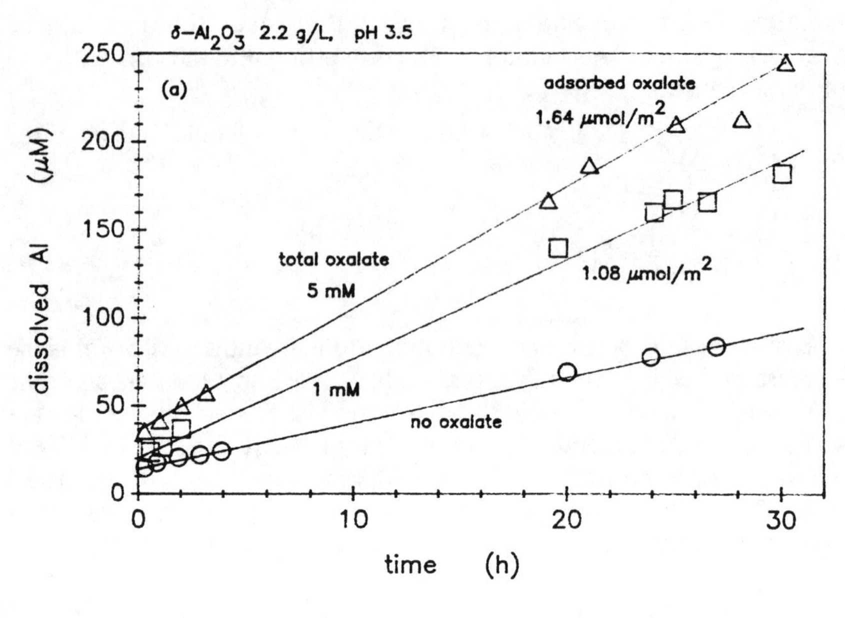

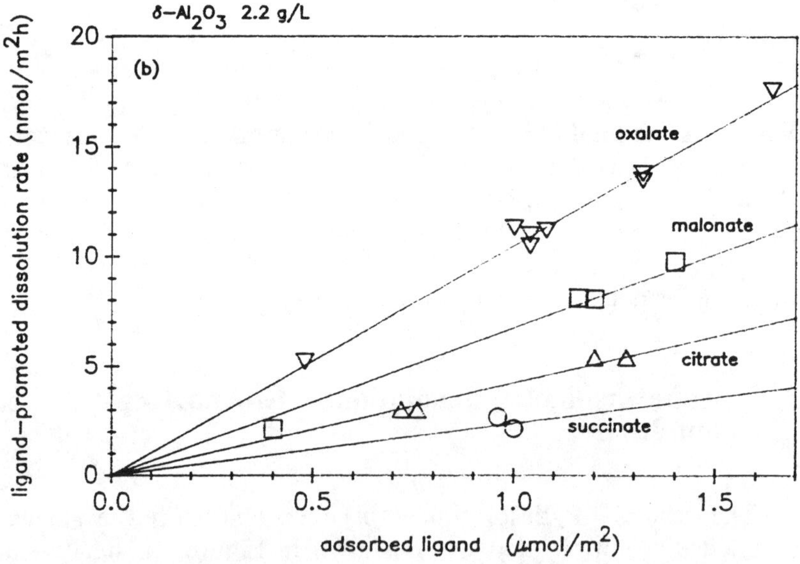

FIGURE 4. ***Ligand-promoted Dissolution of Aluminum Oxide (a) Dissolved Al Concentrations as a Function of Time for Dissolution of δ-Al_2O_3 at pH 3.5 in the Presence and Absence of Oxalate (b) Ligand-promoted Dissolution Rate as a Function of Adsorbed Ligand Concentration (adapted with permission from [25, 26]).***

These observations provide indirect support for the assumptions of the model that the mineral surface is continuously regenerated and that the concentration of the surface complex remains constant over the course of dissolution. These assumptions are also supported by numerical simulations of surface morphologies during dissolution [27]. Recently, direct spectroscopic evidence for the validity of these assumptions has been obtained.

The concentration of surface complexes during dissolution can be monitored by fluorescence spectroscopy in the case of dissolution of δ-Al_2O_3 by a ligand, 8-hydroxyquinoline-5-sulfonate (HQS), which forms fluorescent complexes with Al both in solution and at the oxide-water interface. Photomicrographs of Al oxide aggregates pre-treated with HQS show the intense fluorescence associated with the HQS-surface complex (Figure 5). During HQS-promoted dissolution of δ-Al_2O_3, both the total fluorescence of the suspension and the dissolved (i.e., filterable) fluorescence increase over time (Figure 6). These increases are due to dissolution of the solid and increase in the concentration of the Al-HQS complex in solution. Since the increase in the total fluorescence of the suspension can be entirely accounted for by the increase in dissolved fluorescence, the surface-associated fluorescence and thus the surface HQS concentration must remain constant during (long-term) dissolution of the oxide [28].

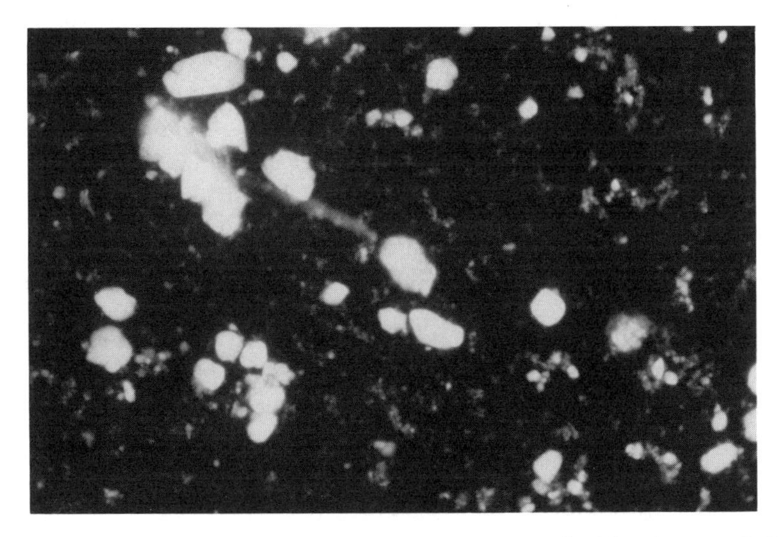

FIGURE 5. Photomicrograph of HQS-treated Aluminum Oxide Aggregates Showing UV-excited Fluorescence of the Particles. Oxide suspensions were equilibrated with HQS, filtered and re-suspended in water.

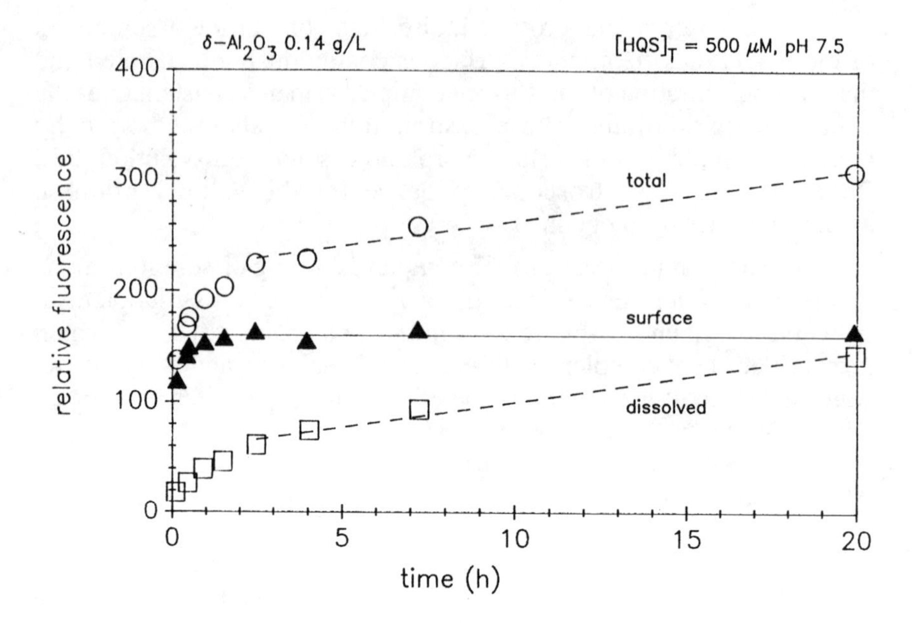

FIGURE 6. ***Fluorescence as a Function of Time During HQS-Promoted Dissolution of Aluminum Oxide. Surface-associated fluorescence (solid symbols) is calculated from the difference between measured total and dissolved fluorescence. The increase in total and dissolved fluorescence over time is a result of dissolution of the oxide. The rate of increase, i.e., the slope of the dashed lines, is proportional to the dissolution rate [28].***

From these experimental observations and the surface-controlled dissolution model, it is possible to identify the surface ligand concentration as a crucial parameter determining the rate of ligand-promoted dissolution. Thus, any extrapolation from laboratory to field conditions must consider the dependence of surface ligand concentrations on both the dissolved ligand concentrations and mineral surface areas.

3.3 Comparison of dissolution and desorption

Transition metals are commonly adsorbed on oxide surfaces as inner-sphere complexes between the metal and surface hydroxyl groups. The formation of inner-sphere complexes is demonstrated by adsorption of (cationic) metals on positively charged oxide surfaces (i.e., overcoming electrostatic effects), by the (minimal) effect of ionic strength on metal adsorption, and by spectroscopic evidence [29-32]. Desorption of the metal requires the dissociation of these inner-sphere complexes, corresponding to the detachment of the surface metal center in dissolution.

Study of dissolution and desorption kinetics in similar model systems allows comparison of these related processes; the surface-controlled model for dissolution may be applied as a theoretical basis for this comparison. Both ligand-promoted dissolution of iron and aluminum oxides and ligand-promoted desorption of aluminum from iron oxide surfaces have been studied with the ligand HQS (see Appendix for experimental conditions).

Dissolution of hematite and δ-Al_2O_3 by HQS at near-neutral pH is shown in Figure 7. After some initial rapid dissolution, linear increases in the dissolved concentrations of Fe or Al with time are observed. The steady state rates of HQS-promoted dissolution (normalized for surface area) are similar for both oxides. To study HQS-promoted desorption of Al from the hematite surface, Al was pre-equilibrated with hematite (at pH 7.5) before the addition of the ligand (see Appendix for experimental conditions). In this mixed Al/hematite system, HQS solubilizes both Al and Fe (Figures 8-10).

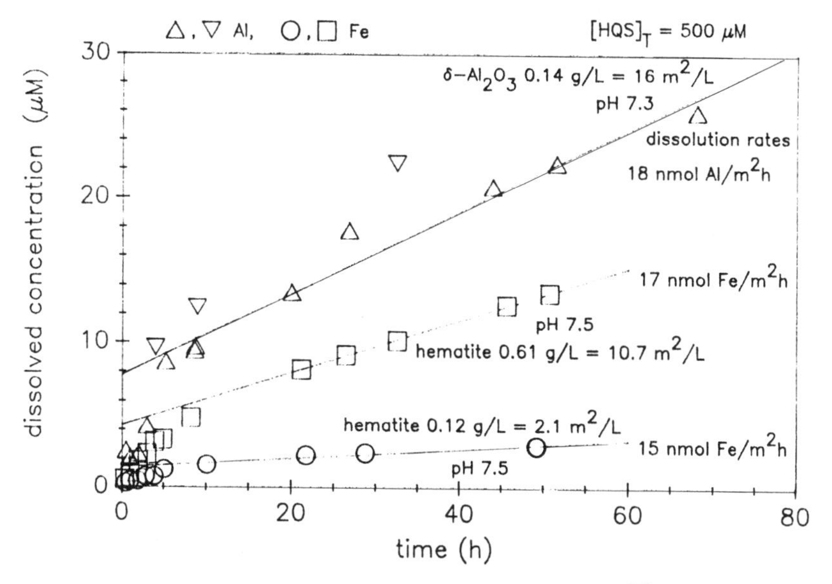

FIGURE 7. ***Concentrations of Dissolved Al (△, ▽) and Fe (o, □) During Dissolution of Hematite and δ-Al_2O_3 by HQS. Solid Lines Indicate Steady-state Dissolution Rates: (▽) Points Omitted in Rate Calculation for δ-Al_2O_3.***

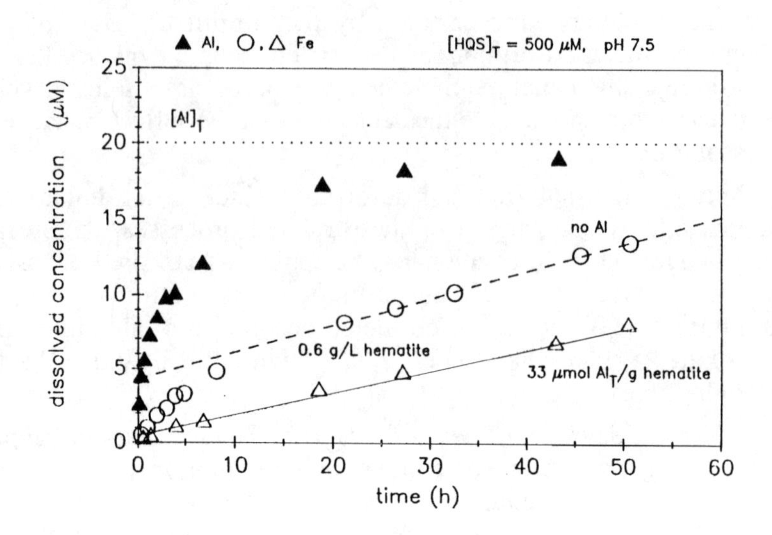

FIGURE 8. Concentrations of Dissolved Al (▲) and Fe (Δ) During Desorption. - Dissolution Experiments. Hematite (0.6 g/L) was pre-equilibrated at pH 7.5 with 20 μM total Al before addition of HQS at time = 0. For comparison, dissolution of pure hematite is also shown (o).

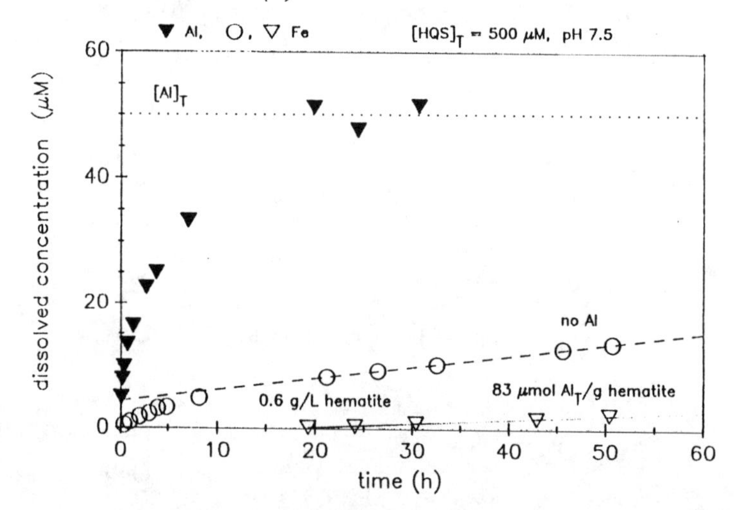

FIGURE 9. Concentrations of Dissolved Al (▼) and Fe (▽) During Desorption.- Dissolution Experiments. Hematite (0.6 g/L) was pre-equilibrated at pH 7.5 with 50 μM total Al before addition of HQS at time = 0. For comparison, dissolution of pure hematite is also shown (o).

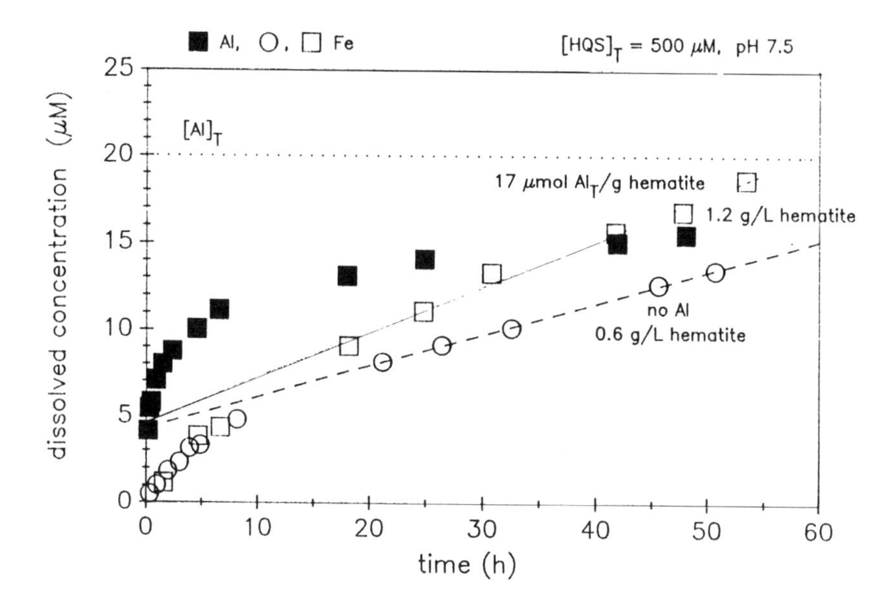

FIGURE 10. Concentrations of Dissolved Al (■) and Fe (□) During Desorption - Dissolution Experiments. Hematite (1.2 g/L) was pre-equilibrated at pH 7.5 with 20 μM total Al before addition of HQS at time = 0. For comparison, dissolution of pure hematite is also shown (o).

Since Al is relatively insoluble at the pH of the Al/hematite pre-equilibration, however, solubilization of Al by HQS in the mixed system probably involves both desorption of Al from the hematite surface and dissolution of amorphous Al hydroxide (precipitated during pre-equilibration). As shown in Figures 8-10, the desorption/dissolution of Al in the mixed system by HQS occurs on roughly the same time scale as the dissolution of pure oxide phases in Figure 7. The sorption/precipitation of Al on the hematite surface protects the oxide surface, to some extent, from attack by the ligand. At a high Al-to-hematite ratio, no hematite dissolution is observed until the Al is solubilized (Figure 9). This preferential reaction of surface Al with HQS is, however, strongly dependent on the Al-to-hematite ratio; at lower Al-to-hematite ratios, concurrent solubilization of Al and Fe is observed (Figures 8 and 10).

These observations are consistent with the operation of similar mechanisms in the dissolution of pure oxide phases and in desorption/dissolution in the mixed system. In ligand-promoted dissolution, detachment of the metal from the oxide crystal lattice is facilitated by surface complex formation between the ligand and the reacting metal center. Thus it is likely that ligand-promoted desorption similarly involves formation of a surface complex between the ligand and the adsorbed metal. Although the adsorption of ligands on a pure oxide surface does not necessarily occur at the same sites as metal adsorption, the inhibition of the initial, rapid dissolution of hematite by sorbed/precipitated Al (Figure 8) indicates that some highly reactive sites may indeed preferentially adsorb both metals and ligands.

It is difficult to extend the model of surface-controlled dissolution quantitatively to describe desorption kinetics because of the depletion of surface-bound Al. That is, the simplifying assumption of steady-state kinetics for dissolution (which depends on regeneration of the oxide surface) cannot be applied in the case of desorption. Nonetheless, the observed similarities in the pure oxide and mixed system suggest that the surface ligand concentration, of central importance in the model of surface-controlled dissolution, may be equally important in the ligand-promoted desorption of metals from oxide surfaces.

3.4 Dissolution of iron oxides by a widely occurring anthropogenic ligand

Ethylenediaminetetraacetic acid (EDTA) is a synthetic complexing agent with a high affinity for many metals. As a result of its widespread use in a range of industrial, pharmaceutical, and agricultural applications and of its resistance to biodegradation [33,34], EDTA is present in groundwaters [1,35,36], sewage effluents [34,35], freshwaters, including drinking water [34,35,37,38], and estuarine waters [39].

EDTA can effect both non-reductive and reductive dissolution of iron oxides [40-45]. Reductive dissolution may also proceed photochemically and result in the decomposition of the ligand [42]. The non-reductive, or ligand-promoted, dissolution of iron oxides by EDTA occurs over a wide pH range (Figure 11a). For both lepidocrocite (Figure 11a) and hematite (Figure 11b), dissolution of the iron oxides (present in large excess of the ligand) is stoichiometric when corrected for the adsorption of EDTA on the oxide surface. That is, the final dissolved Fe(III) concentration is equal to the initial EDTA concentration less the concentration of EDTA adsorbed. In the dissolution of lepidocrocite, the oxide was pre-conditioned with EDTA (as described in the Appendix), and the final dissolved Fe(III) concentration is equal to the concentration of EDTA added during the dissolution experiment [43]. In the dissolution of hematite, the concentration of adsorbed EDTA can be calculated from the discrepancy between maximum

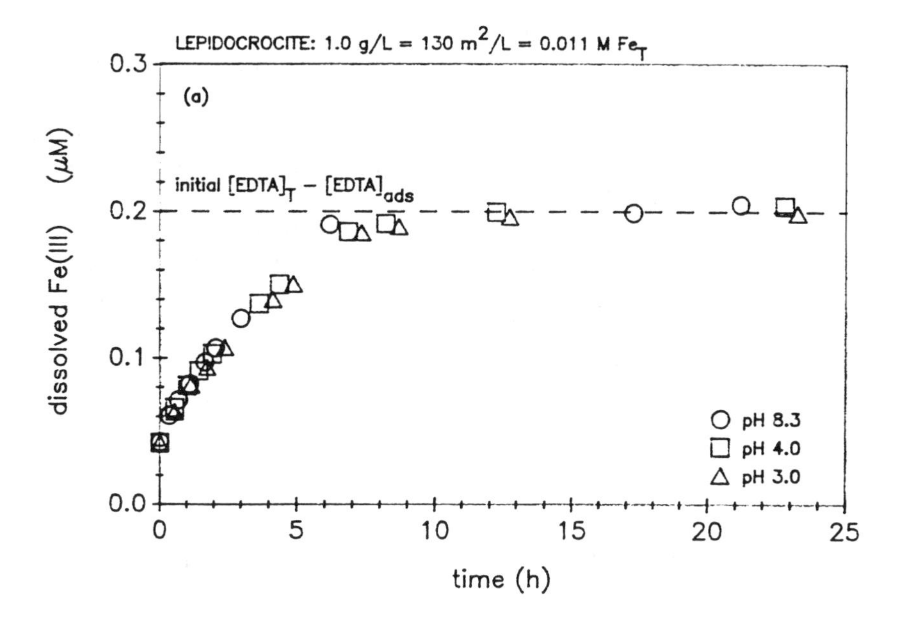

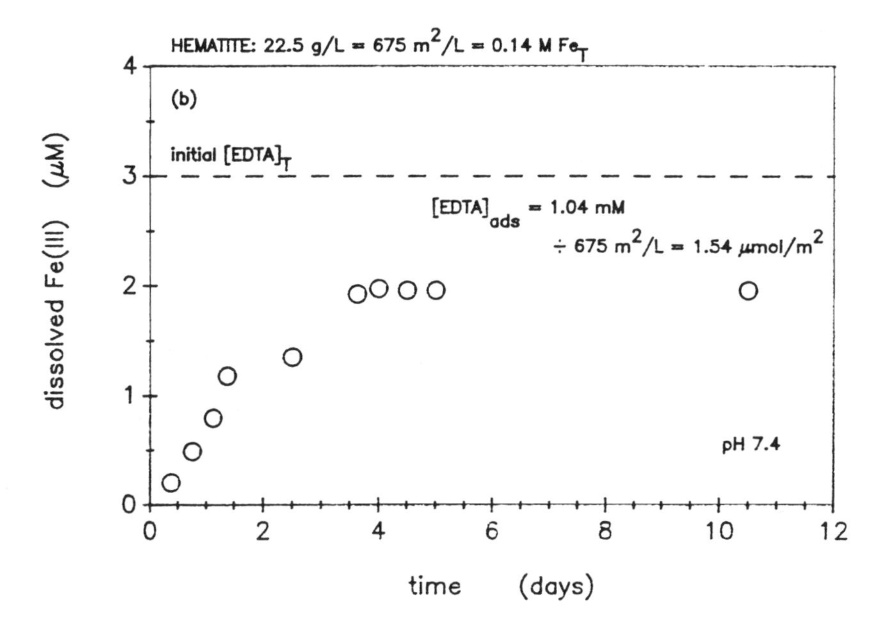

FIGURE 11. Non-reductive Dissolution of Iron Oxides by EDTA (a) Dissolved Fe(III) as a Function of Time for Dissolution of Lepidocrocite [43] and (b) Dissolved Fe(III) as a Function of Time for Dissolution of Hematite (data from [44]).

dissolved Fe(III) concentration and the initial EDTA concentration (Figure 11b); this difference corresponds to an adsorption density of 1.5 μmol EDTA/m^2 [44,45].

These model studies have several important implications for the behavior of EDTA in soils containing a significant fraction of iron oxides. This extrapolation is complicated by two factors, the difference in the EDTA concentrations in the laboratory and field (both its absolute concentration and its concentration relative to the available mineral surface area) and the presence of natural organic matter. Nonetheless, the following processes may be expected, based on laboratory observations, to occur in the subsurface environment: (1) sorption of EDTA onto naturally-occurring iron oxides, (2) EDTA-promoted dissolution of iron oxides and (3) transport of the resulting Fe-EDTA complexes, which do not themselves sorb to iron oxide surfaces.

3.5 Reactions of metal complexes at the mineral-water interface

The above discussion has not explicitly considered the effect of (initial) ligand speciation on the reactions of the ligand at the mineral surface. Ligand-promoted dissolution experiments are usually performed under conditions where the ligand is initially present in its free or protonated form. For strong ligands, however, this is unlikely to be the case in natural waters. Thus the reactions of the complexes of metals and organic ligands (i.e., metal-organic complexes), as well as simply those of the organic ligands, at the mineral-water interface must be considered.

If a metal-organic complex, ML, is introduced into the subsurface environment, the speciation of both the metal and ligand will be influenced by the interaction of these species with mineral surfaces, present in large excess. If the most reactive mineral phase is a naturally-occurring iron oxide, equilibration of the complex with the solid may result in sorption of the metal and/or the ligand to aquifer minerals and dissolution of aquifer minerals, that is

$$\begin{aligned} \equiv FeOH + ML \xrightarrow{H^+} \;& \equiv FeL && \text{(sorption of ligand)} \\ & \equiv FeOM && \text{(sorption of metal)} \\ & FeL_{aq} && \text{(dissolution of aquifer minerals)} \end{aligned} \tag{3}$$

An important question is then the extent to which these reactions are thermodynamically-favorable in a given system. This will depend on parameters such as the composition of the soil or aquifer materials and the chemical composition, particularly pH and redox conditions, of the groundwater.

Even if such reactions are thermodynamically-favorable, the question of the rates of these reactions under natural conditions still remains. This question can only be addressed with some understanding of the reaction mechanisms. Although the mechanisms of the reactions between metal-organic complexes and mineral surfaces have not yet been determined, comparison with complexation reactions in solution and with sorption reactions at mineral surfaces may provide some insight.

The reaction of a metal-organic complex with the surface resulting in metal sorption on the surface

$$\equiv FeOH + ML^{(n-m)+} \leq \equiv FeOM^{(n-1)+} + H^{+} + L^{m-} \tag{4}$$

may be compared with the ligand-exchange reaction in solution

$$ML^{(n-m)+} + Y^{x-} \rightarrow MY^{(n-x)+} + L^{m-} \tag{5}$$

where L and Y are ligands (protonated species omitted for simplicity) and all species are dissolved. Ligand-exchange reactions between dissolved species may involve either dissociation of the initial complex:

$$ML^{(n-m)+} \leq M^{n+} + L^{m-} \tag{6}$$

$$M^{n+} + Y^{x-} \rightarrow MY^{(n-x)+} \tag{7}$$

or direct attack of the incoming ligand on the initial complex:

$$ML^{(n-m)+} + Y^{x-} \leq LMY^{(n-m-x)+} \rightarrow L^{m-} + MY^{(n-x)+} \tag{8}$$

as discussed in References 46 and 47. Analogous reaction pathways may be envisioned for the surface reactions, that is, either

$$ML^{(n-m)+} \leq M^{n+} + L^{m-} \tag{9}$$

$$\equiv \text{FeOH} + \text{M}^{n+} \rightarrow \equiv \text{FeOM}^{(n-1)+} + \text{H}^{+} \quad (10)$$

or

$$\equiv \text{FeOH} + \text{ML}^{(n-m)+} \leq \equiv \text{FeOML}^{(n-m-1)+} + \text{H}^{+}$$
$$\rightarrow \equiv \text{FeOM}^{(n-1)+} + \text{L}^{m-} + \text{H}^{+} \quad (11)$$

In the first case, because of the very large excess of surface binding sites over the free (or protonated) ligand, the rate of the reaction may be effectively limited by the rate of dissociation of the initial complex. In the second case, dissociation of the initial complex may be facilitated by interaction with the surface, i.e., bonding between the surface hydroxyl group and the (complexed) metal destabilizes the metal-organic complex. This is essentially the reverse of the mechanism proposed for ligand-promoted desorption (see Section 3.3) and, as was also suggested for the desorption process, is likely to be strongly influenced by the concentration of the intermediate surface species. That is, the rate of the reaction through the latter pathway is likely to depend on the concentration of adsorbed ML. In the case of EDTA complexes, however, sorption of the metal-EDTA complex on mineral surfaces may not occur to any appreciable extent (as discussed in Section 3.4 above).

Of necessity, this discussion of reactions of metal-organic complexes at mineral surfaces is rather speculative. But on the basis of the above discussion, observations of dissolution and desorption kinetics in laboratory systems, and the model of surface-controlled dissolution, it is possible to develop some ideas of how metal-organic complexes may behave in the subsurface environment and how strong organic complexing agents may influence the transport of metals in the subsurface. In particular, it is likely that the introduction of strong organic complexing agents into an aquifer will result in dissolution of aquifer minerals and/or desorption of trace metals from the surfaces of aquifer minerals. Metal-organic complexes, either formed by such reactions within the aquifer or introduced into the aquifer by infiltration of the pre-formed complex, may be transported with the groundwater if reactions with mineral surfaces (leading to ligand and/or metal sorption) are either thermodynamically- or kinetically-unfavorable. It must be remembered, however, that the rates of these reactions are likely to depend strongly on the extent of sorption of organic ligands or metal-organic complexes on the surface of naturally-occurring minerals, which may be extremely difficult to estimate. In the following sections, field observations of the occurrence and transport of both metals and ligands in river and groundwaters will be compared with the behavior of these species that may be expected based on laboratory experiments.

4. FIELD OBSERVATIONS

Field studies have been conducted to determine the behavior and mobility of metals in the subsurface environment and to elucidate the chemical processes controlling such behavior. In these studies, concentrations of metals and of organic complexing agents, including EDTA, in groundwater and partitioning of radionuclides between groundwater and aquifer materials have been determined. As discussed below, the results of some of these studies suggest that the observed mobility of metals in the subsurface may be due, at least in part, to the transport of metal-organic complexes.

4.1 Chemical composition of groundwater along a river-groundwater infiltration flow path: Glattfelden, Switzerland

The Glatt River, located in northeastern Switzerland, receives considerable inputs (approximately 15-20% of its total discharge) from sewage treatment plants. The infiltration of river water into adjacent aquifers influences the chemical composition of the groundwater; groundwater chemistry along a river-groundwater infiltration flow path has been examined at a study site in Glattfelden [48,49] (see Figure 12).

As a consequence of the inputs of sewage effluents to the Glatt River, the river water contains significant concentrations of EDTA. The concentration of EDTA in the Glatt River is on the order of 0.1 μM though both temporal and spatial variability in its concentration are observed due to variability in the source and possibly to some photochemical removal processes occurring in the river [50]. Similar concentrations of EDTA have been measured in the groundwater at Glattfelden [35,36].

Chemical composition of Glatt River water and of groundwater along the infiltration flow path at Glattfelden have been compared by Jacobs *et al.* [48]. Figure 13 shows the concentrations of redox-active species (oxygen and nitrate) and of metals in the river and groundwater. Groundwater was sampled at wells 2.5, 7, and 100 m from the river; data shown in Figure 13 for the farthest well is from a later sampling date than for the near-field wells to account for the 3-5 week transit time of the water. (Note that mixing with deeper groundwaters also influences groundwater composition at the farthest well; physical mixing, however, is neglected in the following discussion).

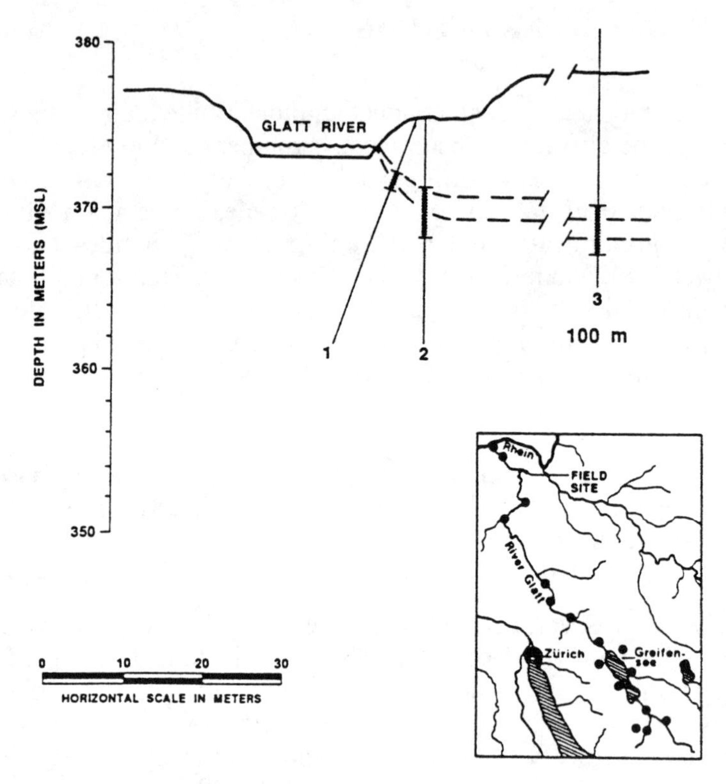

FIGURE 12. Sampling Site at Glattfelden, Switzerland. Closed circles on map indicate location of sewage treatment plants. On well cross-section, hatching indicates screened section of wells and dashed lines correspond to minimum and maximum water table during sampling period (adapted from [48]).

Comparison of the groundwater concentrations of dissolved Cu_T, Zn_T, and Cd_T show some interesting differences in the behavior of these metals. Both Cd and Zn concentrations decrease along the infiltration flow path indicating some removal process, most probably adsorption, within the aquifer. In contrast, the Cu concentrations remain remarkably constant suggesting that Cu is not strongly sorbed to the aquifer materials. Further studies at this site have, in general, confirmed these observations, though stronger adsorption of Zn, particularly in the near field region, and some seasonal variations in Cd concentrations and mobility in groundwater have been observed [51].

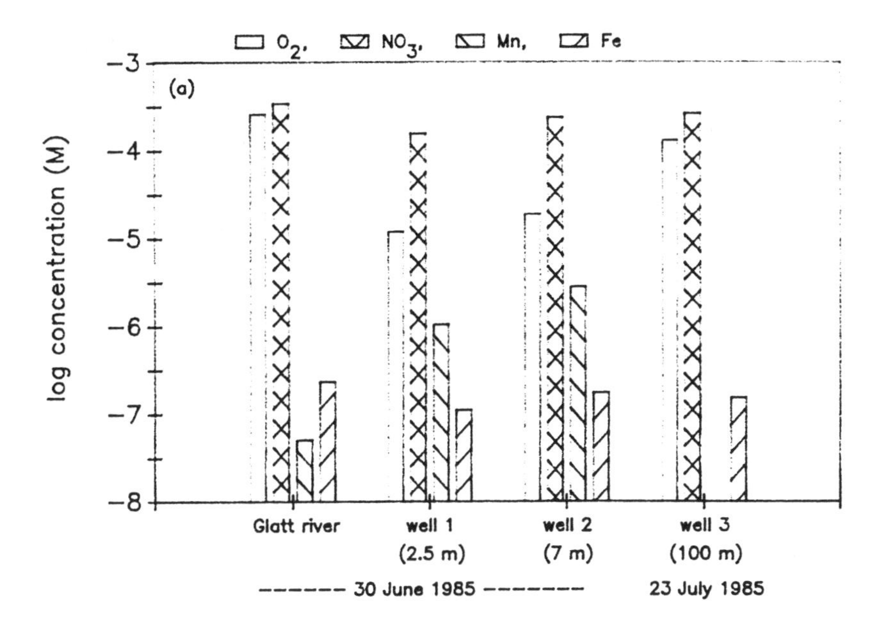

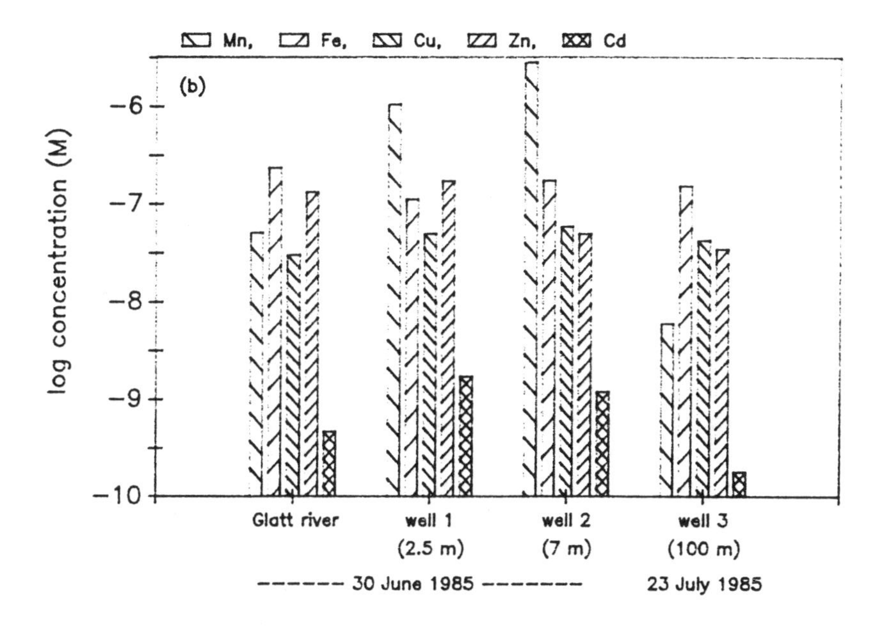

FIGURE 13. Chemical Composition of Glatt River Water and Groundwater at Glattfelden (data from [48]).

Jacobs *et al.* [48] have suggested that the mobility of Cu within the aquifer may be due in part to the transport of Cu-organic complexes, particularly EDTA complexes. They have noted that earlier measurements of EDTA in groundwater at Glattfelden indicated that its concentration, like that of Cu, does not decrease with distance along the flow path [35]. It is then implicit in this explanation of Cu mobility that the Cu-EDTA complex does not react at mineral surfaces within the aquifer and is conservatively transported in the subsurface.

4.2 Transport of radionuclides in groundwater

Disposal of low and intermediate level radioactive wastes in infiltration or seepage pits and trenches has resulted in radionuclide contamination of the subsurface at sites such as Oak Ridge, Tennessee. [1], Maxey Flats, Kentucky [2], and Chalk River, Ontario [3]. The wastes disposed of at these sites are known to have contained organic complexing agents, including EDTA, which were used in the decontamination of nuclear facilities. EDTA has been found in the groundwater at Oak Ridge at approximately 0.3 μM and in trench leachates at Maxey Flats at concentrations up to 1.7 μM [1,2].

Significant migration of radionuclides, such as Co-60 and Pu, from these disposal sites has been observed. This radionuclide mobility has been attributed to the transport of the radionuclides as organic complexes and, at Oak Ridge and Maxey Flats, specifically to the transport of Co-EDTA and Pu-EDTA complexes [1,2]. Again, mobilization of radionuclides as EDTA complexes requires that the complexes not react with mineral surfaces.

5. COMPARISON OF FIELD AND LABORATORY OBSERVATIONS

Laboratory studies of ligand-promoted dissolution and desorption and, in particular, of the (non-reductive) reactions of EDTA with iron oxides may be used as a basis for predicting the behavior of EDTA and metal-EDTA complexes in the subsurface environment. These laboratory studies suggest that EDTA, in its free or protonated form, is likely to adsorb to the surfaces of aquifer minerals, e.g., iron oxides, and to effect the partial dissolution of these minerals. The resulting EDTA complexes, e.g., Fe(III)EDTA, would not be expected to adsorb on aquifer materials and thus could be transported with the groundwater.

Field studies suggest that EDTA is transported conservatively in groundwater [35] and that metals may be transported with little or no retardation, most probably as organic complexes [1-3,48]. In some cases, for Cu at Glattfelden, Co-60 at Oak Ridge, and Pu at Maxey Flats, metal mobility has been attributed to transport of metal-EDTA complexes. Such transport, however, requires that the complexes neither adsorb to aquifer materials nor undergo reactions at the mineral-water interface, which would result in the sorption of both the metals and EDTA to the surface of aquifer materials (present in excess).

That metal-EDTA complexes might not be significantly adsorbed on aquifer materials is consistent with laboratory observations that Fe(III)EDTA complexes are not adsorbed on iron oxides. However, laboratory studies provide little basis for prediction of the extent of reaction of metal-EDTA complexes at mineral surfaces or the rates of such reactions. Clearly, the lack of such information is a serious hindrance to the understanding and prediction of the behavior of metal-organic complexes in the subsurface environment.

6. CONCLUDING REMARKS

Metal transport in the subsurface environment is strongly influenced by chemical processes, such as complexation, sorption, and dissolution. The presence of strong, organic complexing agents can markedly affect the mobility of metals in the subsurface. Laboratory studies of ligand-promoted dissolution and desorption provide some insight into the mechanisms of these reactions. These laboratory studies and the models (such as the surface-controlled dissolution model) developed to describe them form the basis for understanding the behavior of metals and metal-organic complexes in the subsurface environment. In particular, model studies allow the identification of crucial parameters, e.g., the concentration of surface complexes, that determine reaction rates. Field studies provide information on the behavior of chemical species under natural conditions which may be compared with the behavior expected on the basis of model studies. Such comparisons can highlight the limits on extrapolation from laboratory to field conditions and suggest avenues for further research.

Acknowledgments: I would like to thank Prof. Stumm (EAWAG) for his insight, support, and encouragement and for his critical review of this manuscript. I am also grateful to Drs. Zobrist and

Huggenberger (EAWAG) for use of their equipment, to G. Bondietti (EAWAG/ETH) for providing unpublished data, to Dr. Behra (CNRS, Strasbourg) for providing original data for the figure on his work, and to Dr. Ballmer (Carl Zeiss (Schweiz) AG) for providing the photomicrograph shown in Figure 5. This work was supported in part by the Swiss National Science Foundation.

APPENDIX: EXPERIMENTAL SECTION

1. Materials

Hematite was prepared following the method of Matijevic and Scheiner [52] as modified by Penners and Koopal [53]; details of oxide characterization are given in [54]. Lepidocrocite was synthesized as described by Brauer [55]. Degussa aluminum oxide C was pre-treated by washing several times with water, centrifuging the suspension, and discarding the supernatant each time. Analytical grade $NaClO_4 \cdot H_2O$ (Merck) was used as received. 8-Hydroxyquinoline-5-sulfonic acid monohydrate (98%, Aldrich) was recrystallized from a large volume of hot water. Ethylenediaminetetraacetic acid (EDTA) was obtained as a 0.1 M standard solution (Titrisol, Merck). Atomic absorbance standard solutions of Al and Fe (1000 ppm, Baker) were used to prepare calibration standards. All solutions were prepared with doubly deionized water (Barnstead Nanopure).

2. Dissolution and desorption experiments

Dissolution and desorption were studied in batch experiments at constant pH (pH maintained with a pH-stat) at constant ionic strength (in 0.1 M $NaClO_4$ for experiments with HQS and in 0.01 M $NaClO_4$ for experiments with EDTA). For dissolution of pure oxides by HQS, 200 mL of oxide suspensions were equilibrated at the reaction pH and room temperature overnight with stirring, and the reaction was initiated by the addition of HQS at time = 0. Aliquots were removed during the reaction and filtered through 0.01 μm cellulose nitrate (Sartorius) filters. Concentrations of dissolved Al and Fe were determined (as HQS complexes) by absorbance spectrophotometry. Similar procedures were followed in the desorption/ dissolution experiments in the mixed Al/hematite system, except that hematite suspensions were pre-equilibrated with Al (added from an acidic stock solution) and that the concentration of dissolved Al (as the HQS complex) was determined by fluorescence spectrophotometry (cf. ref. 28).

For dissolution of lepidocrocite by EDTA, 150 mL of oxide suspension were pre-conditioned under the same conditions (i.e., of pH and $[EDTA]_T$ and at 25°C) as for the dissolution experiment itself. Suspensions were bubbled with N_2 to exclude CO_2. After 24 h of pre-conditioning, the supernatant solution was removed and the dissolution experiment was initiated (i.e., t = 0 in Figure 11a) by resuspension of the pre-conditioned oxide in fresh reaction media. Aliquots were removed during the dissolution reaction, filtered through 0.1 μm cellulose nitrate (Sartorius) filters, and analyzed for Fe by atomic absorption spectroscopy [43].

REFERENCES

1. Means, J.L., D.A. Crerar and J.O. Duguid, "Migration of Radioactive Wastes: Radionuclide Mobilization by Complexing Agents," *Science* 200:1477-1481 (1978).

2. Cleveland, J.M. and T.F. Rees, "Characterization of Plutonium in Maxey Flats Radioactive Trench Leachates," *Science* 212:1506-1509 (1981).

3. Killey, R.W.D., J.O. McHugh, D.R. Champ, E.L. Cooper and J.L. Young, "Subsurface Cobalt-60 Migration from a Low-Level Waste Disposal Site," *Environ. Sci. Technol.* 18:148-157 (1984).

4. McCarthy, J.F. and J.M. Zachara, "Subsurface Transport of Contaminants," *Environ. Sci. Technol.* 23:496-502 (1989).

5. Evans, L.J., "Chemistry of Metal Retention by Soils," *Environ. Sci. Technol.* 23:1046-1056 (1989).

6. Behra, Ph., "Evidences for the Existence of a Retention Phenomenon During the Migration of a Mercurial Solution Through a Saturated Porous Medium," *Geoderma.* 38:209-222 (1986).

7. Behra, Ph., "How Laboratory Experiments and Modelling Complement Each Other to Identify Physico-Chemical Interaction Mechanisms: Application to the Propagation of a Metal Micropollutant-Mercury," Intl. Conf. on the Impact of Physico-Chemistry on the Study, Design, and Optimization of Processes in Porous Media," (Nancy, France, June 10-12, 1987).

8. Behra, Ph., "Etude du Comportement d'un Micropolluant Métallique – le Mercure – Au Cours de sa Migration à Travers un Milieu Poreux Saturé: Identification Expérimentale des Mécanismes d'échanges et Modélisation des Phénomenès," *Ph.D. Thesis*. Louis Pasteur University, (Strasbourg, France).

9. Jennings, A.A., D.J. Kirkner and T.L. Theis, "Multicomponent Equilibrium Chemistry in Groundwater Quality Models," *Wat. Res. Res.* 18:1089-1096 (1982).

10. Behra, Ph., A. Zysset, L. Sigg, and F. Stauffer, "Modelling of Pollutant Transport in Groundwater: Chemistry as a Key Factor," *EAWAG News* 28/29:6-11 (1990).

11. Cederberg, G.A., R.L. Street, and J.O. Leckie, "A Groundwater Mass Transport and Equilibrium Chemistry Model for Multicomponent Systems," *Wat. Res. Research* 21:1095-1104 (1985).

12. Furrer, G., J Westall and P. Sollins, "The Study of Soil Chemistry Through Quasi-Steady-State Models: 1. Mathematical Definition of Model," *Geochim. Cosmochim. Acta* 53:595-601 (1989).

13. Furrer, G., P. Sollins and J.C. Westall, "The Study of Soil Chemistry Through Quasi-Steady-State Models: II. Acidity of Soil Solution," *Geochim. Cosmochim. Acta* 54:2363-2374 (1980).

14. Kirkner, D.J., A.A. Jennings and T.L. Theis, "Multisolute Mass Transport with Chemical Interaction Kinetics," *J. Hydrol.* 76:107-117 (1985).

15. Theis, T.L., "Reactions and Transport of Trace Metals in Groundwater," in *Metal Speciation: Theory, Analysis, and Application*, J.R. Kramer and H.E. Allen, Eds. pp. 81-98. Lewis Publishers, Inc., (Chelsea, MI 1988).

16. Stone, A.T. and J.J. Morgan, "Reductive Dissolution of Metal Oxides," in *Aquatic Surface Chemistry*, W. Stumm, Ed. pp. 221-254. Wiley-Interscience, (New York, 1987).

17. Hering, J.G. and W. Stumm, "Oxidative and Reductive Dissolution of Minerals," in *Rev. in Mineralogy, Vol. 23, Mineral-Water Interface Geochemistry*, M.F. Hochella and A.F. White, Eds. pp. 427-465. Mineralogical Soc. Am., (Washington, D.C. 1990).

18. Furrer, G. and W. Stumm, "The Coordination Chemistry of Weathering: I. Dissolution Kinetics of δ-Al_2O_3 and BeO," *Geochim. Comochim. Acta* 50:1847-1860 (1986).

19. Zinder, B., G. Furrer and W. Stumm, "The Coordination Chemistry of Weathering: II. Dissolution of Fe(III) Oxides," *Geochim. Cosmochim. Acta* 50:1861-1869 (1986).

20. Wieland, E., B. Wehrli and W. Stumm, "The Coordination Chemistry of Weathering: III. A Generalization on the Dissolution Rates of Minerals," *Geochim. Cosmochim. Acta* 52:1969-1981 (1988).

21. Stumm, W., G. Furrer and B. Kunz, "The Role of Surface Coordination in Precipitation and Dissolution of Mineral Phases," *Croatica Chem. Acta* 56:593-611 (1983).

22. Stumm, W. and G. Furrer, "The Dissolution of Oxides and Aluminum Silicates: Examples of Surface-Coordination-Controlled Kinetics," in *Aquatic Surface Chemistry*, W. Stumm, Ed. pp. 197-219. Wiley-Interscience, (New York, 1987).

23. Petrovic, R., R.A. Berner and M.B. Goldhaber, "Rate Control in Dissolution of Alkali Feldspars - I. Study of Residual Feldspar Grains by X-ray Photoelectron Spectroscopy," *Geochim. Cosmochim. Acta* 40:537-548 (1976).

24. Berner, R.A. and G.R. Holdren, Jr., "Mechanism of Feldspar Weathering - II. Observations of Feldspars From Soils," *Geochim. Cosmochim. Acta* 43:1173-1186 (1979).

25. Furrer, G., "Die Oberflächenkontrollierte Auflösung von Metalloxiden: Ein Koordinationschemicher Ansatz zur Verwitterungskinetik," *Ph.D. Thesis.* Swiss Federal Institute of Technology, (Zurich, Switzerland 1985).

26. Stumm, W., G. Furrer, E. Wieland and B. Zinder, "The Effects of Complex-Forming Ligands on the Dissolution of Oxides and Aluminosilicates," in *The Chemistry of Weathering*, J.I. Drever, Ed. pp. 55-74. D. Reidel Publishing Co., (Dordrecht, 1985).

27. Wehrli, B., "Monte Carlo Simulations of Surface Morphologies During Mineral Dissolution," *J. Coll. Int. Sci.* 132:230-242 (1989).

28. Hering, J.G. and W. Stumm, "Fluorescence Spectroscopic Evidence for Surface Complex Formation at the Mineral-Water Interface: Elucidation of the Mechanism of Ligand-promoted Dissolution," *Langmuir* 7:1567-1570 (1991).

29. Davis, J.A., and D.B. Kent, "Surface Complexation Modeling in Aqueous Geochemistry," in *Rev. in Mineralogy, vol. 23, Mineral-Water Interface Geochemistry*, M.F. Hochella, Jr. and A.F. White, Eds. pp. 177-260. Mineral. Soc. Am., (Washington, D.C. 1990).

30. Sposito, G., "Molecular Models of Ion Adsorption on Mineral Surfaces," in *Rev. in Mineralogy, Vol. 23, Mineral-Water Interface Geochemistry*, M.F. Hochella, Jr. and A.F. White, Eds. pp. 261-279. Mineral. Soc. Am., (Washington, D.C. 1990).

31. Brown,Jr., G.E., "Spectroscopic Studies of Chemisorption Reaction Mechanisms at Oxide-Water Interfaces," in *Rev. in Mineralogy, vol. 23, Mineral-Water Interface Geochemistry*, M.F. Hochella, Jr. and A.F. White, Eds. pp. 309-363. Mineral. Soc. Am., (Washington, D.C. 1990).

32. Zeltner, W.A., E.C. Yost, M.L. Machesky, M.I. Tejedor-Tejedor and M.A. Anderson, "Characterization of Anion Binding on Goethite Using Titration Calorimetry and Cylindrical Internal Reflection-Fourier Transform Infrared Spectroscopy," in *Geochemical Processes at the Mineral-Water Interface*, J.A. Davis and K.F. Hayes, Eds. pp. 142-161. ACS Symposium Ser., 323 (1986).

33. Means, J.L., T. Kucak and D.A. Crerar, "Relative Degradation Rates of NTA, EDTA, and DTPA and Environmental Implications," *Environ. Poll. (Ser. B)* 1:45-60 (1980).

34. Gardiner, J., "Complexation of Trace Metals by Ethylenediaminetetraacetic acid (EDTA) in Natural Waters," *Wat. Res.* 10:507-514 (1976).

35. Giger, W., H. Ponusz, A. Alder, D. Baschnagel, D. Renggli and C. Schaffner, "Untersuchungen Ueber das Umweltverhalten des Phophatersatzstoffes NTA und des Organischen Komplexbildners EDTA.", EAWAG Jahresbericht, (Dubendorf, Switzerland 1987).

36. Giger, W., "Behavior of Organic Micropollutants During Infiltration of River Water into Groundwater: Field Studies," in *Proc. Workshop COST 641*, 16-17 Oct. pp. 33-36. (Chania, Greece 1986).

37. Dietz, F., "Neue Messergebebnisse Ueber die Belastung von Drinkwasser mit EDTA," *GWF Wasser/Abwasser* 128:286-288 (1987).

38. Brauch, H.J. and S. Schullerer, "Verhalten von Ethylenediaminetetraacetat (EDTA) und Nitriloacetat (NTA) bei der Trinkwasseraufbereitung," *Vom Wasser* 69:155-164 (1987).

39. Nirel, P.M.V., (Unpublished results).

40. Torres, R., M.A. Blesa and E. Matijevic, "Interactions of Metal Hydrous Oxides with Chelating Agents. VII. Dissolution of Hematite," *J. Coll. Int. Sci.* 131:567-579 (1989).

41. Rueda, E.H., R.L. Grassi and M.A. Blesa, "Adsorption and Dissolution in the System Geothite/Aqueous EDTA," *J. Coll. Int. Sci.* 106:243-246 (1985).

42. Litter, M.I. and M.A. Blesa, "Photodissolution of Iron Oxides: I. Maghemite in EDTA Solutions," *J. Coll. Int. Sci.* 125:679-687 (1988).

43. Bondietti, G., J. Sinniger and W. Stumm, "The Reactivity of Fe(III) (Hydr)oxides - Effects of Ligands in Inhibiting the Dissolution," *Colloids and Surfaces A-Physico Chemical and Engineering Aspects* 79:157-167 (1993).

44. Chang, H.-C. and E. Matijevic, "Interactions of Metal Hydrous Oxides with Chelating Agents. IV. Dissolution of Hematite," *J. Coll. Int. Sci.* 92:479-488 (1983).

45. Chang, H.-C., T.W. Healy and E. Matijevic, "Interactions of Metal Hydrous Oxides with Chelating Agents. III. Adsorption on Spherical Colloidal Hematite Particles," *J. Coll. Int. Sci.* 92:469-478 (1983).

46. Hering, J.G. and F.M.M. Morel, "The Kinetics of Trace Metal Complexation: Implications for Metal Reactivity in Natural Waters," in *Aquatic Chemical Kinetics*, W. Stumm, Ed. pp. 145-171. Wiley-Interscience, (New York, 1990).

47. Margerum, D.W., G.R. Cayley, D.C. Weatherburn and G.K. Pagenkopf, "Kinetics and Mechanism of Compex Formation and Ligand Exchange," in *Coordination Chemistry, Vol. 2*, A. Martell, Ed. pp. 1-220. ACS Symposium Ser., 174 (Washington, D.C. 1978).

48. Jacobs, L.A., H.R. von Gunten, R. Keil and M. Kuslys, "Geochemical Changes Along a River-Groundwater Infiltration Flow Path: Glattfelden, Switzerland," *Geochim. Cosmochim. Acta* 52:2693-2706.

49. von Gunten, H.R. and T.P. Kull, "Infiltration of Inorganic Compounds From the Glatt River, Switzerland, into a Groundwater Aquifer," *Water, Air, Soil Poll.* 29:333-46 (1986).

50. Giger, W., (Unpublished results).

51. von Gunten, H.R., G. Karametaxas, U. Krahenbuhl, M. Kuslys, R. Giovanoli, E. Hoehn and R. Keil, "Seasonal Biochemical Cycles in Riverborne Groundwater," *Geochim. Cosmochim. Acta* 55:3597-3609 (1991).

52. Matijevic, E. and P. Scheiner, "Ferric Hydrous Oxide Sols III. Preparation of Uniform Particles by Hydrolysis of Fe(III)-chloride, -nitrate, and -perchlorate Solutions," *J. Coll. Int. Sci.* 63:509-524 (1978).

53. Penners, N. and L. Koopal, "Preparation and Optical Properties of Homodisperse Haematite Hydrosols," *Coll. Surf.* 19:337-349 (1986).

54. Banwart, S., S. Davies and W. Stumm, "The Role of Oxalate in Accelerating the Reductive Dissolution of Hematite (α-Fe_2O_3)by Ascorbate," *Coll. Surf.* 39:303-309 (1989).

55. Brauer, G., "Handbuch der Praeparation Anorganischen Chemie," Bd II p.1308. Ferd Enke Verlag, (Stuttgart, Germany 1962).

4

COLLOIDAL TRANSPORT OF METAL CONTAMINANTS IN GROUNDWATER

Liyuan Liang and John F. McCarthy
Environmental Sciences Division
Oak Ridge National Laboratory
Oak Ridge, TN 37830

1. INTRODUCTION

Models of contaminant transport processes typically treat groundwater as a two-phase system with contaminants partitioning between immobile solid constituents and the mobile aqueous phase. However, solid-phase components in the colloidal size range may also be mobile in subsurface environments, although the conditions controlling their mobility are poorly understood. Association of contaminants with mobile colloidal particles may, therefore, enhance the transport of strongly sorbing contaminants [1]. Recent work has attempted to include the role of a colloidal phase in speciation models (COMET; [2]). However, it is more difficult to predict the extent to which colloids will be transported in subsurface environments because little, if any, information is available on the abundance and distribution of colloidal particles in groundwater, or the hydrogeochemical conditions controlling their formation and mobility in subsurface systems. Accurate assessment of current contaminant problems, engineering of containment strategies, and cost-effective remediation approaches are all dependent on a fundamental understanding of transport processes.

This paper outlines (1) the influence of aqueous chemistry on the association of metal contaminants with solid (and colloid) surfaces and (2) the mechanisms governing the formation and the transport of colloids. An example will be discussed on the formation of iron oxide/hydroxide colloids and the transport of organic and inorganic colloids in groundwater systems.

2. TWO PHASE SYSTEM: METAL ADSORPTION ON SOLID SURFACES

The solid-aqueous interface has been recognized as playing an important role in regulating metal concentration in soil environment and natural water systems [3,4]. For fast metal sorption reactions, an equilibrium approach is typically applied to obtain metal speciation in water and on surfaces [5]. A kinetic approach to sorption of metal on solid surfaces has been useful in quantifying reaction rates and delineating how physical chemical processes control metal in a dynamic system. In addition, mechanistic information on metal sorption on surfaces can be derived from a kinetic study [6]. In natural systems, metal sorption can also be influenced by a number of physical chemical processes. For example, ligand-promoted dissolution of metal oxide not only affects the available surface sites for metal sorption but also accounts for increase in the concentration of soluble metal ions [7]. Oxidation/reduction reactions will change the oxidation state of metals, which often determines the affinity of a metal to be sorbed to a surface [8]. Using Cr as an example, Davis *et al.* [9] demonstrated that the removal of dissolved Cr from aqueous phase was a result of irreversible sorption of Cr(III) following the reduction Cr(VI) to Cr(III) in a suboxic sandy aquifer. Redox processes can affect metal sorption through changes in the sorbent as well as the sorbate. For example, addition of a reductant may reduce either the adsorbed metal or the hydrous oxide host phase; the dissolution of the host phase may lead to metal mobilization by lowering the number of available surface sites [10].

Mineral oxides/hydroxides, especially those of Al, Si, and Fe are very important in natural water systems [11]. Therefore, in many metal ion adsorption studies, oxides are used as adsorbents [12,13]. Assuming that metals bind to oxide surface hydroxyl groups analogous to aqueous coordination reactions, general expressions can be written as follows [11,14,15]:

$$\equiv SOH + M^{z+} \rightleftarrows \equiv SOM^{(z-1)+} + H^{+} \qquad {}^{*}K_{1,app} \tag{1}$$

where $\equiv SOH$ denotes a surface site, and M^{z+} represents a metal ion. ${}^{*}K_{1,app}$ and ${}^{*}\beta_{2app}$ are the apparent surface equilibrium constants and are defined as:

$$2(\equiv S-OH)+M^{z+} \rightleftarrows (\equiv SO)_2M^{(z-2)+}+2H^+; \quad *\beta_{2,app} \tag{2}$$

$$^*k_{1,app}=\left\{\equiv SOM^{(z-1)+}\right\}[H^+]/\{\equiv SOH\}\left[M^{z+}\right] \tag{3}$$

$$^*\beta_{2,app}=\left\{(\equiv SO)_2M^{(z-2)+}\right\}[H^+]^2/\{\equiv S_2(OH)_2\}\left[M^{z+}\right] \tag{4}$$

where { } indicate the concentration of surface species and [ ] indicate the concentration of the species in the aqueous phase. Because of the charge characteristics of solid surfaces, the activities of ions at the surface need to be corrected to obtain the intrinsic equilibrium constants. Accordingly, the intrinsic equilibrium constants can be expressed as:

$$^*k_{1,int}={}^*k_{1,app}\exp[(z-1)F\Psi/RT] \tag{5}$$

$$^*\beta_{2,int}={}^*\beta_{2,app}\exp[(z-2)F\Psi/RT] \tag{6}$$

where Ψ is the potential difference between the binding site and the bulk solution. The exponential term accounts for the columbic contribution to the intrinsic equilibrium constants. Because the surface potential can not be determined experimentally, it is generally formulated based on constant capacitance model, diffuse layer model, Stern layer model, or triple layer model [13,16]. Westall and Hohl [17] have compared all these models and found that they are equivalent in predicting metal adsorption on surfaces. Applying mass and charge conservation laws to titration data and correcting for the columbic effect from the charged surface, the intrinsic equilibrium constants for metal ions binding to surface can be determined.

Dzmobak and Morel [18] compiled the existing results of metal sorption on hydrous ferric oxide surfaces and examined these results according to the general theoretical framework of coordination reactions. Using the Gouy-Chapman diffuse layer model, they derived the surface equilibrium constants for various metals on hydrous ferric oxides. They found that even though the approach is somewhat semi-empirical, the modeling results are consistent with the experimental observations.

Furthermore, they summarized the trend of surface complexation (reaction of cations with surface hydroxyl groups) with respect to the aqueous metal ion hydrolysis (reaction of cations with aqueous hydroxyl ions) and found that the equilibrium constants for these reactions are correlated (Figure 1). The correlation between the equilibrium constants of the surface complexation and the aqueous reaction is termed as linear free-energy relationships (LFER) [13], and it is useful in estimating the surface equilibrium constant of a cation in the lack of experimental data. For example, the hydrolysis equilibrium for ferrous ion is:

$$Fe^{2+} + OH^{-} = FeOH^{+} \tag{7}$$

with an equilibrium constant logK = 4.5 (between that of Cd^{2+} and Zn^{2+}). Using the LFER derived in Figure 1, the equilibrium constants for the surface complexation (on the hydrous ferric oxide) as written in Equation (1) is estimated to be $\log{}^{*}K_{1,int} = 0.8$.

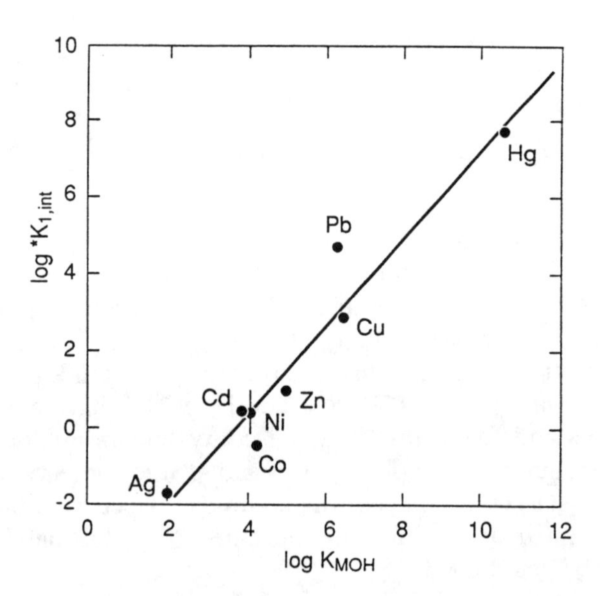

*FIGURE 1. Correlation Between the Equilibrium Constants of the Surface Complexation (log*K1,int) and the First Hydrolysis Constant (log KMOH) for Selected Metal Ions. The Fitted Line Corresponds to log*K1, int = 1.166 log KMOH - 4.374. Dzmobak and Morel [18].*

According to Equations (1) and (2), both pH and surface characteristics of a solid phase are important for metal speciation. In addition, metal speciation will be affected by ionic strength, the type and the concentration of specific ions that may promote or inhibit metal adsorption (such as natural organic matter (NOM)), as well as the redox potential.

For example, sorption of metal on solid surfaces typically shows an "adsorption edge" [12,14,19]. This concept can be illustrated by studies of sorption of Pb onto iron hydrous oxide ([12]; Figure 2).

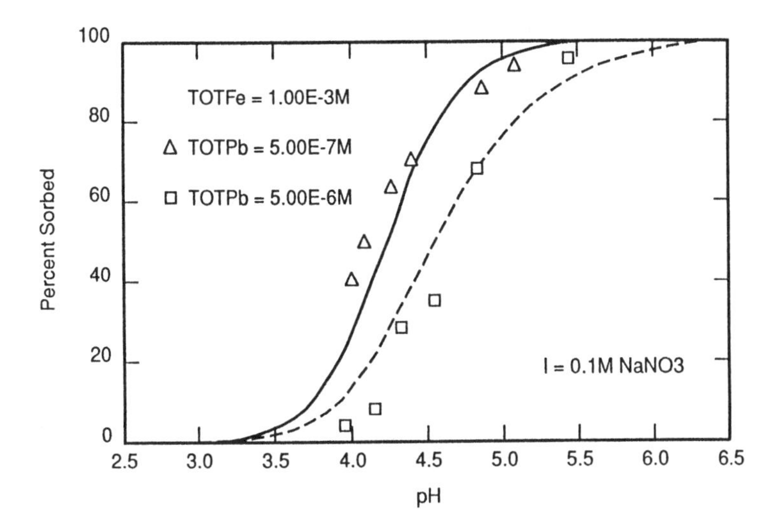

FIGURE 2. ***The pH Dependence of Lead Sorption on Hydrous Ferric Oxide at a Constant Ionic Strength. The Solid Concentration is Expressed as Total Surface Sites, TOTFe. (Δ) represents total lead concentration at 0.5 μM, and (■)) represents total lead at 5 μM. Solid - and dashed lines represent model calculation by Dzmobak and Morel [18]. Adapted from data by Benjamin [12].***

At higher pH (pH>6), Pb is preferentially sorbed on iron oxide, but at pH<3.0, essentially all of the Pb remains in solution. Between pH 3 to 6, there is a sharp increase of Pb adsorption for a small increment of pH (i.e., a "sorption edge"). For different metal ions, this sharp increase in adsorption occurs at different pHs. In general, sorption of ions with a higher affinity for surfaces has the sorption edge at a lower pH than does a weaker binding metal ([14]; Figure 3). Applying the LFER discussed in Dzmobak and Morel [18] to ferrous adsorption on amorphous silica, the Fe^{2+} sorption edge should lie between that of Cu^{2+} and Cd^{2+}. This relationship permits a semi-quantitative prediction of metal sorption under conditions relevant to natural groundwater; for example in an anoxic sandy aquifer, ferrous ions may be expected to be mainly sorbed on the solid material at pH>8.

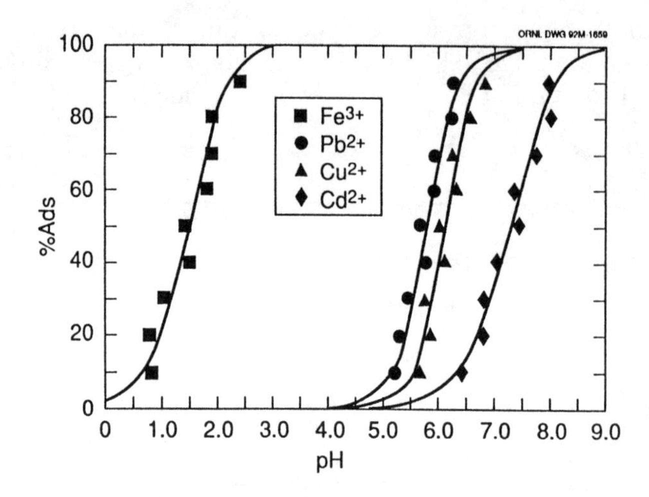

FIGURE 3. ***Adsorption of Fe(III), Cu(II), Cd(II), and Pb(II) on Silica as a Function of pH. Symbols are Actual Data and Solid Lines are Calculated Based on Surface Complexation Equilibrium. Adapted from Schindler et al. [14].***

While an increase in ionic strength usually significantly decreases metal ion activities in solution, alters the oxide surface acid-base speciation, and affects the columbic energies for metal adsorption, the net effect of ionic strength on the overall metal sorption is very small. For example, Hayes and Leckie [20] did not observe any ionic strength effect on the adsorption of Pb^{2+} on a goethite surface (Figure 4). Dzmobak and Morel [18] modeled a similar system (Zn^{2+} adsorption on hydrous ferric oxide) and attributed the lack of increase in metal sorption to the combined effects of a decrease in cation activities and an increase in the diprotonated surface sites as a result of increase in ionic strength. This description is limited to well-defined oxide surfaces and may not be predictive of sorption in natural systems. For example, Shuman (1986) observed that the adsorption of Zn^{2+} on soils varies at different ionic strength.

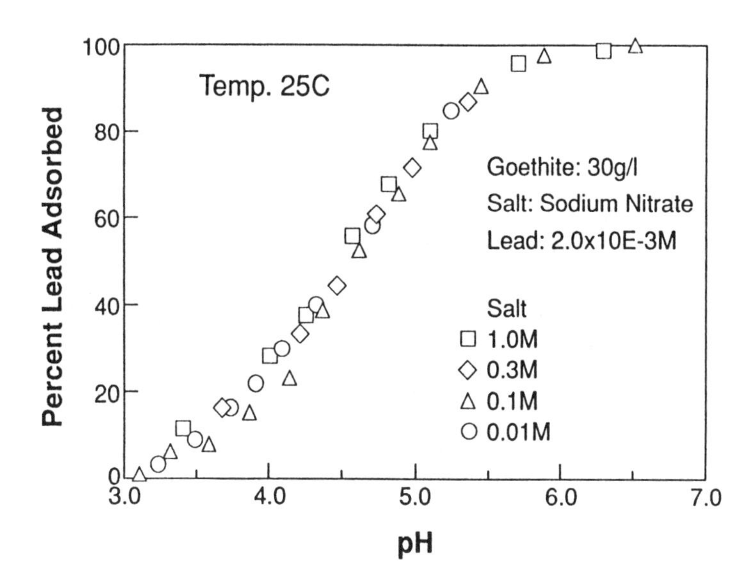

FIGURE 4. ***Effects of Ionic Strength on the Sorption of Lead on Goethite. The total lead ion concentration is 2 mM, iron oxide concentration is 30 g/L and ionic strength varies between 0.01 M to 1.0 M. Results from Hayes and Leckie [20].***

The presence of competing anions (e.g., sulfate, phosphate or NOM) can alter sorption of metals to surfaces. For example, in the presence of NOM, the formation of aqueous complexes will stabilize the metal in solution and effectively decrease the amount of aqueous metal ion available for adsorption [21,22]. In addition, anions such as sulfate and phosphate will compete with metal for surface sites, reducing the metal adsorption on solids [23,24]. The implications of these studies are important, especially in the predictive modeling of contaminant transport in geochemical systems [25].

3. THREE PHASE SYSTEM: ROLE OF A MOBILE COLLOIDAL PHASE IN SUBSURFACE

Models of metal contaminant transport treat groundwater as a two-phase system with contaminants partitioning between immobile solid phase and the mobile aqueous phase. Assuming the aquifer solid phases to have a strong affinity for sorption, these models predict that radionuclides are essentially non-mobile in subsurface [1]. However, there is an increasing body of evidence that components of the solid phase may exist in groundwater in the colloidal (submicron) size range [26]. These colloidal particles have similar composition and surface characteristics to the immobile aquifer solids, but may be mobile within aquifers. These mobile sorbents represent a potentially important vector for transport of metals through the subsurface.

Two lines of evidence suggest that colloids may influence the transport of subsurface contaminants: (1) laboratory column studies demonstrating co-transport of contaminants sorbed to mobile colloids and (2) field studies demonstrating the association of contaminants with natural groundwater colloids. For example, Champ *et al.* [27] observed that when groundwater was pumped through undisturbed aquifer cores, Pu was rapidly transported, and almost 75% of the Pu was associated with colloidal particles. Newman [28] demonstrated that when metal ion was introduced into a soil column, the breakthrough of the metal was associated with the breakthrough of colloids; surface analysis indicated that metals were adsorbed on the surfaces of the colloids. Similarly, the pesticides DDT and paraquat were co-transported with montmorillonite in saturated soil columns [29,30].

Several field studies have demonstrated the association of contaminants with colloidal material from groundwater. At the Nevada Test Site, transition metals and lanthanide radionuclides were associated with inorganic colloids recovered from groundwater 300 m from the nuclear detonation cavity [31]. Pu and Am were associated with siliceous colloids in an alluvial aquifer at Los Alamos National Laboratory [32]. Filterable particles (>0.4 μm) containing radionuclides of Co, Zr, Ru, Cs, and Ce were recovered from contaminant plumes at the Chalk River Nuclear Laboratory [27]. Uranium and daughter species were found associated with iron- and silicon-rich colloids down-gradient from a uranium deposit in Australia [33].

Difficulties in sampling and characterization of colloids in groundwater, and the potential for introduction of artifacts during this process [26,34], has limited our understanding of, and ability to predict, the formation, stability and transport of colloids in the subsurface. Nevertheless, recent research has greatly increased the available data on the nature and abundance of groundwater colloids. Colloidal particles in groundwater may be composed of a variety of materials, including mineral precipitates (notably metal oxides, hydroxides, carbonates, silicates, and phosphates, but also including actinide elements such as uranium, neptunium, plutonium and americium); rock and mineral fragments (including layer silicates, oxides, and other weatherable mineral phases); "biocolloids" (including viruses, bacteria, and fragments of these organisms); microemulsions of nonaqueous phase liquids; and macromolecular components of NOM (NOM covers a wide spectrum of size range and may be considered a molecule at the smaller sizes (<3000 MW) and a colloid at larger sizes).

The very large surface area of colloids (10-500 m^2 g^{-1}) suggests that the role of colloids in sorbing and potentially co-transporting metals could be significant even for relatively low mass concentrations of colloidal particles. The size and concentration of groundwater colloids varies considerably over a range of hydrogeochemical conditions [26]. In general, higher concentrations of colloids are found in disturbed systems. For example, in a range of subsurface environments with stable hydrogeochemistry, including deep fractured granitic systems, sandstones, or shallow sandy aquifers, colloids of 10 nm to 450 nm are present in concentrations of 0.025 to 1.0 mg L^{-1}. Much higher concentrations (20 to 100 mg L^{-1}) are reported in quite similar systems, but in zones with hydrogeochemical perturbations, including changes in temperature, pH, or redox conditions ([26], and references therein).

It is not unreasonable to observe an apparent linkage between geochemical disturbances and increased concentrations of stable colloids in groundwater, based on current understanding of the mechanisms by which colloids can be generated in aquifers. Mechanisms postulated for the generation of colloids include:

> *Dispersion* — Changes in the groundwater chemistry such as a decrease in ionic strength or changes in ionic composition can cause dispersion of colloidal particles and allow them to become mobile in aquifers. For example, introduction of low ionic strength water into a sandy aquifer dispersed submicron-size particles and caused turbidity in wells several hundred meters down-gradient [35].

> *Decementation of Secondary Mineral Phases* — Colloidal particles may be cemented to each other and to larger mineral grains by secondary mineral phases such as oxides and carbonates. Geochemical changes that result in dissolution of these cementing phases can result in release of colloids. For example, reducing conditions in an aquifer beneath a swamp appeared to cause mobilization of clays by dissolving ferric hydroxide coatings binding the clay particles to aquifer solids [36].

> *Geochemical Alteration of Primary Minerals* — In crystalline formations, colloids can be generated by micro-erosion of primary minerals and secondary phase

production [37]. Micro-erosion can result from crushing and resuspension of minerals due to tectonic activity within the formation. Particles can also be produced by the chemical dissolution of primary minerals between less soluble mineral grains (e.g., a biotite particle within a more readily weathered feldspar phase may be freed when the feldspar is altered to clay and subsequently removed). Colloids may also be generated during production and micro-erosion of secondary mineral phases.

Homogeneous Precipitation — Changes in groundwater geochemical conditions such as pH, ionic composition, redox potential, or partial pressures of CO_2 can induce supersaturation and co-precipitation of insoluble colloidal particles. As an example, Gschwend and Reynolds [38] observed precipitation of ferrous phosphate colloids down-gradient of a sewage infiltration site. Many strongly hydrolyzing radionuclides such as uranium also form submicron-sized particles [39]. In the case study discussed later in this paper, we describe the formation of iron oxide colloids in groundwater as a result of changes in dissolved oxygen levels which caused supersaturation with respect to iron oxide (hydroxide).

NOM Effects on Stability — NOM may be a critical factor in maintaining the negative surface potentials of newly formed or dispersed particles and limiting the deposition of colloids to aquifer materials (generally negatively charged). For example, colloids such as iron oxides carry a net, small positive charge near neutral pH [40]; thus they are inherently unstable from colloid stability point-of-view and deposition on aquifers is favored. The association of NOM with iron oxides has been demonstrated in laboratory studies [41] and in surface waters [42-44] to result in an overall negative surface potential at pH 6.5 and to the increase in the colloid stability. Ryan and Gschwend [36] postulated that colloidal hydrous oxides of Fe, Al, and Ti in a coastal sedimentary aquifer were stabilized as suspensions in groundwater by coatings of organic carbon on the inorganic particles.

4. PREDICTING TRANSPORT OF COLLOIDS AND COLLOID-ASSOCIATED METALS

Even if the geochemistry favors the formation of stable colloids, suspended particles will be significant to the transport of metals in groundwater only if they are transported through aquifers. Current approaches to predicting the migration of colloids are based on filtration theory [45]. This theoretical work was initially developed for describing and predicting efficiencies of particle (aqueous or gaseous) removal from a filter bed of solid matrix [46-49].

The theory with regard to the kinetics of deposition of colloidal particles has been conveniently divided into two aspects, transport and attachment [50]. The transport of a colloidal particle from bulk fluid to a collector (e.g., a sand grain of an aquifer) is affected by physical processes such as fluid convection, diffusion of the particle and, for larger particles, gravity and fluid drag forces. The attachment of colloidal particle to the surface of a collector is mainly a physicochemical process and is a function of various forces operative at short distances from the interface, such as electrostatic interaction, van der Waals attraction, hydrodynamic, hydration, hydrophobic interaction, and steric forces.

There are two approaches to model the mass transfer of particles in a porous media. For particles greater than 1 μm, Brownian diffusion is less important, and the Lagrangian or trajectory approach is commonly applied [51]. In this method, the particle travel path toward a collector is followed; the interception of the particle to the collector depends on a balance of torque and forces. These forces include gravity, fluid drag, and interfacial forces [52]. Non-Brownian transport may be of limited significance in most groundwater situations because particles will tend to settle out due to gravity.

For submicron size particles, the Brownian diffusion makes the trajectory method less deterministic and the Eulerian or mass transport approach is generally applied. In this method, the transport of particles toward a collector is determined by solving the convective diffusion equations. In the presence of an external force field, the concentration distribution of Brownian particles over a collector can be described as follows [48]:

$$\frac{\partial C}{\partial t} + \bar{v} \cdot \nabla C \cdot \nabla \left(D \nabla C - \frac{DC}{kT} \bar{F} \right) \qquad (8)$$

where C is the particle number concentration, v is the fluid velocity vector, t is the time, D is the colloid diffusion coefficient, k is the Boltzmann's constant, T is the absolute temperature, and F is the force field vector. Following Levich's (1962) solution to the diffusion equation in the absence of repulsion forces between particles and collectors (i.e., favorable deposition), Yao *et al.* [46] obtained the rate of transport of particles towards a single spherical collector. Incorporating the Happel model (1958) for the effect of neighboring collector in a packed bed, the single collector efficiency is given by:

$$\eta_D = 4.0\, A_s^{1/3}\, Pe^{-2/3} \tag{9}$$

where As depends on the porosity of a packed bed and Pe is dimensionless and depends on the approach velocity of the fluid, the diameter of the collector, and the diffusion coefficient of the particles. Equation (8) accounts for the diffusive transport of particles and is a good approximation for polystyrene latex particles up to 1 μm. A general expression for particle deposition accounting for diffusion, fluid flow and gravity is given by Rajagopalan and Tien [49] and is summarized in a review by O'Melia [45].

The attachment of particles in a porous media depends on the forces acting on particles and a collector at short distance. For example, changes in aqueous chemical conditions can cause a particle to develop negative charges. A negatively charged colloidal particle will be unfavorable to deposition on a negatively charged collector. In the presence of an external force field, the particle deposition efficiency is reduced by a factor, α. This α factor is termed a *sticking coefficient* and can be theoretically modeled by using the analytical solution to Equation (7) given by Spielman and Friedlander [48]. The interaction energy due to the external force field can be evaluated by theory developed by Derjaguin, Landau, Verwey and Overbeek (i.e., DLVO theory).

From an experimental point of view, various physical chemical factors controlling colloidal particle transport in porous media can be evaluated using assumptions of clean bed filtration. Integrating over the depth of the filter column, the colloid concentration versus distance from such experimental setup is given by Yao [53]:

$$\text{In}\left(\frac{C_L}{C_0}\right) = \frac{3}{2}\alpha\eta\ \frac{(1-\varepsilon)}{d_c}L \tag{10}$$

Where C_L is the concentration measured at distance L from the injection point; Co is the initial concentration of colloid injected; d_c is collector diameter and ε is the porosity of the filter bed. Since η can be theoretically modeled by using Equation (9), the only unknown, α, can be determined.

Alternatively, if the sticking coefficient can be modeled from theory, the concentration of colloids at desired distance can be predicted using Equation (9). For example, if a colloidal suspension consists of particles of diameter 100 nm, the effective diffusion coefficient is $2x10^{-12} m^2/sec$. Assuming the sand grains in an aquifer are 0.2-mm in diameter and the porosity is 0.5, for a groundwater flow rate of 3.5 m day^{-1}, the mass transfer rate of the colloid, η, is estimated to be $7x10^{-2}$. If the sticking coefficient is 1.0, only 1% of colloids will be detected at a distance of 3.5 cm; colloid migration under these conditions is quite limited. However, if all parameters remain the same, except that α is reduced from 1.0 to 0.0001, the same percentage of colloids will be detected at a distance of 350 meters (10^4-fold increase in distance). Under such circumstances, colloid migration and the potential for colloid-facilitated transport of contaminants, including metals, can become important.

There are two main difficulties in applying filtration theory to predict colloid transport in the environment:

1. *Theory Limitations* – Filtration theory works well when conditions for attachment are favorable. However, in the presence of a force field, the theoretical predictions are less satisfactory. For example, using quartz as filtration media, Litton and Olson [54] studied the transport of negatively charged polystyrene latex particles. The study showed that at high pH, the experimentally determined sticking coefficient, α, was order of magnitude higher than theoretical prediction (Figure 5). This is partially due to the apparent assumptions within the theory, such as the uniform surface potential and point charges at the interface. In addition, the lack of agreement between theory and experiments has been attributed to the inadequate description of the particle-particle interaction energies. For example, theory has yet to consider ion relaxation kinetics in the double layer, surface roughness of colloid and collectors, and ion interactions within a diffuse layer [50].

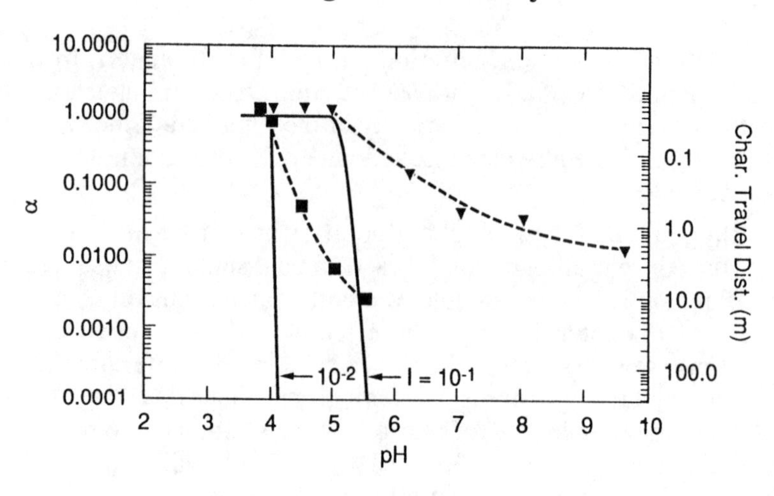

FIGURE 5. ***Attachment Efficiencies and Characteristic Travel Distance (in meters) vs pH for 0.245μm Carboxyl Latex Spheres in Quartz (0.2-mm) at Varying Ionic Strength: (▼) 0.1 M (■) 0.01 M. Solid lines are theoretical predictions. The theoretical prediction agrees with experimental results at low pH when attachment is favorable (α = 1). At high pH, theory underestimates the attachment efficiencies. Adapted from Litton and Olson [54].***

2. *System Limitations* – Some fundamental assumptions of filtration theory simply do not apply in many natural groundwater systems. For example, filtration theory assumes a uniform isotropic porous media; however, in soils and aquifers, water often flows preferentially through channels and secondary pore structure rather than through the intergranular pore space. In laboratory columns, the organic colloid, blue dextran (2×10^6 dalton molecular weight) eluted from the column faster than ^{3}H-water, presumably by being forced through larger pores and excluded from the smaller pores [55,56]. Likewise, Harvey *et al.* [57] observed that bacteria injected into the Cape Cod aquifer eluted slightly ahead of the nonreactive tracer. Preferential flow paths such as fractures, solution channels, or soil macropores can greatly enhance transport and reduce retention of particles. Smith *et al.* [58] recovered 22 to 79% of bacteria injected onto intact soil columns, but only 0.2 to 7% of the bacteria were recovered if the columns were prepared from

mixed, repacked soil; the results suggested that flow through macropores present in the intact columns bypassed the adsorptive surfaces of the porous media. The role of fractures on particle transport was examined by Toran and Palumbo [59] who simulated fractures by inserting small (0.2- or 1-mm diameter) tubes into a sand column. The "fractures" significantly increased the transport (decreased retention) of latex microspheres and bacteria.

In summary, there appears to be a preponderance of evidence that groundwater colloids are able to bind contaminants and may, in at least some circumstances, be mobile in aquifers. Unfortunately, much of the information specific to groundwater colloids is more anecdotal than systematic, and is insufficient to reliably evaluate the significance of colloids as a transport vector or to develop a predictive capability. Additional progress in this field will benefit from controlled field research, supported by laboratory studies elucidating the fundamental processes underlying the field observation of mobile colloids.

5. A CASE STUDY

In the summer of 1990, a field experiment involving injection of NOM was conducted in a sandy aquifer in order to improve capabilities to predict the subsurface transport of contaminants that sorb to mobile NOM. The research also sought to understand the role of NOM in mobilizing solid-phase iron in the aquifer, as well as its role in stabilizing iron oxide colloids. The discussion of this study will be limited to those aspects related to the formation and stability of iron oxide colloids. The results of the study of NOM transport and of complexation of iron by mobile NOM are described in McCarthy *et al.* [60] and Liang *et al.* [61].

The field experiment was conducted in a shallow, sandy, coastal plain aquifer at Hobcaw Field within the Baruch Forest Science Institute of Clemson University located in Georgetown, South Carolina. The aquifer media was approximately 90-95% sand, with 3-10% clay, 0.3-4.1 mg g^{-1} total iron, and 0.03-0.05% organic carbon. The experiment involved injection of 80,000 L water containing high levels of dissolved organic carbon (66 mg carbon L^{-1}); the injection solution also contained 2 mg L^{-1} dissolved oxygen (DO). Iron dynamics were followed for 2 weeks in three saturated horizons at sampling wells located 1.5 m and 3 m from the injection well. The initial oxidation/reduction potential of the aquifer favored Fe(II) in the iron-rich groundwater, and the redox potential was expected to increase as oxygen-rich water was introduced into the groundwater. The changing redox potentials were hypothesized to result in a decrease in Fe(II) and an increase Fe(III) due to oxidation, with the Fe(III) mostly in the ferric oxide/hydroxide colloidal fraction; that is, the ferric fraction would be found predominantly in sizes >0.1 μm. Furthermore, the ferric oxide/hydroxide colloids were expected to have a negative surface charge even in the pH range of the groundwater (6.0–7.2) due to NOM adsorption.

It was observed that in the zone within the aquifer where DO substantially increased from 0.1 to 1.5 mg L^{-1}, turbidity increased by a factor of 10 (Figure 6). A comparison of ferric/ferrous concentration ratios and size distribution at the beginning and the end of injection is shown in Figure 7. Fe(III) increased more than 20-fold over the course of the injection, while Fe(II) was reduced to one-tenth of its initial value during the same period. At the end of the injection, Fe(II) remained in the dissolved form (passing through 3000 mol. wt. hollow fiber filter), but 73% of Fe(III) was large enough to be retained on a 0.1-μm filter. The NOM (measured as total organic carbon in a groundwater sample) was largely in the <3000 mol. wt. size range, but about 5% of organic carbon was also retained by a 0.1 μm filter.

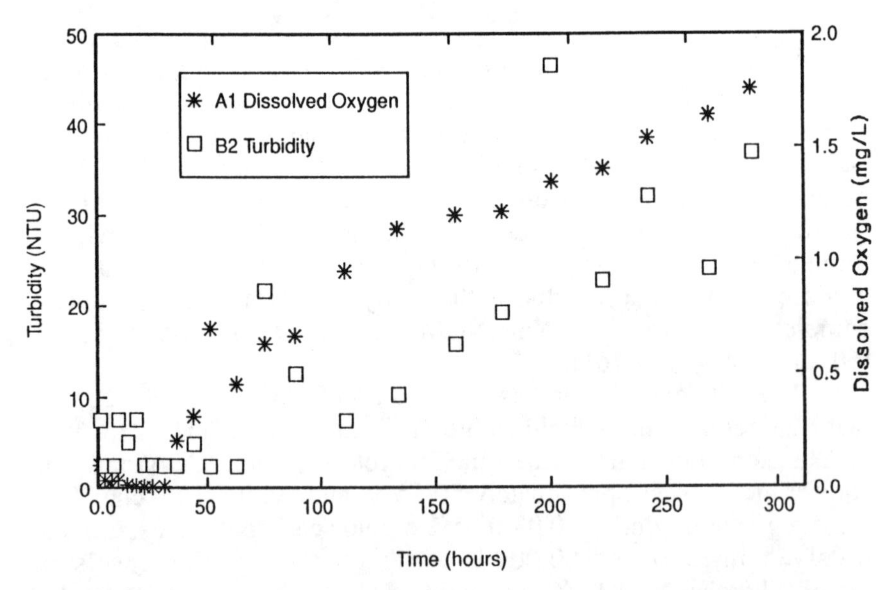

FIGURE 6. ***Changes in Dissolved Oxygen and Turbidity in a Sampling Well during the Injection of NOM-rich Water Containing 2 mg L^{-1} Dissolved Oxygen. Adapted from Liang et al. [61].***

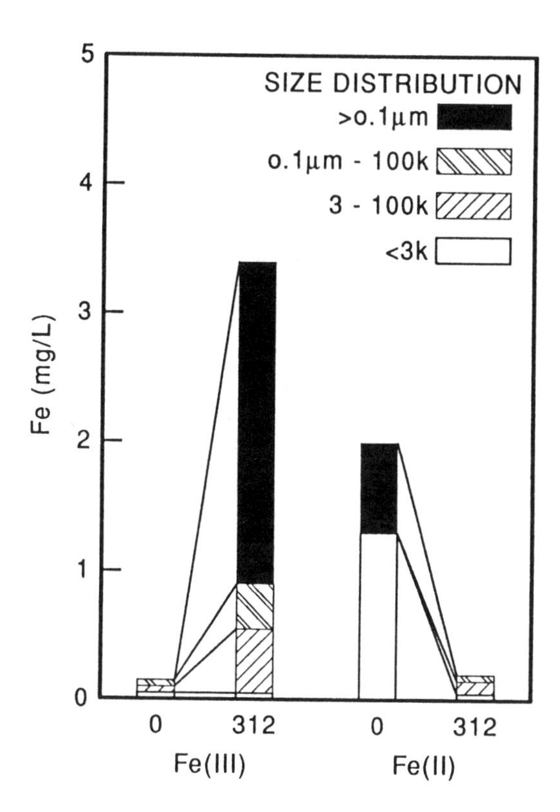

FIGURE 7. Changes in Concentration and Size of Fe(II) and Fe(III) during the Injection of Water with High NOM and DO Content. "0" indicates the starting time of injection. "312" indicates the elapsed time (in hours) since the injection began. Adapted from Liang et al. [61].

The electrophoretic mobilities of groundwater particles in the oxygenated zone showed negative surface potentials (-20 mv). The number of groundwater particles increased substantially, as indicated by photon correlation spectroscopy (PCS); scattered light counts-per-second (CPS) increased from 8400 CPS at the beginning to 148,000 CPS at the end of the injection. The sizes of the groundwater particles fell into two ranges: one, about 0.2 μm; the other, >3 μm. This bimodal distribution is in good agreement with observations under a scanning electron microscopy (SEM), which showed individual spherical particles of 0.2 μm and large aggregates of these particles. X-ray analysis of SEM samples indicated the presence of iron in the colloidal particles. The oxygenation product of the same groundwater was analyzed. These colloids had similar size and shape, by SEM, as those sampled from the wells during the injection experiment; X-ray diffraction showed amorphous structure.

It is understood that in the presence of oxygen, the Fe(II) is rapidly oxidized, and the half-life is on the order of minutes, as demonstrated in laboratory studies [62]. The disappearance of Fe(II) and the production of Fe(III) during the injection can be interpreted as a rapid oxygenation process.

Because Fe(III) hydroxides/oxides are sparingly soluble, precipitation of Fe(III) will follow oxidation. The observations from SEM and PCS support the existence of colloidal groundwater particles. These colloids are most likely iron colloids because most of the Fe(III) is in the size range >0.1 μm; colloids of this size range are very effective in scattering light and influencing turbidity.

At pH 6.0-7.2, most iron hydroxide/oxide particles have a positive or neutral charge [40]. The consistently negative surface potential from electrophoretic mobility measurement suggests that iron colloids are associated with anions. Fourier Transform Infrared Spectroscopy analysis on the groundwater showed a strong spectrum for $-COO^-$ (Dr. N. Marley, Argonne National Laboratory, personal communication). Carboxylate bonding to iron oxide has been established by Peck *et al.* [63]; hence, the association of iron colloids with organic anions may result in the negative surface potential of particles.

Results also suggest that the stable NOM-coated iron colloids were transported through the aquifer. In the field study, the injected NOM moved rapidly through the aquifer; however, differences were observed in the mobility of different size components of NOM. A small amount of large NOM (>0.1 μm in size, by filtration) was rapidly transported. It is postulated that this NOM was sorbed to, and co-transported with, iron colloids. In support of this, it was noted that the increase in Fe(III) was twice as great as the decrease of Fe(II) in the sampling well at the end of the injection, suggesting that transport of iron colloids may be occurring.

In this study, we have demonstrated *in situ* iron oxide colloid formation as a result of an oxidation process. Furthermore, the interaction of inorganic colloids and natural organic matter can be of importance in the transport of the colloids. These results highlight the point that changes in geochemical parameters may cause the formation of mobile sorbents and thereby affect the transport of metals in subsurface.

6. CONCLUDING REMARKS

The motivation for this discussion of groundwater colloids centers on their potential role in migration of highly adsorbed contaminants, including many metals and radionuclides. However, evidence for enhanced transport of contaminants as a result of sorption and co-transport on mobile colloids is largely circumstantial. Several studies have demonstrated that colloids are present in groundwater and that their abundance is promoted by geohydrochemical perturbations [64]. In a few instances, contaminants associated with colloidal particles have been recovered from monitoring wells much further from the input source than models and laboratory sorption experiments predicted. Unfortunately, there is no clear demonstration that the contaminants migrated by attaching to mobile colloids; solute transport through preferential flow paths and subsequent attachment to particles around the sampling well cannot be unequivocally discounted. There are a few examples that clearly demonstrate transport of (uncontaminated) colloidal particles (mostly biocolloids) from a clearly identified source to a monitoring well [65]. Sorption experiments (including a limited number of studies using natural groundwater colloids) have demonstrated that colloids are capable of associating with many metals and radionuclides [32,34,66,67]. However, information on the affinity, kinetics and reversibility, which is critical to evaluating the significance of colloids to contaminant transport, is largely unavailable for groundwater particles.

Acknowledgments: This research was sponsored by the Subsurface Science Program, Environmental Sciences Division, U..S. Department of Energy under contract DE-AC05-84OR21400 with Martin Marietta Energy Systems, Inc. Publication No. 4117, Environmental Sciences Division, Oak Ridge National Laboratory.

REFERENCES

1. McCarthy, J.F., and J.M. Zachara, "Subsurface Transport of Contaminants," *Environ. Sci. Technol.* 23:496-503 (1989).

2. Mills, W.B., S. Liu and F.K. Fong, "Literature Review and Model (COMET) for Colloid/Metals Transport in Porous Media," *Groundwater* 29:199-208 (1991).

3. Bolt, G.H. Ed., *Soil Chemistry, Vol. B: Physico-Chemical Models* (Elsevier, Amsterdam).

4. Honeyman, B.D. and P.H. Santschi, "Metals in Aquatic Systems," *Environ. Sci. Technol.* 22:862-871 (1988).

5. Page, A.L., personal communication (1992).

6. Sparks D.L., "Kinetics of Metal Sorption Reactions," in *Metal Speciation and Contamination of Soil.* H.E. Allen, C.P. Huang, G.W. Bailey and A.R. Bowers, Eds. pp. 35-58. Lewis Publishers, (Chelsea, MI 1994).

7. Hering, J.G., "Implications of Complexation, Sorption and Dissolution Kinetics for Metal Transport in Soils," in *Metal Speciation and Contamination of Soil.* H.E. Allen, C.P. Huang, G.W. Bailey and A.R. Bowers, Eds. pp. 59-86. Lewis Publishers, (Chelsea, MI 1994).

8. Stucki, J.W., G.W. Bailey and H. Gan, "Redox Reactions in Phyllosilicates and Their Effects on Metal Transport," in *Metal Speciation and Contamination of Soil.* H.E. Allen, C.P. Huang, G.W. Bailey and A.R. Bowers, Eds. pp. 113-182. Lewis Publishers, (Chelsea, MI 1994).

9. Davis, J.A., D.B. Kent, B.A. Rea, S.P. Garabedian and L.C.D. Anderson, "Effect of the Geochemical Environment on Heavy-Metal Transport in Ground Water," in *Water Resources Investigations Report*, G.E. Mallard and D.A. Aronson, Eds. pp. 91-4034. (Monterey, CA 1991).

10. Stone, A.T., "Reductive Mobilization of Oxide-Bound Metals: The Role of Reductant Capacity and Reductant Reactivity in Determining Mobilization Rates in Soils and Sediments," *DOE/ER*/60946 pp. (1990).

11. Stumm W. and J.J. Morgan, *Aquatic Chemistry*, Wiley & Sons, (New York, 1981).

12. Benjamin, M.M., *Effects of Competing Metals and Complexing Ligands on Trace Metal Adsorption at the Oxide/Solution Interface,. Ph.D. Thesis.* Stanford University, (Stanford, CA 1978).

13. Sigg, L. and W. Stumm, "The Interaction of Anions and Weak Acids with the Hydrous Goethite (-FeOOH) Surface," *Colloid and Surfaces* 2:101-117 (1981).

14. Schindler, P.W., F. Furst, R. Dick and P.U. Wolf, "Ligand Properties of Surface Silanol Groups I: Surface Complex Formation with Fe^{3+}, Cu^{2+}, and Pb^{2+}," *J. Colloid Interf. Sci.* 55:469-475 (1976).

15. Morel, F.M.M., *Principles of Aquatic Chemistry*, Wiley & Sons, (New York, 1987).

16. Davis, J.A. and J.O. Leckie, "Surface Ionization and Complexation at the Oxide/Water Interface: II. Surface Properties of Amorphous Iron Oxyhydroxide and Adsorption of Metal Ions," *J.Colloid Interf. Sci.* 67:90-107 (1978).

17. Westall, J.C. and H. Hohl, "A Comparison of Electrostatic Models for the Oxide/Solution Interface," *Advan. Colloid Interf. Sci.* 12:265-294 (1980).

18. Dzmobak, D.A. and F.M.M. Morel, *Surface Complexation Modeling - Hydrous Ferric Oxide*, Wiley & Sons, (New York, 1990).

19. Hayes, K.F. and J.O. Leckie, "Mechanisms of Lead Ion Adsorption at the Goethite-Water Interface" in *Geochemical Processes at Mineral Surfaces. ACS Symp. Ser.* 323, J.A. Davis and K.F. Hayes, Eds. pp.114-141. Am. Chem. Soc., (Washington, D.C. 1986).

20. Hayes, K.F. and J.O. Leckie, "Modeling Ionic Strength Effects on Cation Adsorption at Hydrous Oxide/Solution Interfaces," *J.Colloid Interf. Sci.* 115:564-572 (1987).

21. Lion, L.W., R.S. Altmann and J.O. Leckie, "Trace-metal Adsorption Characteristics of Estuarine Particulate Matter: Evaluation of Contributions of Fe/Mn Oxide and Organic Surface Coatings," *Environ. Sci. Technol.* 16:660-666 (1982).

22. Davis, J.A., "Complexation of Trace Metals by Adsorbed Natural Organic Matter," *Geochim. Cosmochim. Acta* 48:679-691 (1984).

23. Zachara, J.M., C.C. Ainsworth, C.E. Cowan and C.T. Resch, "Adsorption of Chromate by Subsurface Soil Horizons," *Soil Sci. Soc. Am. J.* 53:418-428 (1989).

24. Neal, R.H., G. Sposito, K.M. Holtzclaw and S.J. Traina, "Selenite Adsorption on Alluvial Soils: II Solution Composition Effects," *Soil Sci. Soc. Am. J.* 51:1165-1169 (1987).

25. Davis, J.A. and D.B. Kent, "Surface Complexation Modeling in Aqueous Geochemistry," in *Mineral-Water Interface Geochemistry, Reviews in Mineralogy (23). Mineralog. Soc. Am.* M.F. Hochella and A.F. White, Eds. pp. 177-260. (Washington, D.C. 1990).

26. McCarthy, J.F. and C. Degueldre, "Sampling and Characterization of Colloids and Particles in Groundwater for Studying Their Role in the Subsurface Transport of Contaminants," in *Environmental Particles.* J. Buffle and H. Van Leeuwen, Eds. pp. 247-315. Lewis Publishers, (Chelsea, MI 1992).

27. Champ, D.R., W.P. Merritt and J.L. Young, "Potential for the Rapid Transport of Plutonium in Groundwater as Demonstrated by Core Column Studies," *Scientific Basis or Radioactivity Waste Management* V:745-754 (1982).

28. Newman, M.E., "Effects of Alterations in Groundwater Chemistry on the Mobilization and Transport of Colloids," *Ph.D. Thesis.* Clemson University, (Clemson, SC 1990).

29. Vinten, A.J.A., B. Yaron and P.H. Nye, "Vertical Transport of Pesticides Into Soil When Adsorbed on Suspended Particles," *J. Agric. Food Chem.* 31:662-665 (1983).

30. Vinten, A.J.A. and P.H. Nye, "Transport and Deposition of Dilute Colloidal Suspensions in Soils," *J. Soil Sci.* 36:531-541 (1985).

31. Buddemeier, R.W. and J.R. Hunt, "Transport of Colloidal Contaminants in Groundwater: Radionuclide Migration at the Nevada Test Site," *Appl. Geochem.* 3:535-548 (1988).

32. Penrose, W.R., W.L. Polzer, E.H. Essington, D.M. Nelson and K.A. Orlandini, "Mobility of Plutonium and Americium Through a Shallow Aquifer in a Semiarid Region," *Environ. Sci. Technol.* 24:228-234 (1990).

33. Short, S.A., R.T. Lowson and J. Ellis, "$^{234}U/^{238}U$ and $^{230}Th/^{234}U$ Activity Ratios in the Colloidal Phases of Aquifers in Lateritic Weathered Zones," *Geochim. Cosmochim. Acta* 55:2555-2563 (1988).

34. Backhus, D.A., "Colloids in Groundwater: Laboratory and Field Studies of Their Influence on Hydrophobic Organic Contaminants," *Ph.D. Thesis*. Massachusetts Institute of Technology, (Cambridge, MA 1990).

35. Nightingale, H.I. and W.C. Bianchi, "Groundwater Turbidity Resulting From Artificial Recharge," *Groundwater* 15:146-152 (1977).

36. Ryan,Jr., J.N., and P.M. Gschwend, "Colloid Mobilization In Two Atlantic Coastal Plain Aquifers," *Water Resour. Res.* 26:307-322 (1990).

37. Drever, J.I., Ed.,. *The Chemistry of Weathering*, D. Reidel Publishing Co., (Dordrecht, Holland 1985).

38. Gschwend, P.M. and M.D. Reynolds, "Monodisperse Ferrous Phosphate Colloids in an Anoxic Groundwater Flume," *J.Contam. Hydrol.* 1:309-327 (1987).

39. Ho, C.H. and N.H. Miller, "Formation of Uranium Oxide Sols in Bicarbonate Solutions," *J. Colloid Interf. Sci.* 113:232-240. (1986).

40. James, R.O. and G.A. Parks, "Characterization of Aqueous Colloids by Their Electrical Double-layer and Intrinsic Surface Chemical Properties," in *Surface and Colloid Science, Volume 12*. E. Matijevic Ed. pp. 119-216. (Plenum, New York 1982).

41. Laboratory Studies and Implications For Natural Systems. *Aquatic Sci.* 52:32-55.

42. Hunter, K.A. and P.S. Liss, "The Surface Charge of Suspended Particles in Estuarine and Coastal Waters," *Nature* 282:823-825 (1979).

43. Tipping, E., "The Adsorption of Aquatic Humic Substances by Iron Oxides," *Geochim. Cosmochim. Acta.* 45:191-199 (1981).

44. Davis, J.A., "Adsorption of Natural Dissolved Organic Matter at the Oxide/water Interface," *Geochim. Cosmochim. Acta* 46:2381-2393 (1982).

45. O'Melia, C.R., "Kinetics of Colloid Chemical Processes in Aquatic Systems," in *Aquatic Chemical Kinetics*, W. Stumm, Ed. pp. 447-474. Wiley & Sons, (New York, 1990).

46. Yao, K.M., M.T. Habibian and C.R. O'Melia, "Water and Wastewater Filtration: Concepts and Applications," *Environ. Sci. Technol.* 5:1105-1112 (1971).

47. Ruckenstein, E. and D.C. Prieve, "Rate of Deposition of Brownian Particles Under the Action of London and Double Layer Forces," *J. Chem. Soc. Far. Trans. II* 69:1522-1536 (1973).

48. Spielman, L.A. and S.K. Friedlander, "Role of Electric Double Layer in Particle Deposition by Convective Diffusion," *J. Colloid Interf. Sci.* 46:22-31 (1974).

49. Rajagopalan, R. and C. Tien, "The Theory of Deep Bed Filtration," in *Progress in Filtration and Separation Vol. 1*. R.J. Wakeman, Ed. pp. 179-269. (Elsevier, Amsterdam 1979).

50. O'Melia, C.R., "Particle-Particle Interactions," in *Aquatic Surface Chemistry*, W. Stumm Ed. pp. 385-404. Wiley & Sons, (New York, 1987).

51. Vaidyanathan, R., "Colloid Deposition in Granular Media Under Unfavorable Surface Conditions," *Ph.D. Thesis*. Syracuse University, (Syracuse, New York 1987).

52. Rajagopalan, R. and C. Tien, "Trajectory Analysis of Deep-bed Filtration With the Sphere-in-cell Porous Media Model". *Am. Ins. Chem. Eng*. 22:523-533 (1976).

53. Yao, K.M., "Influence of Suspended Particle Size on the Transport Aspect of Water Filtration," *Ph.D. Thesis*. University of North Carolina, (Chapel Hill, 1968).

54. Litton, G.M. and T.M. Olson, "Colloid Deposition in Silica Bed Media and Artifacts Related to Collector Surface Preparation Methods," *Environ. Sci. Technol.* 27:185-193 (1993).

55. Enfield C.G. and G. Bengtsson, "Macromolecular Transport of Hydrophobic Contaminants in Aqueous Environments," *Groundwater* 26:64-70 (1988).

56. Enfield, C.G., G. Bengtsson and R. Lindqvist, "Influence of Macromolecules on Chemical Transport," *Environ. Sci. Technol.* 23:1278-1286 (1989).

57. Harvey, R.W., L.H. George, R.L. Smith and D.R. LeBlanc, "Transport of Fluorescent Microsphere and Indigenous Bacteria Through A Sandy Aquifer: Results of Natural- and Forced-Gradient Tracer Experiments," *Environ. Sci. Technol.* 23:51-56 (1989).

58. Smith, M.S., G.W. Thomas, R.E. White and D. Ritonga, "Transport of Escherichia-coli Through Intact and Disturbed Soil Columns," *J. Environ. Qual.* 14:87-91 (1985).

59. Toran, L. and A.V. Palumbo, "Colloid Transport Through Fractured and Unfractured Laboratory Sand Columns," *Journal Contaminant Hydrology* 9:289-303 (1992).

60. McCarthy, J.F., L. Liang, P.M. Jardine and T.M. Williams, "Mobility of Natural Organic Matter Injected Into A Sandy Aquifer," in *Manipulation of Groundwater Colloids for Environmental Restoration*, J.F. McCarthy and F.J. Wobber, Eds. pp. 35-39. Lewis Publishers, (Chelsea, MI 1992).

61. Liang, L., J.F. McCarthy, T.M. Williams and L. Jolley, "Iron Dynamics During Injection of Natural Organic Matter in a Sandy Aquifer," in *Manipulation of Groundwater Colloids for Environmental Restoration*, J.F. McCarthy and F.J. Wobber, Eds. pp. 263-267. Lewis Publishers, (Chelsea, MI 1992).

62. Sung, W. and J.J. Morgan, "Kinetics and Product of Ferrous Iron Oxygenation in Aqueous Systems," *Environ. Sci. Technol.* 14:561-568 (1980).

63. Peck, A.S., L.H. Raby and M.E. Wadsworth, "An Infrared Study of the Flotation of Hematite with Oleic Acid and Sodium Oleate". *Trans. AIME* pp. 235:301 (1966).

64. Gschwend, P.M., "Geochemical Factors Influencing Colloids in Waste Environments," in *Concepts in Manipulation of Groundwater Colloids for Environmental Restoration.* J.F. McCarthy and F.J. Wobber, Eds. pp. Lewis Publishers, Inc., (Chelsea, MI 1992).

65. Keswick, B.H., D.S. Wang and C.P. Gerba, "The Use of Microorganisms as Groundwater Tracers: A Review," *Groundwater*, 20:142-149 (1982).

66. Higgo, J.J.W., "Review of Sorption Data Applicable to the Geological Environments of Interest For the Deep Disposal of ILW and LLW in the UK," *Safety Studies Nirex Radioactive Waste Disposal.* NSS/R-162. British Geological Survey, (Keyworth, Nottingham 1988).

67. Vilks, P. and C. Degueldre, "Sorption Behavior of ^{85}Sr, ^{131}I, and ^{137}Cs on Grimsel Colloids," *Appl. Geochem.* 2:620-655 (1991).

5

REDOX REACTIONS IN PHYLLOSILICATES AND THEIR EFFECTS ON METAL TRANSPORT

Joseph W. Stucki[1], *George W. Bailey*[2] *and Huamin Gan*[1]
[1]Agronomy Department
University of Illinois
Urbana, IL 61801
[2]Environmental Research Lab
U. S. Environmental Protection Agency
Athens, GA 30613

1. INTRODUCTION

The risk that metal cations pose to the global ecosystem depends largely on their activity in porous media, which can be calculated only if the true exposure, i.e., the fate and transport, of the cations is known. Exposure is determined by the bioavailability, which in turn depends on the speciation of chemical compounds in the system. The fate and transport of redox-active pollutants in porous media are governed by solubility and adsorption processes, but modifications also occur due to redox reactions in the solid-liquid interface. These effects are governed by the thermodynamic energy of redox couples, and by the rates at which reactions proceed. Solid surfaces in the media create a unique chemical environment that influences both the energetics and the kinetics of reactions among chemical components.

Knowledge of the solution concentration or activity alone is insufficient to provide a complete model of exposure because, in this interfacial region, two phenomena occur that are absent in solution. First, the solid surfaces become an active participant, as both reactant and product, in the chemical reactions. Second, chemical constituents come under the influence of van der Waals, electrostatic, hydration, and possibly other forces that alter their total potential energy or reactivity. Accurate definition of speciation thus requires an understanding of the processes and forces that are operating in the interface, in terms of both the participants and the rates of reaction. When this understanding is achieved, the opportunity to predict the behavior of percolating ions in porous media will be greatly enhanced.

In this chapter, we will review the general principles of redox reactions in soils and sediments, and provide a summary of equilibrium thermodynamics, concepts of solid state electron transfer, and acid-base phenomena. We also will present the known effects of changes in redox conditions on the behavior and properties of phyllosilicate minerals. Several different mechanisms have been proposed for the reduction or oxidation of structural iron in phyllosilicates. These will be discussed along with recent advances and evidence supporting the current hypotheses, giving an assessment of the potential importance of redox reactions in phyllosilicates in modifying or predicting the fate and transport of hazardous metal species in the vadose zone.

2. PHYLLOSILICATE STRUCTURES AND COMPOSITION

The purpose of this section is to summarize briefly the principal crystal structural components of the various phyllosilicates. For detailed descriptions of phyllosilicate crystal structures, chemical compositions, and occurrence, the reader is referred to the classic works of Grim [1,2], Brindley and Brown [3], and others (Ross and Hendricks [4]; Kerr [5]; Brindley and MacEwan [6]; Weaver and Pollard [7]). The crystal structures of these minerals are comprised of two basic units -- the Si tetrahedron, consisting of a Si^{4+} ion surrounded by four O^{2-} ions in tetrahedral coordination, and the Al octahedron, consisting of an Al^{3+} ion surrounded by four O^{2-} and two OH^- ions in octahedral coordination (Figure 1). These structural units combine to form, respectively, tetrahedral and octahedral sheets by adjacent Si tetrahedra sharing O at their basal corners, and by Al octahedra sharing O and OH at their edges. These sheets, in turn, are joined together to form the clay mineral layer by sharing the apical oxygens of the tetrahedral sheet (Figure 2). The clays are classified (Martin *et al.* [8]; Table 1) according to layer type and charge per formula unit. Electrical neutrality requires four positive charges for each tetrahedral site (satisfied with one Si^{4+} ion), and six for every three octahedral sites (satisfied in dioctahedral clays with one vacancy and two trivalent ions such as Al^{3+} or Fe^{3+}, and in trioctahedral clays with three divalent ions such as Mg^{2+} or Fe^{2+}). Hydroxide ions are distributed in the octahedral sites either *cis* or *trans* to one another, with the *cis* sites being twice as prevalent as the *trans* sites (Figure 3).

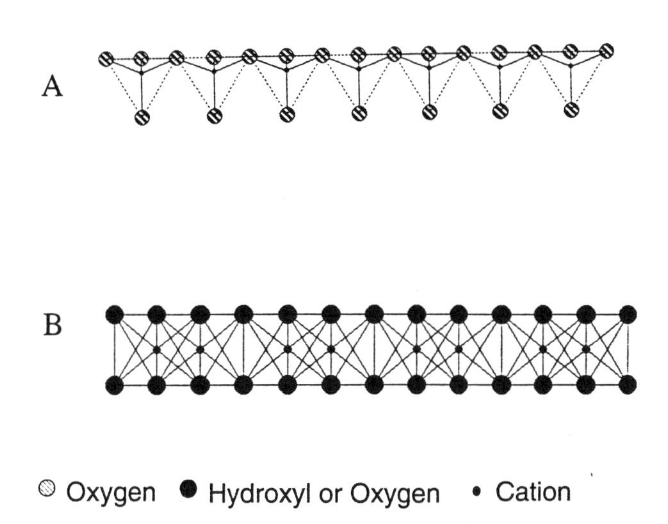

FIGURE 1. Schematic Illustration of (a) the Tetrahedral Sheet, and (b) the Octahedral Sheet.

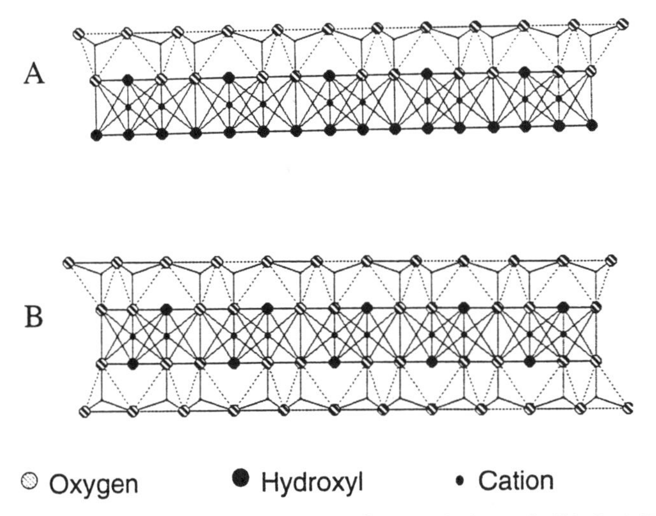

FIGURE 2. Schematic Illustration of a (a) 1:1, and (b) 2:1 Lattice Organization.

Table 1. Classification of Planar Hydrous Phyllosilicates [Ref. 8].

Layer type	Interlayer material[1]	Group	Octahedral Character	Species
1:1	None or H_2O only (x~0)	Serpentine-kaolin	Trioctahedral	Lizardite, berthierine, amesite, cronstedtite, neopuite, kellyite, fraipontite, brindleyite
			Dioctahedral	Kaolinite, dickite, nacrite, halloysite (planar)
			Di-trioctahedral	Odinite
2:1	None (x~0)	Talc-pyrophyllite	Trioctahedral	Talc, willemseite, kerolite, pimelite
			Dioctahedral	Pyrophyllite, ferripyrophyllite
	Hydrated exchangeable cations (x~0)	Smectite	Trioctahedral	Saponite, hectorite, sauconite, stevensite, swinefordite
			Dioctahedral	Montmorillonite, beidellite, nontronite, volkonskoite
	Hydrated exchangeable cations (x~0.6-0.9)	Vermiculite	Trioctahedral	Trioctahedral vermiculite
			Dioctahedral	Dioctahedral vermiculite
	Non-hydrated monovalent cations (x~0.6-1.0)	True (flexible) mica	Trioctahedral	Biotite, phlogopite, lepidolite, etc.
			Dioctahedral	Muscovite, ilite, glauconite, celadonite, paragonite, etc.
	Non-hydrated divalent cations (x~1.8-2.0)	Brittle mica	Trioctahedral	Clintonite, kinoshitalite, bityite, anadite
			Dioctahedral	Margarite
	Hydroxide sheet (x = variable)	Chlorite	Trioctahedral	Clinochlore, chamosite, pennantite, nimite, baileychlore
			Dioctahedral	Donbassite
			Di-trioctahedral	Cookeite, tosudite
2:1	Regularly interstratified (x = variable)		Trioctahedral	Corrensite, aliettite, hydrobiotite, kulkeite
			Dioctahedral	Rectorite, tosudite

[1]x is net layer charge per formula unit.

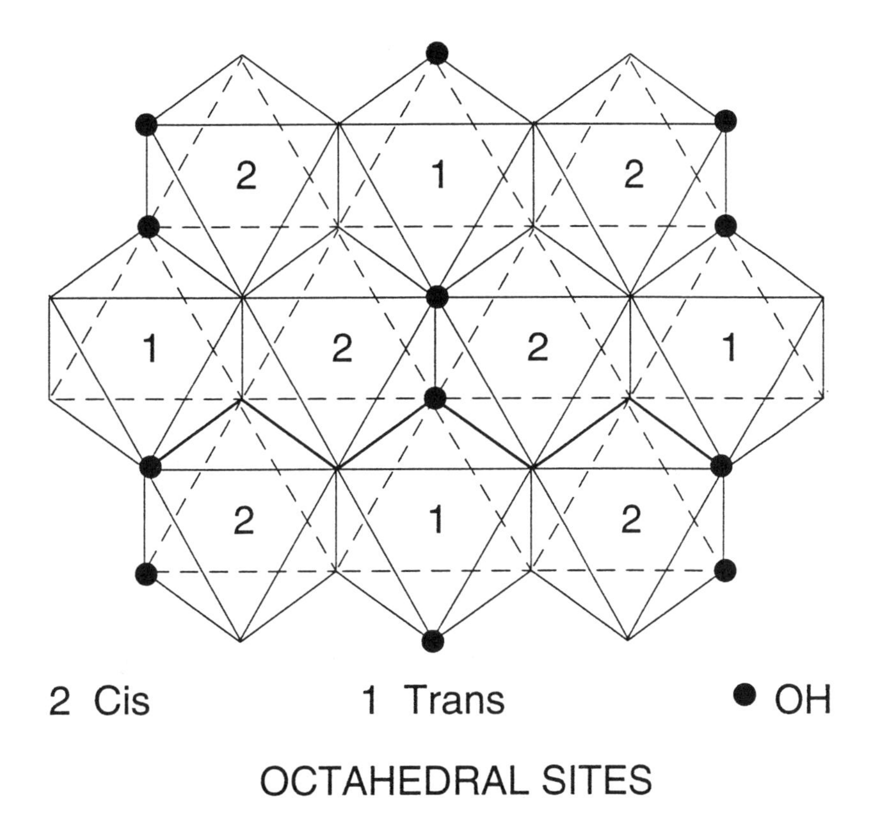

FIGURE 3. ***Projection in the ab Plane of the Position of the trans and cis Sites and OH Groups (after Bonnin et al., [54]).***

Minerals in the Kaolinite-Serpentine group are of 1:1 layer type in which the individual layers consist of one tetrahedral and one octahedral sheet, and the layer charge, ν, is very low ($\nu \approx 0$). The basal oxygen surface of the octahedral sheet is fully protonated.

All other phyllosilicates except the chlorites, including talc-pyrophyllite, smectite, mica, and brittle mica, are of the 2:1 layer type -- the layers consist of one octahedral and two tetrahedral sheets. The layer charge ranges from very low ($\nu = 0$) to very high ($\nu = 2$) per formula unit.

The chlorite group of minerals are classified as 2:1:1, with each layer consisting of a 2:1 layer as in the smectites and an additional octahedral sheet occupying the interlayer position. Consequently, two types of octahedral sheets exist within each chlorite layer. These sheets may be either di- or trioctahedral, and can be mixed within the same

2:1:1 layer. When the interlayer octahedral sheet is dioctahedral, it is termed gibbsitic because of high Al content; when trioctahedral, it is termed brucitic because of high Mg content. Iron in both the di- and trivalent states may exist in either of the octahedral sheets.

Phyllosilicates vary widely in structural Fe content, but virtually all contain some structural Fe. Among them are those commonly referred to as illite [9,10], glauconite [11-16], celadonite [17], vermiculite [18-20], chlorite [21-23], mica [24], fibrous phyllosilicates [25,26], and smectite [27]. Iron also plays a significant role in the kaolinite minerals [28].

For smectites and micas, possible combinations of compositions with respect to Fe, Mg, Al, Li, and Si in the octahedral and tetrahedral sheets are represented by the three-dimensional parallellograms shown in Figure 4 for di- and trioctahedral structures. So-called end-members of the dioctahedral smectites having no octahedral charge are pyrophyllite, beidellite, nontronite, and ferripyrophyllite [29,30]. Corresponding trioctahedral smectite end-members are talc, saponite, lembergite, and "Fe^{2+}-talc," respectively. No official name has been assigned to the Fe^{2+} analogue of talc, hence, the generic designation in quotations is used here [31-35]. No names have been assigned to the smectite end-member compositions with a net octahedral charge, except montmorillonite and hectorite, which respectively are the di- and trioctahedrally-charged species with, ideally, no tetrahedral or Fe substitution.

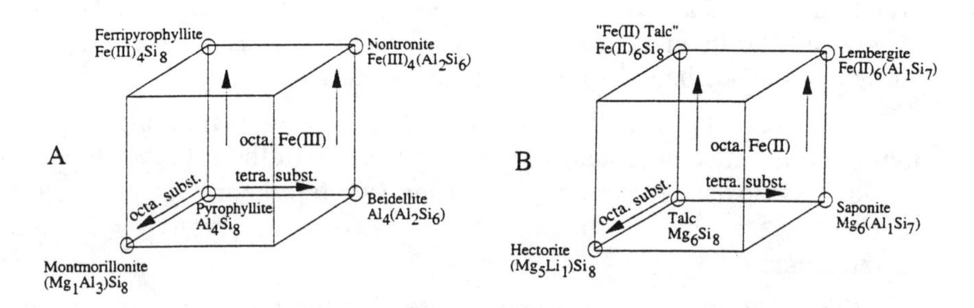

FIGURE 4. Composition and Classification of Known (●) end-Members of (a) Dioctahedral and (b) Trioctahedral Smectites. (O = theoretical, but unknown end-members).

For mica having no octahedral charge, the zero-Fe dioctahedral end-member is muscovite, in which 25% of the Si is replaced by Al in the tetrahedral sheet. The trioctahedral end-members are phlogopite (only Mg^{2+} in the octahedral sheet) and annite (only Fe^{2+} in the octahedral sheet).

Complete homovalent replacement of octahedral Al or Mg, e.g., Fe^{3+} for Al^{3+} or Fe^{2+} for Mg^{2+}, is possible theoretically. But heterovalent substitution, e.g., Mg^{2+} for Al^{3+} or Al^{3+} for Si^{4+}, is limited as indicated by the approximate stoichiometry of the end-member formulas (Figure 4). The choice of octahedral cation need not be limited to Al, Fe, Li, or Mg, as Cr, Cu, Ni, and Zn are known also to exist [36,37]. Chromium-rich dioctahedral smectite has been given the status of its own species name, volkonskoite [38-40], and sauconite is the name for Zn-rich smectite [41], testifying to the fact that such structural ions are present in rather large quantities in some minerals. Nickel-bearing smectites are also rather common in some geographical areas [42-45]. Invariably, Fe is also present.

Few if any phyllosilicates are ideal end-members, but end-member names often are used for minerals having compositions near that of the end-member. None is completely devoid of structural Fe, so the actual compositions of virtually all known smectites and micas fall inside the parallellograms (Figure 4) rather than on any of the surfaces or edges. Hectorite is probably the most notable example of an "Fe-free" smectite, but even it contains sufficient structural Fe to be detected by electron spin resonance [46-48].

The precise location of Fe in phyllosilicate structures has been the subject of much interest. In particular, efforts have focussed on the existence of Fe^{3+} in the tetrahedral sheet [49-58], and on its distribution between the *cis* and *trans* octahedral sites [26,50,53,55,56,59-67]. Ferrous iron is limited to octahedral sites. Dioctahedral phyllosilicates usually contain very little Fe^{2+}, whereas the trioctahedral species can be very rich in Fe^{2+} as well as Fe^{3+}.

The substitution of Fe for other ions in the smectite crystal structure is sufficient to invoke numerous changes in clay properties and behavior. Of equal or greater significance are the further changes that may occur, sometimes reversibly, upon *in situ* alteration of the oxidation state of the Fe. This ability to change its properties *in situ* sets Fe apart from other common constituents of clay minerals, and creates a new dimension to its effects. The properties of smectites known to be affected by structural Fe are summarized in Table 2. This summary, however, excludes possible effects on catalytic properties. Since most of these properties are intricately woven together with the macroscopic chemical and physical behavior of soils and sediments, the impact of Fe is great. In the last section of this chapter, the effects of Fe on various of these properties will be reviewed, including the effects of changes in oxidation state.

Table 2. The Effects of Structural Fe on Smectite Properties [Ref. 27].

	Change in Property Due to+		
	Fe(III)		Fe(II)
Property	Octa	Tetra	Octa
		Dioctahedral	
Layer Charge/Cation Exchange Capacity	0	+	+
Swelling Water	-	?	-
Crystallographic *b*-dimension	+	+	+
Surface Area	?	?	-
Color	yellow-brown	?	blue, green, blue-grey
Chemical and Thermal Stability	-	-	-
		Trioctahedral	
Surface Charge/Cation Exchange Capacity	0	+	+
Swelling Water	?	?	?
Crystallographic *b*-dimension	+	+	+
Surface Area	?	?	?
Color	yellow brown	?	green blue
Chemical and Thermal Capacity	-	?	+

+Symbols signify positive (+), negative (-), and no (0) correlation between the property and Fe(II). ? = unknown.

3. THERMODYNAMIC PRINCIPLES OF REDOX REACTIONS

Oxidation and reduction reactions in soils and clay minerals can be properly interpreted only if the thermodynamic principles underlying such reactions are clearly set forth and understood. Reactions representing these processes, such as those summarized in Table 3, are often presented without details regarding their origin or assumptions, expecting the reader to already have the background for extracting or deducing this information. As a result, their utility is limited and they become susceptible to misapplication. Following is a brief, but complete presentation of the thermodynamic principles of chemical redox reactions, beginning with first principles. A first glance at the number of equations may be somewhat discouraging, but the reader is encouraged to study the following treatment. It will yield a rich harvest of understanding.

Table 3. Selected Environmentally Significant Half-cell Reactions on Various Scales.

Reduction Half-Cell Reaction	$\mathcal{E}°$ (V)	pe°	Electron Free Energy Kcal/mole e^-
$O_{(g)} + 2H^+ + 2e^- = H_2O$	2.421	41.0	27.9
$FeO_4^- + 8H^+ + 3e^- = Fe^{3+} + 4H_2O$	2.200	37.3	16.9
$O_3 + 2H^+ + 2e^- = O_2 + H_2O$	2.076	35.2	23.9
$Ag^{2+} + e^- = Ag^+$	1.980	33.6	45.7
$H_2O_2 + 2H^+ + 2e^- = 2H_2O$	1.776	30.1	20.5
$N_2O + 2H^+ + 2e^- = N_2 + H_2O$	1.776	29.9	20.4
$Au^+ + e^- = Au$	1.692	28.7	39.0
$2NO + 2H^+ + 2e^- = N_2O + H_2O$	1.591	27.0	18.4
$Mn^{3+} + e^- = Mn^{2+}$	1.542	26.1	35.6

Reduction Half-Cell Reaction	$\mathcal{E}^\circ$ (V)	pe°	Electron Free Energy Kcal/mole e⁻
$Au^{3+} + 3e^- = Au$	1.498	25.4	11.5
$HO_2 + H^+ + e^- = H_2O_2$	1.495	25.3	34.5
$PbO_2 + 4H^+ + 2e^- = Pb^{2+} + 2H_2O$	1.455	24.7	16.8
$Au^{3+} + 2e^- = Au+$	1.401	23.8	32.3
$Cl_{2(g)} + 2e^- = 2Cl^-$	1.358	23.0	15.7
$2HNO_2 + 4H^+ + 4e^- = N_2O + 2H_2O$	1.297	22.0	7.47
$O_3 + H_2O + 2e^- = O_2 + 2OH^-$	1.240	21.0	14.3
$Cr_2O_7^{2-} + 14H^+ + 6e^- = 2Cr^{3+} + 7H_2O$	1.232	20.9	4.73
$O_2 + 4H^+ + 4e^- = 2H_2O$	1.229	20.8	7.08
$MnO_2 + 4H^+ + 2e^- = Mn^{2+} + 2H_2O$	1.224	20.8	14.1
$[Fe(phen)_3]^{3+} + e^- = [Fe(phen)_3]^{2+}$	1.147	19.4	26.5
$Pu^{5+} + e^- = Pu^{4+}$	1.099	18.6	25.3
$N_2O_4 + 4H^+ + 4e^- = 2NO + 2H_2O$	1.035	17.5	5.97
$HNO_2 + H^+ + e^- = NO + H_2O$	0.983	16.7	22.7
$NO_3^- + 3H^+ + 2e^- = HNO_2 + H_2O$	0.934	15.8	10.8
$2Hg^{2+} + 2e^- = Hg_2^{2+}$	0.920	15.6	10.6
$N_2O_4 + 2e^- = 2NO_2^-$	0.867	14.7	10.0
$Hg^{2+} + 2e^- = 2Hg$	0.851	14.4	9.81
$Ag^+ + e^- = Ag$	0.800	13.6	18.4
$Hg_2^{2+} + 2e^- = 2Hg$	0.797	13.5	9.18
$Fe^{3+} + e^- = Fe^{2+}$	0.771	13.1	17.7
$2NO + H_2O + 2e^- = N_2O + 2OH^-$	0.760	12.9	8.76
$O_2 + 2H^+ + 2e^- = H_2O_2$	0.695	11.8	8.01
$UO_2^{2+} + 4H^+ + e^- = U^{4+} + 2H_2O$	0.612	10.4	14.1
$MnO_4^- + 2H_2O + 3e^- = MnO_2 + 4OH^-$	0.595	10.1	4.57
$Cu^+ + e^- = Cu$	0.521	8.83	12.0
$Cu^{2+} + 2e^- = Cu$	0.342	5.80	3.94

Reduction Half-Cell Reaction	$\mathcal{E}^\circ$ (V)	pe°	Electron Free Energy Kcal/mole e⁻
$UO_2^{2+} + 4H^+ + 2e^- = U^{4+} + 2H_2O$	0.327	5.54	3.77
$HAsO_2 = 3H+ + 3e^- = As + 2H_2O$	0.248	4.20	1.91
$As_2O_3 + 6H^+ + 6e^- = 2As + 3H_2O$	0.234	3.97	0.90
$Cu^{2+} + e^- = Cu^+$	0.153	2.59	3.53
$Sn^{4+} + 2e^- = Sn^{2+}$	0.151	2.56	1.74
$2NO_2^- + 3H_2O + 4e^- = N_2O + 6OH^-$	0.150	2.54	0.87
$S + 2H^+ + 2e^- = H_2S_{(aq)}$	0.142	2.41	1.64
$N_2 + 2H_2O + 6H^+ + 6e^- = 2NH_4OH$	0.092	1.56	0.35
$UO_2^{2+} + e^- = UO_2^+$	0.062	1.05	1.43
$NO_3^- + H_2O + 2e^- = NO_2^- + 2OH^-$	0.010	0.17	0.12
$2H^+ + 3e^- = H_{2(g)}$	0.000	0.00	0.00
$Fe^{3+} + 3e^- = Fe$	-0.037	-0.63	-0.28
$O_2 + H_2O + 2e^- = HO_2^- + OH^-$	-0.076	-1.29	-0.88
$Pb^{2+} + 2e^- = Pb$	-0.126	-2.14	-1.45
$Sn^{2+} + 2e^- = Sn$	-0.138	-2.34	-1.59
$O_2 + 2H_2O + 2e^- = H_2O_2 + 2OH^-$	-0.146	-2.48	-1.68
$2NO_2^- + 2H_2O + 4e^- = N_2O_2^- + 4OH^-$	-0.180	-3.05	-1.04
$Ni^{2+} + 2e^- = Ni$	-0.257	-4.36	-2.97
$Co^{2+} + 2e^- = Co$	-0.280	-4.75	-3.23
$SeO_3^{2-} + 3H_2O + 4e^- = Se + 6OH^-$	-0.366	-6.21	-2.11
$Se + 2H^+ + 2e^- = H_2Se(aq)$	-0.399	-6.77	-4.60
$Cd^{2+} + 2e^- = Cd$	-0.403	-6.83	-4.65
$2S + 2e^- = S_2^{2-}$	-0.428	-7.26	-4.93
$Fe^{2+} + 2e^- = Fe$	-0.447	-7.58	-5.15
$NO_2^- + H_2O + 3e^- = NO + 2OH^-$	-0.460	-7.80	-10.6
$S + 2e^- = S^{2-}$	-0.476	-8.07	-5.49
$S + H_2O + 2e^- = HS^- + OH^-$	-0.478	-8.10	-5.51

Reduction Half-Cell Reaction	$\mathcal{E}^\circ$ (V)	pe°	Electron Free Energy Kcal/mole e⁻
$TiO_2 + 4H^+ + 2e^- = Ti^{2+} + 2H_2O$	-0.502	-8.51	-5.80
$Fe(OH)_3 + e^- = Fe(OH)_2 + OH^-$	-0.560	-9.49	-12.9
$As + 3H^+ + 3e^- = AsH_3$	-0.608	-10.3	-4.67
$AsO_2^- + 2H_2O + 3e^- = As + 4OH^-$	-0.680	-11.5	-5.23
$AsO_2^{3-} + 2H_2O + 2e^- = AsO_2^- + 4OH^-$	-0.710	-12.0	-8.19
$Cr^{3+} + 3e^- = Cr$	-0.744	-12.6	-5.71
$Zn^{2+} + 2e^- = Zn$	-0.762	-12.9	-8.79
$2H_2O + 2e^- = H_2 + 2OH^-$	-0.828	-14.0	-9.55
$2NO_3^- + 2H_2O + 2e^- = N_2O_4 + 4OH^-$	-0.850	-14.4	-9.80
$Se + 2e^- = Se^{2-}$	-0.924	-15.7	-10.7
$Mn^{2+} + 2e^- = Mn$	-1.185	-20.1	-13.7
$UO_2^{2+} + 4H^+ + 6e^- = U + 2H_2O$	-1.444	-24.5	-5.55
$Ti^{2+} + 2e^- = Ti$	-1.630	-27.6	-18.8
$Al^{3+} + 3e^- = Al$	-1.662	-28.2	-12.8
$SiO_3^{2-} + 3H_2O + 4e^- = Si + 6OH^-$	-1.697	-28.8	-9.78
$U^{3+} + 3e^- = U$	-1.798	-30.5	-13.8
$Pu^{3+} + 3e^- = Pu$	-2.031	-34.4	-15.6
$H_2 + 2e^- = 2H-$	-2.230	-37.8	-25.7
$Mg^{2+} + 2e^- = Mg$	-2.372	-40.2	-27.4
$Na^+ + 2e^- = Na$	-2.710	-45.9	-62.5
$Ca^{2+} + 2e^- = Ca$	-2.868	-48.6	-33.1
$Sr^{2+} + 2e^- = Sr$	-2.890	-49.0	-33.3
$Ba^{2+} + 2e^- = Ba$	-2.912	-49.4	-33.6
$K^+ + e^- = K$	-2.931	-49.7	-67.6
$Li^+ + e^- = Li$	-3.040	-51.5	-70.1

(After Vanysek, [139])

3.1 Definitions

Before proceeding with a presentation of thermodynamic principles, however, a few definitions are necessary. Throughout this discussion reference is made to a system. This is defined as a region in space, such as a soil pedon, enclosed by an imaginary boundary that separates the system from its surroundings. If matter is allowed to cross the boundary, the system is open; when no matter crosses, the system is closed. If heat flow across the boundary is precluded, the system is adiabatic, otherwise it is non-adiabatic. When the transfer of neither heat nor matter is permitted, the system is isolated. A soil system usually must be considered to be open and non-adiabatic.

3.2 Equations of state

The First Law of chemical thermodynamics is based on three important observations arising from the experience of mankind:

1. Every chemical system in the universe contains an internal energy, E;
2. This energy can neither be created nor destroyed; and
3. Only changes in E, rather than its absolute value, can be measured.

If a change in the energy state of a system occurs, one or more of the following events has ensued: work has been done on or by the system, by or on its surroundings, matter has been added to or subtracted from the system, or heat has transferred between the system and the surroundings. These processes are summarized in the formal mathematical expression for the First Law of Thermodynamics, *viz.*,

$$dE = TdS - PdV + \sum_i \left(\frac{\partial E}{\partial n_i} \right)_{S,V,n} dn_i \qquad (1)$$

(energy) = (heat) - (work) + (matter)

where TdS represents the change in heat, with T being the absolute temperature and S being the entropy (the relationship between heat and entropy comes from the Second Law of Thermodynamics and will be used here without providing the formal proof); PdV is the mechanical work, with P being the pressure exerted on the system and V being volume; n_i is the number of moles of each chemical species in the system, where i corresponds to the respective species; and $(\partial E/\partial n_i)$ represents the change in internal energy of the system that occurs when an infinitessimal amount of species i is added to the system. The terms within the Σ account for changes in composition. If the system is closed, by definition no matter can cross the boundary between the system and its surroundings and theΣ term is zero; if open, it is non-zero.

The internal energy is a function of the state of the system, or a state function. Other state functions also exist. One of them is the entropy, S, which has already been introduced. The others are the enthalpy, H, Helmholtz free energy, F, and the Gibbs free energy, G. The term "free energy" originates in the concept that the capacity of the system to do work on its surroundings, e.g., expand its volume and thereby compress the volume of the surroundings, is determined by the amount of energy possessed by the system that is extra or free to be dissipated or utilized. Mathematical expressions for these state functions are given, by definition, as follows:

$$H = E + PV \tag{2}$$

$$F = E - TS \tag{3}$$

$$G = E + PV - TS \tag{4}$$

Because measurements must be directed only to changes in energy or in the state of the system, the differential forms of the equations of state (Equations 2-4) are of interest and are written

$$dH = dE + PdV + VdP \tag{5}$$

$$dF = dE - TdS - SdT \tag{6}$$

$$dG = dE + PdV + VdP - TdS - SdT \tag{7}$$

Notice that Equations 5 through 7 all have the term dE, which is given by Equation 1. Substituting Equation 1 into Equations 5 through 7 yields

$$dH = TdS + VdP + \sum_i \left(\frac{\partial E}{\partial n_i} \right)_{S,V,n} dn_i \tag{8}$$

$$dF = -SdT - PdV + \sum_i \left(\frac{\partial E}{\partial n_i} \right)_{S,V,n} dn_i \tag{9}$$

$$dG = -SdT + VdP + \sum_i \left(\frac{\partial E}{\partial n_i} \right)_{S,V,n} dn_i \tag{10}$$

By examining Equations 8 through 10, one can see that each state function, namely, H, F, or G, depends on certain characteristic variables that are independent of one another. For H these are S, P, and n_i; for F they are T, V, and n_i; and for G they are T, P, and n_i.. Because the differential equations of these state functions are exact or perfect differentials (Low [68]), Equations 8 through 10 also can be written in the form

$$dH = \left(\frac{\partial H}{\partial S} \right)_{P,n} dS + \left(\frac{\partial H}{\partial P} \right)_{S,n} dP + \sum_i \left(\frac{\partial H}{\partial n_i} \right)_{S,P,n_{j \neq i}} dn_i \tag{11}$$

$$dF = \left(\frac{\partial F}{\partial T} \right)_{V,n} dT + \left(\frac{\partial F}{\partial V} \right)_{T,n} dV + \sum_i \left(\frac{\partial F}{\partial n_i} \right)_{T,V,n_{j \neq i}} dn_i \tag{12}$$

$$dG = \left(\frac{\partial G}{\partial T} \right)_{P,n} dT + \left(\frac{\partial G}{\partial P} \right)_{T,n} dP + \sum_i \left(\frac{\partial G}{\partial n_i} \right)_{T,P,n_{j \neq i}} dn_i \tag{13}$$

3.3 Partial molar Gibbs free energy and total potential

Comparing Equations 8 through 10 with Equations 11 through 13 reveals the following relationships

$$\left(\frac{\partial H}{\partial S} \right)_{P,n} = T; \qquad \left(\frac{\partial H}{\partial P} \right)_{S,n} = V$$

$$\left(\frac{\partial F}{\partial T}\right)_{V,n} = -S; \qquad \left(\frac{\partial F}{\partial V}\right)_{T,n} = -P$$

$$\left(\frac{\partial G}{\partial T}\right)_{P,n} = -S; \qquad \left(\frac{\partial G}{\partial P}\right)_{T,n} = V \tag{14}$$

and from the Σ term in Equations 11 through 13

$$\left(\frac{\partial H}{\partial n_i}\right)_{S,P,n_{j\neq i}} = \left(\frac{\partial F}{\partial n_i}\right)_{T,V,n_{j\neq i}} = \left(\frac{\partial G}{\partial n_i}\right)_{T,P,n_{j\neq i}} = \left(\frac{\partial E}{\partial n_i}\right)_{S,V,n_{j\neq i}} = \phi_i \tag{15}$$

where ϕ_i is the total potential of the respective chemical species in the system.

The total potential was originally defined for soil and clay systems by Low [68], patterned after the thermodynamic methods of Gibbs [69], and was developed specifically to account for electrical and other force fields emanating from the surfaces of soil colloids that are not present in the types of solutions usually considered by the classical physical chemist. Commonly, the intrinsic chemical potential, μ_i, is used in Equation 15 instead of the total potential, ϕ_i ; but doing so introduces an unnecessary limitation that later greatly impedes an intuitive understanding of thermodynamics when external force fields besides pressure (e.g., electrical, magnetic, and gravitational forces) are present. Gibbs [69] clearly understood this when he defined the chemical potential because he indicated that the value of μ_i would vary with such force fields. The relationship between total potential and intrinsic chemical potential was given by Low [68] where θ_i is the potential energy of component i due to its interaction with external force fields.

$$\phi_i = \mu_i + \theta_i \tag{16}$$

Equations 8 through 10 can thus be rewritten so that ϕ_i replaces the partial differential under the Σ sign. For the Gibbs free energy, for example, the equation is

$$dG = -SdT + VdP + \sum_i \phi_i dn_i \tag{17}$$

From Equations 13 and 15, however, notice also that

$$\phi_i = \left(\frac{\partial G}{\partial n_i} \right)_{T,P,n_{j \neq i}} \tag{18}$$

To reach the final form of the differential equation for the Gibbs free energy, the definition of a partial molar quantity, $\overline{U}_i$, must be introduced. The partial molar quantity due to constituent i in the system is defined as the change in magnitude of a given property of the system when one mole of that constituent i is added to an infinite copy of the system, i.e., when the addition of the constituent does not change the overall composition of the system. For an easily understood example, consider the partial molar volume of water. This can be viewed as the volume change by which the addition of one mole of water (18 g) increases the overall volume of the ocean. The quantity of water added is finite and measureable, but it leaves the overall composition of the ocean virtually unchanged. Mathematically, the partial molar quantity is given by

$$\overline{U}_i = \left(\frac{\partial U}{\partial n_i} \right)_{T,P,n_{j \neq i}} \tag{19}$$

where U represents any thermodynamic variable or state function. The partial molar Gibbs free energy is of particular interest, and is given by

$$\overline{G}_i = \left(\frac{\partial G}{\partial n_i} \right)_{T,P,n_{j \neq i}} \tag{20}$$

Because $\overline{G}_i$ thus exactly equals the total potential of the constituent, as can be seen by comparing Equations 18 and 20, then the partial differentials with respect to n_i in Equations 13 and 15, and the term ϕ_i in Equation 17 can be replaced with $\overline{G}_i$ giving

$$\phi_i = \overline{G}_i \tag{21}$$

and

$$dG = -SdT + VdP + \sum_i \overline{G}_i dn_i \tag{22}$$

Equation 22 is the fundamental, complete expression for the change in Gibbs free energy associated with any process in the universe. All forces and parameters are implicit in the variables given.

The partial molar Gibbs free energy of any chemical species is given by its concentration or activity. Proof of this is found by recalling the ideal gas law, PV = nRT, where R is the universal gas constant and n is the total number of moles in the system, i.e., $n = \Sigma n_i$. The partial pressure (P_i) of the system due to species i is found from the relation

$$P = \frac{P_i}{x_i} \tag{23}$$

where $x_i = n_i / \Sigma n_i$ is the mole fraction of species i in the system. The partial molar volume for any of the species i is thus obtained by solving the ideal gas law for V, then differentiating with respect to n_i, giving

$$\overline{V}_i = \left(\frac{\partial V}{\partial n_i}\right)_{T,P,n_{j \neq i}} = \frac{\partial}{\partial n_i}\left(\frac{nRT}{P}\right) = \frac{RT}{P} = x_i \frac{RT}{P_i} \tag{24}$$

But notice from Equation 22 that

$$\left(\frac{\partial G}{\partial P}\right)_{T,n_i} = V \tag{25}$$

and upon differentiating Equation 25 with respect to n_i,

$$\left(\frac{\partial}{\partial n_i}\left(\frac{\partial G}{\partial P}\right)\right)_{T,P,n_{j \neq i}} = \left(\frac{\partial V}{n_i}\right)_{T,P,n_{j \neq i}} = \overline{V}_i \tag{26}$$

Because the order of differentiation is immaterial, Equation 26 is the same as

$$\left(\frac{\partial}{\partial P}\left(\frac{\partial G}{\partial n_i}\right)\right)_{T,n} = \left(\frac{\partial \overline{G}_i}{\partial P}\right)_{T,n} = \overline{V}_i \tag{27}$$

After substituting Equation 24 into Equation 27 for $\overline{V}_i$, the result is

$$\left(\frac{\partial \overline{G}_i}{\partial P}\right)_{T,n} = \overline{V}_i = \frac{RT}{P} = x_i \frac{RT}{P_i} \tag{28}$$

Now multiply both sides by dP to obtain

$$\left(d\overline{G}_i\right)_{T,n} = x_i \frac{RT}{P_i} dP \tag{29}$$

Differentiating Equation 23 at constant composition reveals that

$$dP = \frac{1}{x_i} dP_i \tag{30}$$

Substituting this result into Equation 29 yields

$$\left(d\overline{G}_i\right)_{T,n} = RT \frac{dP_i}{P_i} = RT\, d(\ln P_i) \tag{31}$$

Now integrate both sides of Equation 31 between the limits of any two energy states of the system, namely, 1 and 2, to obtain

$$\overline{G}_{i2} - \overline{G}_{i1} = RT \ln\left(P_{i2} / P_{i1}\right) \tag{32}$$

If state 1 is chosen to be a standard or reference state where the pressure due to species i equals exactly 1, then Equation 32 reduces to

$$\overline{G}_i - \overline{G}_i^\circ(T) = RT \ln(P_i) \tag{33}$$

This is the basic form of the expression for the partial molar Gibbs free energy and applies to a gaseous system behaving ideally in the absence of external force fields. With these conditions applied, it also equals the intrinsic chemical potential, and can be written

$$\mu_i = \mu_i^\circ + RT \ln P_i \tag{34}$$

Regardless of whether external force fields are present, the standard state can always be chosen as one in which these forces are absent and thus $\overline{G}_i^o \equiv \mu_i^o$. Equations 33 and 34 can also be expressed in terms of other concentration units, namely, molarity (c_i) or mole fraction (x_i). This is accomplished by recalling the relation

$$P_i = c_i RT = x_i P \tag{35}$$

then substituting for P_i in Equation 33, giving

$$\overline{G}_i - \overline{G}^{\circ}_{i(T)} = RT(\ln P_i) = RT(\ln c_i) + RT(\ln RT) = RT(\ln x_i) + RT(\ln P) \tag{36}$$

Because RT and P are constant parameters, these terms may be incorporated into the term representing the standard state, to give

$$\overline{G}_{P_i} = \overline{G}^{\circ}_{i(T)} + RT(\ln P_i) \tag{37}$$

$$\overline{G}_{c_i} = \overline{G}^{\circ}_{i(T)} + RT(\ln RT) + RT(\ln c_i) = \overline{G}^{\circ}_{c_i(T)} + RT(\ln c_i) \tag{38}$$

$$\overline{G}_{x_i} = \overline{G}^{\circ}_{i(T)} + RT(\ln P) + RT(\ln x_i) = \overline{G}^{\circ}_{x_i(T,P)} + RT(\ln x_i) \tag{39}$$

Notice that in each case the standard state depends parametrically on T, but when the units are mole fraction, it also depends on P.

Remember that the relations given in Equations 37 through 39 were derived for an ideal gas, but they may be applied to solutions by recalling Raoult's Law and Henry's Law, summarized as follows,

Raoult's Law: $$P_o = P_o^* x_o \tag{40}$$

Henry's Law: $$P_i = k_i x_i \tag{41}$$

where P_o and x_o are the partial vapor pressure and mole fraction of the solvent in dilute solution, respectively, and P_o^* is the vapor pressure of the pure solvent; P_i is the vapor pressure of solute i above its dilute solution, x_i is the mole fraction of solute i in dilute solution, and k_i is a proportionality constant that depends on temperature and total pressure. Substituting these expressions for x_i in Equation 39 yields

$$\overline{G}_O = \overline{G}^{\circ}_{x_i(T,P)} - RT(\ln P^{*}_O) + RT(\ln P_O) = \overline{G}^{\circ}_{O(T,P)} + RT(\ln P_O) \quad (42)$$

$$\overline{G}_i = G^{\circ}_{x_i(T,P)} - RT(\ln k_i) + RT(\ln x_i) = \overline{G}^{\circ}_{i(T,P)} + RT(\ln x_i) \quad (43)$$

Similar substitutions can be made if units of molarity or molality in solution are desired by recalling the relations between these units and mole fractions. The value of $\overline{\mu}^{\circ}_i$, while still a constant after substitution, will be different depending on the choice of concentration units and on whether the species is in the vapor or solution phase.

3.4 Activity

Lewis [70] recognized that the true chemical reactivity of a species often differs from that predicted by the concentration, so he introduced the concept of activity. He defined activity as a correction to the concentration, using an activity coefficient, *viz.*, $a_i = \gamma_i c_i$. He then kept the form of the equation for the intrinsic chemical potential (Equation 34), giving

$$\mu_i = \mu^{\circ}_i + RT \ln \gamma_i c_i = \mu^{\circ}_i + RT \ln a_i \quad (44)$$

And in keeping with the concept of total potential, Low [68] introduced the concept of total activity, $\overline{a}_i$, giving by analogy

$$\overline{G}_i = \overline{G}^{\circ}_i + RT \ln(\overline{\gamma}_i c_i) = \overline{G}^{\circ}_i + RT \ln \overline{a}_i \quad (45)$$

Recalling Equation 16 and substituting Equation 44 for μ_i gives the relationship between activity and total activity, *viz.*,

$$\overline{a}_i = a_i \exp\left(\frac{\theta_i}{RT}\right) \quad (46)$$

and the activity coefficient relative to the total activity coefficient is

$$\overline{\gamma}_i = \gamma_i \exp\left(\frac{\theta_i}{RT}\right) \quad (47)$$

3.5 Chemical equilibrium

Attention is now turned to the principles of equilibrium. Equilibrium is defined as that state or condition when the system is at rest and can do no work on its surroundings. Work must be exerted upon the system to move it from the equilibrium state. Consider the following chemical process

$$a\,\mathrm{A} + b\,\mathrm{B} = c\,\mathrm{C} + d\,\mathrm{D} \tag{48}$$

where the upper case letters represent certain chemical species or reagents that react with one another to form products and the lower case letters represent the number of moles of each participant in the reaction. If equilibrium or a steady state exists, the reaction can be rearranged to

$$a\,\mathrm{A} + b\,\mathrm{B} - c\,\mathrm{C} - d\,\mathrm{D} = 0 \tag{49}$$

A short-handed notation for Equation 49 is

$$\sum_{\mathrm{i}} \nu_{\mathrm{i}} \mathrm{A}_{\mathrm{i}} = 0 \tag{50}$$

where ν_i are the stoichiometry coefficients and A_i are the reactants and products. Note that some stoichiometry numbers are positive and others negative. This is taken into account in the values for the ν_i terms.

The reaction is allowed to proceed until equilibrium or a steady state is reached, then the system is maintained at constant T and P. The equilibrium state, therefore, is characterized by rewriting Equation 22 as

$$d\overline{G}_{(T,P)} = \sum_{\mathrm{i}} \overline{G}_{\mathrm{i}} dn_{\mathrm{i}} = 0 \tag{51}$$

where the d denotes a virtual or infinitessimal displacement of the reaction equilibrium in which changes in the various n_i must occur synchronously according to the chemical reaction. For example, in the reaction

$$PCl_5 = PCl_3 + Cl_2 \tag{52}$$

one can envision a synchronous change in the number of moles of PCl_3 and Cl_2 in such a way as to maintain the condition of Equation 51. It is convenient to introduce the quantity λ to represent the infinitessimal or virtual (which also implies reversible) unit of advancement of the chemical reaction, $\sum_i \nu_i A_i = 0$, which leads to

$$dn_i = \nu_i d\lambda \tag{53}$$

and after substituting this expression for dn_i into Equation 51 the free energy change is given by

$$d\overline{G}_{(T,P)} = \sum_i \overline{G}_i \nu_i d\lambda_{(T,P)} \tag{54}$$

or

$$\Delta G_{(T,P)} = \left(\frac{\partial \overline{G}}{\partial \lambda}\right)_{T,P} = \sum_i \nu_i \overline{G}_{i(T,P)} \tag{55}$$

The term ΔG in this case is the free energy change that occurs when the reaction is reversibly advanced at constant T and P by Δn moles of component i in an infinite copy of the system, and does not necessarily represent the total change in Gibbs free energy for the total reaction. Equation 55 applies whether the system is held by external constraints in a steady state condition (could be far removed from equilibrium) in which the mole numbers n_i remain invariant at arbitrarily prescribed values, or whether equilibrium prevails where n_i assume appropriate equilibrium values.

Now substitute into Equation 55 the expression for $\overline{G}_i$ from Equation 45, giving

$$\Delta G_{(T,P)} = \sum_i \nu_i \overline{G}_i^o + RT \sum_i \nu_i \ln \bar{a}_i = 0 \tag{56}$$

If the reaction has reached equilibrium, then

$$\Delta G_{(T,P)} = \sum_i \nu_i \overline{G}_i^o + RT \left(\sum_i \nu_i \ln \bar{a}_i \right)_{eq} = 0 \tag{57}$$

or after rearranging

$$-\frac{\sum_i \nu_i \overline{G}_i^o}{RT} = \left(\sum_i \nu_i \ln \bar{a}_i \right)_{eq} \tag{58}$$

Because all terms on the left side of Equation 58 are constant, the values of $\bar{a}_i$ are unique for the equilibrium case and are represented by the equilibrium constant, K_{eq}, according to the following expressions

$$\ln K_{eq} = -\frac{\sum_i \nu_i \overline{G}_i^{\circ}}{RT} = \left(\sum_i \nu_i \ln \bar{a}_i \right)_{eq} \tag{59}$$

These expressions give the true, complete thermodynamic equilibrium constant for any process in any system in the universe. Hence, Equations 22, 59, and 60 are the mathematical expressions giving the complete thermodynamic energy and equilibrium conditions for any chemical reaction.

$$K_{eq} = \exp\left(-\frac{\sum_i \nu_i \overline{G}_i^{\circ}}{RT} \right) = \prod_i \left(\bar{a}_i^{\nu_i} \right)_{eq} = \left(\frac{\bar{a}_C^{c} \bar{a}_D^{d}}{\bar{a}_A^{a} \bar{a}_B^{b}} \right)_{eq} \tag{60}$$

3.6 Applications to redox reactions

Oxidation and reduction (redox) is the terminology used to describe the chemical process that changes the electrical charge of a chemical element or compound. The change in charge occurs when electrons are transferred from one species to another and are nothing more than a chemical reaction as shown in Equation 48. Redox processes, however, traditionally are represented by dividing the reaction into two parts: one part is comprised of the species which gives up electrons, denoted the reductant; and the other part consists of the species which accepts the electron, or the oxidant. By convention, both processes are written independently in the form

$$M^{m+} + \alpha e^- = M^{(m-\alpha)+} \tag{61}$$

where M is the species of valence m^+ that is reduced by accepting α electrons, e^-. Equation 61 is termed a half-cell reaction because it represents only half of the total reaction. Like all reactions, a change in free energy is associated with the half-cell reaction, and, if at equilibrium, a unique balance in the stoichiometry coefficients will exist. An equilibrium constant will also exist according to Equations 59 and 60. Because redox reactions involve the transfer of electrons, they also produce an electromotive force (emf) that can be measured with a suitable electrode and volt meter. The emf is related to the total partial Gibbs free energy of the reactants and products by the expression [71]

$$\Delta\overline{G} = -n\,\mathcal{F}\,\mathcal{E} \tag{62}$$

where n is the number of moles of electrons involved in the reaction, $\mathcal{F}$ is the Faraday (equal to 96,500 coulombs per electron), and $\mathcal{E}$ is the electrode potential. The electrode potential of the standard state, $\mathcal{E}^\circ$, by analogy is given by

$$\Delta\overline{G}^\circ = -n\,\mathcal{F}\,\mathcal{E}^\circ \tag{63}$$

If the reaction occurs at constant temperature and pressure, these equations can be combined with Equation 56 and written

$$-n\,\mathcal{F}\,\mathcal{E} = -n\,\mathcal{F}\,\mathcal{E}^\circ + RT\sum_i \nu_i \ln\bar{a}_i \tag{64}$$

Then assuming combining Equations 59 and 64 yields $\Delta G^\circ = \sum_i \nu_i \overline{G}_i^o$.

$$\ln K_{eq} = \frac{n\,\mathcal{F}\,\mathcal{E}^\circ}{RT} = -\frac{\sum_i \nu_i \overline{G}_i^o}{RT} = \left(\sum_i \nu_i \ln\bar{a}_i\right)_{eq} \tag{65}$$

By knowing the value of $\mathcal{E}^\circ$ for each half-cell reaction, the equilibrium constant for the half-cell reaction can be calculated and the direction and thermodynamic extent of the reaction between any two half-cell reactions can be determined.

To illustrate how redox half-cell reactions are properly coupled, consider the reaction of the half-cell given in Equation 61 with another half cell, giving

$$M^{m+} + \alpha e^- = M^{m-\alpha} \qquad \mathcal{E}^\circ = x \tag{66}$$

$$N^{n+} + \beta e^- = N^{n-\beta} \qquad \mathcal{E}^\circ = y$$

Because the reactions are written in the standard format, the values of $\mathcal{E}^\circ$ are termed standard electrode potentials and for many elements and compounds are available in handbooks. (See Table 3 for a summary of environmentally important electrode potentials.) If the reactions had been written with the electrons on the right side, these values would have been termed oxidation potentials and would be reversed in sign. The preferred format is as written. All half-cell potentials reported in this way are derived relative to the standard hydrogen electrode (SHE) that is assigned a potential of zero volts at all temperatures.

Three steps are required to construct the proper stoichiometry and potentials for the full-cell reactions: (1) multiply both equations (except the value of $\mathcal{E}°$) by an appropriate factor so that both half-cell reactions have the same number of electrons; (2) subtract the lower from the upper equation, including values for $\mathcal{E}°$; and (3) make all terms positive by transposing those that are negative to the opposite side of the equation. For the reactions given in Equation 66, the result is

$$\begin{array}{ll} \beta\, M^{m+} + \alpha\beta\, e^- = \beta\, M^{m-\alpha\beta} & \mathcal{E}° = x \\ \alpha\, N^{n+} + \alpha\beta\, e^- = \alpha\, N^{n-\alpha\beta} & \mathcal{E}° = y \\ \hline \beta\, M^{m+} + \alpha\, N^{n-\alpha\beta} = \beta\, M^{m-\alpha\beta} + \alpha\, N^{n+} & \mathcal{E}° = x - y \end{array} \qquad (67)$$

If the resulting value of $\mathcal{E}°$ is greater than zero, the reaction will be favored as written; if less than zero, the reaction will proceed in the opposite direction to the way it is written. The more positive the value of $\mathcal{E}°$ for the whole cell reaction, the more the reaction is favored as written (see Equations 60 and 65 also).

The simple algebraic addition of values for $\mathcal{E}°$ may at first glance appear to be wrong given the fact that partial molar Gibbs free energies, not electrode potentials, are the additive quantities. One can prove from Equation 63, however, that by normalizing the number of electrons in each half-cell reaction the cell potential is indeed the algebraic sum of the half-cell potentials as written in Equation 67.

Because chemical species, including electrons, move in response to total potential gradients, the measurement of emf by an electrode reflects the total potential, not simply the concentration or solution activity, of the species (according to Equation 62). This concept is essential when interpreting electrode measurements in soils and clays. If an electrode measurement is used as a measure of the concentration, for example, one explicitly assumes that the species is behaving ideally in solution and that no external force fields are present. In a soil system, these assumptions likely are seldom valid. This applies to any electrode measurement, including those by pH and specific ion electrodes. When a pH electrode is placed in a soil suspension, the electrode will reflect the total hydrogen activity in the system, including the influence of electrostatic potentials at clay mineral surfaces. One must use caution, therefore, when converting the electrode value to either concentration or solution activity.

3.7 Different scales for reporting redox potentials

Several different scales have been used to represent electron behavior in a chemical reaction. That involving the standard electrode potential, $\mathcal{E}°$, has already been presented. Another is the electron activity, which is an index for the quantity of electrons in a porous media system and it denotes the contribution of electrons to the partial molar Gibbs free energy, $\Delta\overline{G}$. As in the case of hydrogen ion activity, the electron activity, $\bar{a}_{e^-}$, is sometimes expressed as a logarithmic function for which the notation pe is used just as pH, is used for $\bar{a}_{H^+}$ *viz.*,

$$pe = -\log \bar{a}_{e^-} \tag{68}$$

The standard value of pe, denoted pe°, for any half-cell reaction is reported numerically as relative to the electrode potential given by the hydrogen electrode and, thus, may be either positive or negative. It can be calculated from the whole-cell $\mathcal{E}°$ using Equation 67 if the temperature and equilibrium total activities of the other species in the half-cell reaction are known. The mathematically necessary condition, however, is that the ratio of these activities equal 1. A large negative value for pe indicates high electron activity and reducing conditions, while a large positive value indicates low electron activity and strongly oxidizing conditions.

Another scale is Eh, which is related to the fundamental thermodynamic relationships given above by

$$Eh = \mathcal{E}° + 0.0591n\frac{\bar{a}_{oxidant}}{\bar{a}_{reductant}} - 0.059\frac{m}{n}pH \tag{69}$$

where $\mathcal{E}°$ is the standard electrode potential (volts); m and n, the number of H^+ and e^- used in the reaction, respectively; and $\bar{a}_i$, the total activity of the oxidant or reductant. At 25°C, the relationship between pe and Eh reduces to

$$pe = \frac{Eh}{0.059} \tag{70}$$

A third is the Lindsay-Sadiq "pe + pH" scale. Because the standard hydrogen electrode is the reference for both the pe and the pH scales, a new convention and log of the redox equilibrium constant are zero for the half-cells, $2H^+(aq) + 2e^- = H_2(g)$ therefore pe + pH = -1/2 $\log(pH_2)$. Lindsay and Sadiq [72] found that many redox reactions in soils occur at fixed pe + pH levels. They proposed the use of the combined term "pe + pH" as a redox parameter. At the O_2/H_2O stability line with P_{O_2} = 1 atm. pe + pH = 20.8 at 25 °C and at the H_2/H_2O stability with pH_2 = 1 atm., pe + pH = 0.

The Gurney electron free energy relationship is the fourth scale. pe is a measure of the partial molar free energy of a redox reaction because pe and ΔG are related through constants (see above developments). Eh also is another measure of the free energy associated with electron transfer. The free energy change associated with the series of half-cell reactions of environmentally significant metals can be seen in Table 3. The corresponding Eh values and pe units are also present on the scale. This reduction sequence is from strongest oxidant first to the weakest last, and corresponds to increasing positive $\Delta G / n$ values.

Developing, calculating, and interpreting half-cell and electrochemical reactions can be confusing, particularly with regards to the algebraic sign involved. A sign convention and guidelines have been developed and implemented [73-75].

4. HETEROGENEOUS ELECTRON-TRANSFER REACTIONS

Heterogenous electron transfer reactions involve systems consisting of at least two different phases. For example, this may involve a mineral (phyllosilicate, sulfide, or oxide which may be a semiconductor), a humic substance, or a microbial cell wall and a donor or acceptor in solution. The electron is delivered to or withdrawn from the solid phase that serves as an electron sink or source. The redox reaction can take place in a micro-heterogeneous system. Most biological electron transfer reactions are micro-heterogeneous in nature. Such semiconductors as Fe_2O_3 and TiO_2 occur in nature as minerals and may be present in colloidal suspensions. Redox, catalytic, and photocatalytic reactions involving these types of surfaces have been observed.

A common denominator among all of these systems is the presence of a phase boundary that generally is electrically charged. An electrical potential difference is established across the interface that influences, in a very decisive manner, electron transfer processes. Control of the thermodynamics and kinetics of heterogeneous redox reactions by the electrical field present at the phase boundary makes heterogeneous reactions unique with respect to the homogeneous analogs.

The colloidal dimensions of the dispersed phase must be taken into consideration with respect to micro-heterogeneous systems. Redox processes may take place only within ensembles or restricted space, and no interactions exist between the ensembles that are present in different host aggregates or in space domains that remain isolated on the time scale of the reaction. Under such circumstances, the rate laws derived for homogeneous solution do not give the appropriate description of the reaction dynamics associated with colloidal phase or interfacial systems. New concepts are required in order to account for the restricted size or availability of the reaction space. Where the location of the donor or acceptor is restricted to the surface of the colloidal particle, the reduction in dimensionality of the reaction species has to be taken into account.

4.1 Electron transfer in the solid phase

In a multiphasic redox reaction, at least three processes must take place: (1) diffusion (transport) of the oxidant or reductant to the solid surface, (2) electron transfer from the surface to the site of the redox reaction, and (3) redox reaction resulting in a change in oxidation number of one of the half-cell couples. The solid-state transport of a reaction product, e.g., a hydroxyl ion or water, and the subsequent diffusion of this product into the bulk solution could be considered the fourth and fifth processes, respectively, in the multi-phasic redox reaction. The following development for an environmental system parallels that of Gratzel [76] for a pure semiconductor system.

Electron transfer in the solid state manifests itself as electrical conductivity; the so-called hopping mechanism is one that has been set forth to describe conductivity in solids. With this mechanism, an electron localized on a particular atom or molecule can be transferred to a neighboring site by a thermally activated process analogous to thermally activated intramolecular transfer. For the case at hand, we will consider that the minerals, humic substances, and microbial cell walls/envelopes act as semiconductors.

The electrical conductance of a semiconductor type material can be derived as follows

The resistance, R, of a semiconductor is defined by Ohm's law

$$i = \Delta\Phi R^{-1} \tag{71}$$

where i is the current and $\Delta\Phi$ is the applied potential difference. The conductivity, σ, of the material is defined by

$$\sigma = l / AR \tag{72}$$

where l is the length of the sample and A is the cross sectional area. The current density, J, is then given by

$$J = i / A = \sigma\, \Delta\Phi = \sigma\, E \tag{73}$$

where E is the field strength, which in this case is uniform. A more general expression applicable to non-uniform fields is

$$J = -\sigma\, \Delta\Phi = \sigma\, E \tag{74}$$

where J and E are the corresponding vectors. For isotropic materials, σ is a scaler quantity; for nonisotropic materials, it is a tensor.

If the current is viewed as the flow of charged particles, the criterion for the applicability of Ohm's law is that the average velocity, $\{v\}$, of such particles is proportional to the field strength, E. We define mobility, μ, by the equation

$$\{v\} = \mu E \tag{75}$$

whence

$$\sigma = nqu \tag{76}$$

where q is the charge per particle and n the particle "density", i.e., the number of charged particles per unit volume. More generally, if different types of charge carriers operate in the same material, then

$$\sigma = \Sigma\sigma_i = \Sigma(n_i q_i u_i) \tag{77}$$

and the transport number, t, for each type is defined as the fraction of total current, i.e.,

$$t_i = n_i q_i u_i / nqu \tag{78}$$

In terms of the molar concentration {X} of the charge carrier, the n in the equation becomes

$$\sigma = L\{X\}qu \tag{79}$$

Before discussing electron transfer in the solid state, let us first consider the energetics of the interfacial redox process. First, consider a Two-Phase System, I and II, where an electron is transferred from a donor, D, to an acceptor, A, located in two different phases:

$$(D)_I + (A)_{II} = \left(D^+\right)_I + \left(A^-\right)_{II} \tag{80}$$

Let Φ_I be the Galvanic potential of Phase I and Φ_{II} that of Phase II. If this is the only external force field operating on the electron, then the thermodynamic driving force for the electron transfer is given by

$$\Delta\overline{G} = \Sigma\nu_i\overline{\mu}_i = \overline{\mu}_{D^+} + \overline{\mu}_{A^-} - \overline{\mu}_A - \overline{\mu}_D \tag{81}$$

where $\overline{\mu}_i$ is the electrochemical potential of the species i, as defined by the relationship

$$\overline{\mu}_i = \mu_i + Z_i\mathcal{F}\,\Phi \tag{82}$$

where μ_i is the chemical potential, Z_i is the charge of the species, $\mathcal{F}$ is Faraday's constant (95,600 coulombs per charge), and Φ is the electrical potential of the phase in which it is located. Electrons will move from the donor across the interface to the acceptor until equilibrium is reached, where $\Delta G = 0$; equilibrium being reached when

$$\overline{\mu}_{D^+} - \overline{\mu}_D = \overline{\mu}_A - \overline{\mu}_{A^-} \tag{83}$$

The potential difference between the two phases at equilibrium is

$$\left(\Phi_{II} - \Phi_I\right) = (1/\mathcal{F})\left(\mu_{D^+} + \mu_{A^-} - \mu_D - \mu_A\right) \tag{84}$$

If the solid and the liquid phases behave as ideal solutions, concentrations can be used instead of activities and we obtain

$$\left(\Phi_{II}-\Phi_{I}\right)_{eq}=\left(\overset{\circ}{\mu}_{D^+}+\overset{\circ}{\mu}_{A^-}-\overset{\circ}{\mu}_{D}-\overset{\circ}{\mu}_{A}\right)+\frac{RT}{\mathcal{F}}\ln\left(\frac{C_{A^-}C_{D^+}}{C_{A}C_{D}}\right) \quad (85)$$

Now let us take a specific example where we want to oxidize Fe^{2+} to Fe^{3+} in a mineral like biotite, and reduce species A in solution. Considering the biotite to be the reducing agent, and thus the source of electrons, we can write the half-cell reaction

$$\left(ne^{-}\right)_{solid}+A_{solution}=A^{n^-}_{solution} \quad (86)$$

At equilibrium, the condition is

$$n\left(\bar{\mu}_{e^-}\right)_{solid}=\bar{\mu}_{A^{n-}}-\bar{\mu}_{A} \quad (87)$$

where

$$\left(\bar{\mu}_{e^-}\right)_{solid}=\left(\bar{\mu}_{e^-}\right)_{solid}-\mathcal{F}\Phi_{solid} \quad (88)$$

is the electrochemical potential of the electron in the solid. In a solid the distribution of electrons over the available states follows from Fermi statistics, with the average occupancy, f, of a given energy level, E, being given by

$$f(E)=\left(\exp\left(\frac{E-E_f}{kT}\right)+1\right)^{-1} \quad (89)$$

where E_f is the kinetic Fermi energy of the electron. This energy corresponds to a state that, on the average, is occupied by 0.5 electrons. E_f is expressed on an energy scale whose zero point is the bottom of the conduction band of the semiconductor solid. From a simple quantum mechanical model for T = 0,

$$E_f=\left(h^2/8\pi^2 m_e\right)\left(3\pi^2 n_e\right)^{2/3} \quad (90)$$

where m_e is the mass of the electron and n_e the number of conduction band electrons per unit volume of solid. We must distinguish E_f from the electrochemical potential, $\bar{\mu}_{e^-}$, which corresponds to the free energy that is gained upon transferring the electron from a vacuum (where it is assumed to be at rest) into the Fermi level of the semiconductor solid. $\bar{\mu}_{e^-}$, in contrast to E_f, is a function of the galvanic or inner potential of the solid conductor.

The difference of $\bar{\mu}_{A^-}$ - $\bar{\mu}_A$ as given above is frequently referred to as the Fermi level of the redox electrolyte. Therefore, the $\mathcal{E}_f$ (redox) can be given by

$$\mathcal{E}_f(\text{redox}) = \frac{\mu_{A^{n-}} - \mu_A}{n\mathcal{F}} - \Phi_{\text{solution}} \tag{91}$$

or under standard conditions we have:

$$\mathcal{E}_f^\circ(\text{redox}) = \frac{\mu^\circ_{A^{n-}} - \mu^\circ_A}{n\mathcal{F}} - \Phi_{\text{solution}} \tag{92}$$

Now let us establish a relation between $\mathcal{E}_f$ (redox) and the standard potential of the D/D^{n-} redox couple. By convention, the standard hydrogen electrode (SHE) is taken as the reference. Under this convention, the redox potential $\mathcal{E}^\circ_{SHE}(D/D^{n-})$ reflects the Gibbs free energy, ΔG°, for the solution reaction

$$\frac{n}{2}H_2 + A = A^{n-} + nH^+ \tag{93}$$

where, under standard conditions, recall Equation 63, *viz.*,

$$\Delta G^\circ = -n\mathcal{F}\,\mathcal{E}^\circ_{SHE} \tag{63}$$

To use the vacuum or absolute potential scale rather than the SHE scale consider the following

$$ne^-_{vac} + A = A^{n-} \tag{94}$$

for which we can obtain

$$\Delta G^\circ = \mu^\circ_{A^{n-}} - \mu^\circ_A - n\mathcal{F}\,\mathcal{E}^\circ_{vac} \tag{95}$$

This assumes that the chemical potential, μ, of the electron at rest in a vacuum at a finite distance from the solid surface is zero. It follows from above that

$$\mathcal{E}_f^\circ(\text{redox}) = -\mathcal{E}^\circ_{vac} - \mathcal{E}^\circ_{SHE} - \Phi_{\text{solution}} \tag{96}$$

Relating $\mathcal{E}^{\circ}_{vac}$ to $\mathcal{E}^{\circ}_{SHE}$ is easily done since Equation 94 is obtained from Equation 93 by adding the reaction

$$ne^- + (nH^+)_{water} = (\frac{n}{2}H_2)_{water} \quad (97)$$

for which the statnard free energy has been found to be -n(4.5 ± 0.3)eV [77]. This relationship shows that the Fermi level of the electron in solution is not identical with the redox potential given on the vacuum scale. To convert the redox potential, E_f, measured against the SHE, use the relationship

$$\mathcal{E}^{\circ}_{vac} = \mathcal{E}^{\circ}_{SHE} + 4.5 \quad (98)$$

5. PROCESSES AFFECTING METAL OXIDATION STATES

5.1 Complexation and solubility

Complexation by both organic and inorganic ligands plays a vital role in the movement, mobility, and bioavailabilty of transition and other heavy metals to biota in soils and sediments through influencing their solubility and sorption to both immobile mineral and organic surfaces in soil. For example, reduction of Cr(VI) to Cr(III) causes a change in species, i.e., from an anion (CrO_4^{2-}) to a cation (Cr^{3+}), and thus its sorptive characteristics. The complex equilibria involved in metal speciation in soils can be seen in Figure 5. Metal complexation by organic and inorganic ligands also can affect the oxidation state of metals, thereby exerting a profound influence on their behavior, including solubility and potential for binding with minerals and humic and fulvic acids through ion exchange and specific adsorption mechanisms. Metal complexation by organic ligands can also influence the rate and equilibria of their redox reactions. Likewise, metal complexes affect the energetics of the oxidation-reduction potential of a metal couple. This can be seen as follows for the reduction half-cell reaction for the Fe^{2+}/Fe^{3+} couple and the Fe phenathroline (Fe-phen) complex:

$$Fe^{3+}_{(aq)} + e^- = Fe^{2+}_{(aq)} \qquad \mathcal{E}^{\circ}=0.77V \quad (99)$$

$$Fe(phen)_3^{3+} + e^- = Fe(phen)_3^{2+} \quad E°=1.147V$$

At higher metal ion concentrations in the soil solution, precipitation can be a major immobilization process for the metal, but is a function of the nature and concentration of the metal and anion species (e.g., CO_3^{2-}, PO_4^{3-}, S^{2-}, SO_4^{2-} and OH^-). For the CO_3^{2-} case, pH and CO_2 govern the levels of the concentration of CO_3^{2-} available for reaction and precipitation with a heavy metal ion. The pe determines the speciation of sulfur in the soil system. Table 3 shows that sulfur forms a variety of species depending on the activity of electrons present, i.e., the pe value, but the pH also exerts a great influence on the reaction. Co-precipitation also influences heavy metal activity in the soil solution.

Reaction of either the mineral phase or the aqueous ion species with a soluble complexing agent can: (1) solubilize the metal ion; (2) increase its mobility in the soil solution; (3) stabilize a certain valence state; (4) alter its bioavailability; and (5) oxidize or reduce the metal species, i.e., change its oxidation number/valence charge.

5.2 Effect of organic ligand and pH on valence-state stabilization

The bioavailabilty of Fe as a nutrient or reactant in a redox reaction depends on the rate of Fe^{2+} oxidation and the stabilization of the resultant Fe^{3+} species. The very low solubilities of the various mineral phases of species is well known and indicates the importance on the availability and transport of the oxidized Fe form. Theis and Singer [78] reported that a wide variety of model humic substances, e.g., citric acid, phenol, vanillin, pyrogallol, and tannic acid were capable of reducing Fe^{3+} to Fe^{2+} when the pH was below 3. Theis and Singer [78] also reported on the effect of Fe^{2+} oxidation at a constant oxygen partial pressure and found that organic compounds that model humic substances exhibited a wide range of effects on the rate of the oxidation reaction. Tannic acid, gallic acid, and pyrogallol completely inhibited the oxygenation of Fe^{2+} over a 1 hr period. These compounds, therefore, stabilized reduced Fe and thus kept it in a soluble form. Presumably this is due to the formation of a chelate that made the Fe^{2+} species unavailable for oxidation, at least in the short run. Glutamic acid, tartaric acid, and glutamine significantly reduced the rate of the oxidation reaction without causing total retardation. Histidine, phenol, resorcinol, vanillin, vanillic acid, and syringic acid had no observable effect on the rate of Fe^{2+} oxidation.

METAL BEHAVIOR IN THE TERRESTIAL ENVIRONMENT

METAL-PORE WATER EQUILIBRIA		IMMOBILE SOLID PHASES	TRANSPORT PATHWAYS
$[M(OH)_y(H_2O)_{n-y}L_j]^{z-(n-y)}$ ⇅ ($-L_j$ / $+L_j$) $[M(OH)_y(H_2O)_{n-y}]^{z-y}$ ⇅ ($-H^+$ / $+H^+$) $[M(H_2O)_n]^{z+}$ ⇅ ($-L_j$ / $+L_j$) ML_j; $[M(H_2O)_n]^{z+}$ ⇌ MA_{solid} ($+A$ / $-A$); MA_{solid} ⇅ $M^{z-m}A_{solid}$ ($-me^-$ / $+me^-$); $[M(H_2O)_n]^{z+}$ ⇌ $[M(H_2O)_n]^{z-m}$ ($+me^-$ / $-me^-$)	ADSORPTION PROCESSES $\xrightarrow{K}$	• Minerals - Silicates - oxides - Carbonates - Sulfides • Humic Acid • Microorganisms - Bacteria - Fungi - Algae	SURFACE RUNOFF - water - particulates ↘ TO SURFACE WATER SUBSURFACE INTERFLOW ↗ PERCOLATION TO GROUNDWATER ↓

LEGEND

Chemical Processes

$+L_j$ = Complexation

$-H^+$ = Hydrolysis

$+A^-$ = Precipitation

$-A^-$ = Dissolution

$-me^-$ = Oxidation

$+me$ = Reduction

$+M^+$ = same equilibrium components as pore water

Z = valence

m,n,x,y = stochiometric coefficients

* = Treat as $[M(H_2O)_n]^{z+}$ species, process-wise

$K = \sum_{i=1}^{j} K_j$ equilibrium constants

M = metal

$L = \sum_{i=1}^{j} L_j$ ligands

A = anion, e.g., SO^{2-}, CO_3^{2-}, SO_4^{2-}, PO_4^{3-}

5.3 Acid-base complex formation

From Figure 5 we see the possibility of several competing reactions for the metal ion present in the pore water, including (1) hydration, (2) hydrolysis, (3) complex formation, and (4) oxidation-reduction (both in the liquid and in the solid phase). These four reactions can be grouped into two elementary processes: generalized acid-base reactions (encompasses the first three reactions), and oxidation-reduction reactions. The meaning of the term generalized acid-base reactions is derived from the definition of acids and bases by G.N. Lewis [70]. A base has the ability to partially donate a pair of electrons to form a coordinate or dative covalent bond, while an acid has the ability to accept at least one pair of electrons from a base. A base, therefore, is an atom, ion, or molecule that has at least one pair of valence electrons that has not already been shared to form a covalent bond. An acid is an ion, atom, or molecule where at least one atom has a vacant orbital in which a pair of electrons can be accommodated. A Lewis base also is a ligand (l) and, in organic chemistry terminology, a nucleophile or an electron donor, while a Lewis acid is an electrophile or an electron acceptor.

All metal ions (M^{z+}) are Lewis acids, and normally are coordinated to one or more Lewis bases (L^{1-}) simultaneously to form inner-sphere complexes of the form $M(L)_n^{z-n\times1}$, where n is the number of ligands and also is the coordination number of the metal ion in the complex. The resulting complex may be a neutral molecule like, Fe_2O_3, or it may be positively or negatively charged depending on the value of l. For example, the hydrated metal complex $M(H_2O)_6^{z+}$ carries the positive charge of the metal because the ligand (H_2O) is neutral l = 0; whereas the hexacyano complex $Fe(CN)_6^{3-}$ is negative because the ligand is negative (l = -1). Anions like COO^-, OH^- and SH^- are Lewis bases and may be important constituents of humic and fulvic acids. Other types of acid-base complexes are charge transfer complexes, free atoms, and free radicals acting as Lewis acids and forming complexes with a variety of bases. Complexes of free radicals have yet to be isolated but they exert a great impact on the reactivity of the molecule.

Metal reactions with ligands (whether with water molecules, ligands present in the structure of humic substances, oxygen, or hydroxyl ions at the edge of phyllosilicates) are interactions that can be generalized as those of acids and bases. What is needed are rules to predict the stability of the metal-ligand reaction. Pearson [79-85] suggested the "Principle of Hard and Soft Acids and Bases" (HSAB), which states, "Hard acids prefer to coordinate, bind, or react with hard bases, and soft acids prefer to coordinate, bind, or react with soft bases." This principle has direct application to the formation of metal complexes with ligands present in humic substances and with other types or sources of ligands. The kinetic analog of the HSAB principle is that hard acids react faster with hard bases and that soft acids react faster with soft bases.

Hard and soft acid properties are shown in Table 4, while hard and soft base properties are give in Table 5. A hard acid is one that has small size, high charge, high electropositivity, low polarizability, a few and not easily excitable outer electrons on donor atoms, and forms ionic/electrostatic bonds; a soft acid has the converse properties. Strictly speaking, it is the acceptor atom of the acid that has these properties. A soft base is large in size, its electrons are easily distorted, polarized, and accessed or removed, and it forms covalent π-type bonds. A hard base is the converse, i.e., one that holds on to its electrons much more tightly.

A corollary to the HSAB principle stated above was given by Arhland *et al.* [86]. It states that hard acids coordinate best to the lightest atom of a family of elements and soft acids coordinate best to the heaviest atom of the same family. Further, as illustrated in Table 6, hard acids bind best to the least polarizable (hardest) atom of the family while soft acids bind to a more polarizable atom of a family of elements. However, soft acids do not form their most stable complexes with the most polarizable atoms, the reason being that some very soft atoms are very weak Lewis bases towards all acids.

Table 7 shows a number of representative Lewis acids and bases that are classified as being either hard, soft, or borderline. Note that many of the divalent transition metals are classified as borderline between hard and soft acids. Whether they act as hard or strong acids will be influenced by the solution environment. Water is a very hard solvent, either as an acid or as a base. It strongly solvates both small anions and small, highly charged cations.

In most cases, hard acids and hard bases are held together by ionic or polar bonds, whereas most soft acids and bases are held together by largely covalent bonds. π bonding centers are also important in soft acid-base reactions. Most soft bases are π bond donors because they have filled orbitals that can donate π electrons. Soft acids have certain empty outer orbitals that can accept electrons from soft bases. π bonds are not involved in hard base-hard acid reactions because of the highly ionic character of the outer orbitals in these species.

Two properties are needed to estimate the stability of a metal-ligand complex: (1) intrinsic strengths of the respective acid and base; and (2) the hardness or softness of the acid and the base. Numerical estimates of the softness parameters are needed to predict the stability of potential metal-ligand complexes. Two approaches to these numerical calculations of softness parameters have been used: the Klopman quantum mechanical perturbation theory [87,87], and the Misono dual parameter scale for calculating softness, assuming that softness is a consequence of a metal ion forming a dative π bond [89].

Table 4. Hard and Soft Acid Properties*

Hard	Property	Soft
Low	Polarizability	High
High	Electropositivity	Low
Large	Positive charge or oxidation state	Small
Small	Size	Large
Ionic, electrostatic	Bond type	Covalent
Few and not easily excited	Outer electrons on donor atoms	Several easily excited

**After Williams [140]*

Table 5. Hard and Soft Base Properties*

Hard	Property	Soft
Low	Polarizability	High
High	Electronegativity	Low
Large	Negative charge	Small
Small	Size	Large
Ionic, electrostatic	Types of bond usually associated with the base	Covalent Pi
High energy and inaccessible	Available empty orbital or donor atom	Low lying and accessible

**After Williams [140]*

In Klopman's approach, emphasis is placed on charge- and frontier-controlled effects. The frontier orbitals are the highest-occupied orbital of the Lewis base and the lowest empty orbital of the Lewis acid.

Table 6. Lewis Acid Preferences*

Hard Acid Preference For Ligand Donor Atoms							
N	>	>	P	>	As	>	Sb
O	>	>	S	>	Se	>	Te
F	>	>	Cl	>	Br	>	I
Soft Acid Preference For Ligand Donor Atoms							
N	<	<	P	>	As	>	Sb
O	<	<	S	<	Se	<	Te
F	<	<	Cl	<	Br	<	I

**After Pearson [85]*

Table 7a. Classification of Lewis Acids and Bases*

Hard Acids										
Inorganic								Organic		
Charge										
0	1+	2+	3+	4+	5+	6+	7+	0	1+	2+
BF_3	H	Be	Al	Si	I	Cr	I	CO_2	RPO_2	$(CH_3)_2$
$AlCl_3$	Li	Mg	Sc	Ti	--	--	Cl	$BeMe_2$	$ROPO_2$	--
AlH_3	Na	Ca	Ga	Zr	--	--	--	$B(OR)_3$	RSO_2	--
SO_3	K	Sr	In	Th	--	--	--	$Al(CH_3)_3$	$ROSO_2$	--
HX†	--	Mn	La	U	--	--	--	--	$(RO)_3P$	--
--	--	UO_2	N	Pu	--	--	--	--	RCO	--
--	--	VO	Gd	Ce	--	--	--	--	NC	--
--	--	--	Lu	Hf	--	--	--	--	--	--
--	--	--	Cr	--	--	--	--	--	--	--
--	--	--	Co	--	--	--	--	--	--	--
--	--	--	Fe	--	--	--	--	--	--	--
--	--	--	As	--	--	--	--	--	--	--
--	--	--	Mo	--	--	--	--	--	--	--

† (hydrogen-bonding molecules)

**After Pearson, [85].*

Table 7b. Classification of Lewis Acids and Bases*

Soft Acids							
Inorganic						Organic	
Charge							
0	1+	2+	3+	4+	3-	0	1+
BH_3	Cu	Pd	Tl	Pt	$Co(CN)_5$	$Tl(CH_3)_3$	I
$GaCl_3$	Ag	Cd	--	Te	--	$Ga(CH_3)_3$	H
GaI_3	Au	Pt	--	--	--	ICN	--
$InCl_3$	Tl	Hg	--	--	--	Trinitrobenzene	--
Br_2	Hg	--	--	--	--	Chloranil,quinones	RO
O	I	--	--	--	--	Tetracyanoethylene	RS
Cl	--	--	--	--	--	CH_2, carbenes	RSe
Br	--	--	--	--	--	--	RTe
I	--	--	--	--	--	--	--
N	--	--	--	--	--	--	--
M° (metal atoms)	--	--	--	--	--	--	--
Bulk metals	--	--	--	--	--	--	--

**After Pearson [85]*

When the energy difference of these orbitals is very large, little electron transfer occurs and a charge-controlled interaction results. Ionic forces dominate in this type of complex. When frontier orbitals are of similar energy, a strong electron transfer occurs from the Lewis base to the Lewis acid. This is a frontier-controlled reaction and binding forces are mainly covalent. Hard acid-hard base reactions result in charge-controlled type reactions, whereas soft-soft interactions are frontier-controlled. Klopman [88] used ionization potential, electron affinities, ion sizes, and hydration energies (Born equation) to calculate a set of orbital electronegativity values (eV) which are, in essence, softness values for the metal ions. Tables 8 and 9 summarize the softness calculations for major cations (Lewis acids) and anions (Lewis bases), respectively. Numbers show very good correlation with the known behavior of each ion as a Lewis acid or base.

Table 7c. Classification of Lewis Acids and Bases*

Hard Bases							Soft Bases			
Inorganic				Organic			Inorganic		Organic	
Charge										
0	1-	2-	3-	0	1-	2-	1-	2-	0	1-
H_2O	OH	SO_4	PO_4	ROH	CH_3CO_2	CO_3	I	S_2O_3	R_2S	RS
Cl	F	--	--	RNH_2	RO	--	H	--	RSH	SCN
NH_3	ClO_4	--	--	--	R_2O	--	--	--	R_3P	CN
N_2H_4	NO_3	--	--	--	--	--	--	--	R_3As	R
--	--	--	--	--	--	--	--	--	$(RO)_3P$	--
--	--	--	--	--	--	--	--	--	RNC	--
--	--	--	--	--	--	--	--	--	CO	--
--	--	--	--	--	--	--	--	--	C_2H_4	--
--	--	--	--	--	--	--	--	--	C_6H_6	--

**After Pearson [85]*

Orbital electronegativities consist of two parts: (1) the energies of the frontier orbitals themselves in an average bonding condition, and (2) changes in the solvation energies that accompany covalent bond formation or electron transfer. The dielectric constant plays an important role in solvation energy calculations. Because of this, one would expect that all cations would become softer acids in less polar solvents, i.e., solvents having lower dielectric constants. Conversely, the softest anions, i.e., the softest Lewis bases in solution, become harder bases in solvents of lower and lower dielectric constant.

In Misono's approach, a softness parameter is calculated as a function of the ionization potential, ionic radius, and charge of the metal ion. Table 10 gives this softness value as a function of oxidation state/valence charge for a variety of metal ions. Note the effect of oxidation state on the softness of the metal cation. Also note the relatively soft nature of the Lewis acids Ag^{+}, Pb^{2+}, and Cr^{3+}. These would be expected to react with softer Lewis bases than ions such as Al^{3+}, alkalis, and the alkaline earths.

Table 7d. Classification of Lewis Acids and Bases*

Borderline Lewis Acids and Bases							
Acids					Bases		
Inorganic			Organic		Inorganic		Organic
Charge							
0	2+	3+	0	1+	1-	2-	0
SO_2	Fe	Sb	$B(CH_3)_3$	R_3C	N_3	S_2O_3	$C_6H_5NH_2$
Cl	Co	Bi	--	C_6H_5	Br	--	C_5H_5N
NH_3	Ni	Rh	--	--	NO_2	--	--
N_2H_4	Cu	Ir	--	--	--	--	--
--	Zn	--	--	--	--	--	--
--	Pb	--	--	--	--	--	--
--	Sn	--	--	--	--	--	--
--	NO		--	--	--	--	--
--	Ru	--	--	--	--	--	--
--	Os	--	--	--	--	--	--

**After Pearson [85]*

The reader is cautioned to remember that the HSAB principle still is qualitative and is used to indicate trends, not strict quantitative assessments.

6. REDOX PHENOMENA IN PHYLLOSILICATES

The most common transition metal in phyllosilicates is Fe, which may exist in either the tri- or divalent positive form in the octahedral sheet, or in the trivalent form in the tetrahedral sheet. Little is known regarding the redox behavior of tetrahedral Fe, but many studies have been conducted to understand the oxidation and reduction of octahedral Fe. Such processes are common in soils as a result of biological activity, alternate wetting and drying events, and redox-sensitive chemicals in the soil solution. They are responsible for many changes in the physical and chemical properties of soils [90-94] and clays [95], including weathering [96,97], swelling in water [98], electrical charge [99,118], magnetic exchange interactions within the clay crystal [66,67], and surface area [100]. Effects of Fe on physical chemical properties of clays have been reviewed by Stucki [27], Stucki and Lear [95], Jepson [101], and Scott and Amonette [24].

Although the thermodynamic properties of clay-water systems have been investigated extensively by Babcock [102], Low and co-workers [68,103-108], and Sposito [109], the thermodynamic properties of redox reactions in clays have yet to be explored in depth. Foster [110] and Stucki *et al.* [98] observed significant effects of Fe oxidation state on the swelling pressure (π) of smectites, but no thermodynamic calculations were made. Future studies need to be directed to determining the redox potentials of transition metals bound in the crystal structures of the phyllosilicates.

The oxidation state of structural Fe in smectites significantly alters the short-range forces next to clay layers, but has little or no influence on long-range forces (Wu *et al.* [111]). Stucki *et al.* [98] observed profound effects of Fe^{2+} on the swellability of four different smectites, indicating that the water holding capacity of these clays is decreased greatly by the reduction of structural Fe. Wu *et al.* [111] measured the Bragg diffraction of clay-water gels at various water and Fe^{2+} contents, and noted that the peaks from fully expanded layers were centered at rather uniform d-spacings which depended on π but not on Fe^{2+} content, whereas the partially expanded spacing did follow the Fe^{2+} content. All clay layers appeared to be either partially or completely expanded in these gel systems. The effect of Fe^{2+} on swelling must occur, therefore, because of changes in the relative numbers of partially versus fully expandable layers. Further evidence for this hypothesis was found by Lear and Stucki [100] who, using samples prepared the same as those studied by Wu *et al.* [111], observed that a lower specific surface area accompanied Fe reduction, indicating a rather high degree of layer collapse had occurred. The reduction of structural Fe^{3+} to Fe^{2+} in Na smectites also increases the amount of non-exchangeable Na^+ [99]; K^+ [112,113]; Ca^{2+}, Cu^{2+}, and Zn^{2+} [113] by undried smectite gels.

From these studies one could easily infer that some of the layers in the reduced clay gel become completely collapsed (10 Å). Reduction also caused the smectite particles to become more consolidated, consisting of as many as 40 layers and of limited lateral extent; whereas, oxidized clay particles consist of long, open networks of ribbons from 1 to 6 layers thick [114]. Studies by electron diffraction revealed that the

Table 8. Calculated Orbital Electronegativity (E_n) of Cations in Water (ϵ = 80) and Their Softness Character*

Cation	IP[a] (eV)	EA[b] (eV)	Orbital Energy (eV)	r+0.82[c] (A)	Desolvation Energy (eV)	E_n (eV)	Softness Character
Al (3+)	28.44	18.82	26.04	1.33	32.05	6.01	↑
La (3+)	19.17	11.43	17.24	1.96	21.75	4.51	
Ca (2+)	11.87	6.11	10.43	1.81	12.76	2.33	
Fe (3+)	30.64	15.96	26.97	1.46	29.19	2.22	HARD
Sr (2+)	11.03	5.69	9.69	1.94	11.90	2.21	
Cr (3+)	30.95	16.49	27.33	1.45	29.39	2.06	
Ba (2+)	10.00	5.21	8.80	2.16	10.69	1.89	
Cr (2+)	15.01	7.28	13.08	1.65	13.99	0.91	↓
Fe (2+)	16.18	7.90	14.11	1.56	14.80	0.69	Borderline
Mn (2+)	15.64	7.43	13.59	1.62	14.25	0.66	↑
Co (2+)	16.49	8.42	14.47	1.54	14.99	0.52	
Li (+)	5.39	0.82	4.25	1.50	4.74	0.49	
H (+)	13.60	0.75	10.38	--	10.80	0.42	
Ni (2+)	17.11	8.67	15.00	1.51	15.29	0.29	↓
Na (+)	5.14	0.47	3.97	1.79	3.97	0.00	SOFT
Pb (2+)	14.96	0.37	11.31	2.14	10.79	-0.52	↑
Cu (2+)	17.57	9.05	15.44	1.54	14.99	-0.55	
Zn (2+)	17.96	9.39	15.82	1.56	14.80	-1.02	
Cd (2+)	16.90	8.99	14.93	1.79	12.89	-2.04	
Cu (+)	7.72	2.00	6.29	1.78	3.99	-2.30	
Ag (+)	7.57	2.20	6.23	2.08	3.41	-2.82	
Au (+)	9.22	2.70	7.59	2.19	3.24	-4.35	↓

* *After Klopman [88]*
a *Ionization Potential*
b *Electron Affinity*
c *Radius*

Table 9. Calculated Orbital Electronegativity (E_n) of Anions in Water (ϵ = 80) and Their Softness Character*

	IP[a]	EA[b]	Orbital Energy	Radius	Desolvation Energy	Electronegativity E_n	Softness Character
X^z	eV	eV	eV	Å	eV	eV	
P	17.42	3.48	6.96	1.36	5.22	-12.18	↑
H_2O	25.4	12.6	15.8	1.40	-5.07	-10.73	
H	13.10	2.8	5.38	1.40	5.07	-10.45	HARD
Cl	13.01	3.69	6.02	1.81	3.92	-9.94	
Br	11.84	3.49	5.58	1.95	3.64	-9.22	Borderline
CN	14.6	3.2	6.05	2.60	2.73	-8.78	
SH	11.1	2.6	4.73	1.84	9.86	-8.59	SOFT
I	10.45	3.21	5.02	2.16	3.29	-8.31	↓
H	13.6	0.75	3.96	2.08	3.41	-7.37	

* *After Klopman [88]*

a *Ionization Potential*

b *Electron Affinity*

Table 10. Misono's Metal Ion Softness* Values+

Charge					
1+		2+		3+	
Metal Ion	Softness	Metal Ion	Softness	Metal Ion	Softness
H	a	Mg	0.87	Al	0.70
Li	0.36	Ca	1.62	In	2.24
K	0.92	Sr	2.08	Fe	2.37
Na	0.93	Zn	2.34	La	2.45
Rb	2.27[b]	Ba	2.62	Co	2.56
Cs	2.73	Ni	2.82	Cr	2.70
Cu	3.45	Cu	2.89	Ti	3.23
Ag	3.99	Co	2.96		
		Mn	3.03		
		Cd	3.04		
		Fe	3.09		
		Sn	3.17		
		Pb	3.58		
		Hg	4.24		

\+ *After Sullivan [141]*
* *Softness increases as values increase.*
a *Hydrogen ion softness cannot be calculated by Misono's softness equation.*
b *This value was calculated by the author using Misono's softness equation.*

forces operating between reduced clay layers are sufficient to cause the layers to stack one upon the other in a much more ordered manner than when the Fe is in the oxidized state, again providing evidence for stronger short-range forces between clay layers when the structural Fe is in the reduced state [114].

The mechanism for redox reactions in smectite clay minerals has been under investigation for many years[1] and continues to be the subject of considerable effort in the authors' laboratories. The mechanism proposed for the reduction of nontronite by H_2 gas at 350 °C [114] is that H_2 enters the crystal structure of the clay, protonates the apical O that is shared by two octahedral cations and one tetrahedral Si, and protonates the structural OH group to form a structural H_2O. The H_2O then diffuses out of the structure, leaving a defective octahedral site. The net result of this process is that no change in structural OH content occurs, but two different types of OH exist in the structure, namely, those in their original positions that were not involved in reduction (assuming some sites are occupied by ions other than Fe or that some Fe remains unreduced) and those at tetrahedral apices. If two Fe ions exist in adjacent sites and both are reduced, then, by this mechanism, a coordination number as low as four would be expected. Although five-coordinate Fe in the octahedral sheet has been suggested in other reducing systems [99,116,117], further studies are necessary to show evidence of four-coordinate Fe in the reduced structure. The presence of two different OH environments also requires confirmation. Perhaps this mechanism precludes the reduction of Fe in adjacent sites and places an upper limit on the extent of reduction possible.

The hypothesis proposed by Stucki and Lear [95] is that the redox mechanism depends on the chemical nature of the reducing agent, but reduction by sodium dithionite ($Na_2S_2O_4$) proceeds according to the following reactions

$$m\left(Fe^{3+}\right)_x + (m-a)Z^{q-} + ae^- = m\left(Fe^{2+}\right)_x + (m-a)Z^{1-q} \tag{100}$$

$$2r\left(OH^-\right)_x = r\left(O^{2-}\right)_x + rH_2O \tag{101}$$

$$r\left(O^{2-}\right)_x + r\left(H^+\right)_s = r\left(OH^-\right)_x \tag{102}$$

---

[1] See 1988 review by Stucki, [27].

where subscripts x and s denote clay and solution phases, respectively; Z is an unidentified electron donor located within the clay crystal; e^- represents a reducing agent in the solution surrounding the clay crystal; and m, r, and a are stoichiometry coefficients. According to this hypothesis, the structural Fe is reduced partially by internal and partially by external reducing agents, and the reduction is accompanied by dehydroxylation and reprotonation reactions. Lear and Stucki [118] determined that two of the stoichiometry coefficients are linearly related, *viz.*, r = 0.32m, but the value of a has yet to be determined experimentally. Komadel *et al.* [119] found that virtually all Fe^{3+} in the clay crystal can be reversibly reduced to Fe^{2+} with $Na_2S_2O_4$, indicating that the reaction represented by Equation 100 can be taken to completion.

The proposed overall reaction then is

$$m\left(Fe^{3+}\right)_x + 2r\left(OH^-\right)_x + r\left(H^+\right)_s + (m-a)Z^{1-q} + ae^- =$$

$$m\left(Fe^{2+}\right)_x + r\left(OH^-\right)_x + (m-a)Z^{1-q} + rH_2O \tag{103}$$

Equations 100 and 103 are in the form of a redox half-cell reaction and can be written more generally as

$$Clay^{ox} + X + e^- = Clay^{red} + Y \tag{104}$$

where X and Y represent non-clay reactants and products such as H^+ and H_2O in Equation 103.

The redox status of soils has been studied by many workers [90,109,120-122], and general electrochemical principles for measuring Eh and electrode potentials have been developed [90,109]. For the clay redox system of interest to the present discussion (Equation 104), the electrode potential, $\mathcal{E}_{1/2}$, of the half-cell reaction given by Equation [103] at constant temperature and pressure is given by

$$\mathcal{E}_{1/2} = -\frac{\Delta\overline{G}}{n\mathcal{F}} = -\frac{1}{z\mathcal{F}}\left(\overline{G}_c^{red} - \overline{G}_c^{ox} + \overline{G}_Y - \overline{G}_X - \overline{G}_{e^-}\right) =$$

$$\mathcal{E}^\circ_{1/2} - \frac{RT}{n\mathcal{F}}\ln\frac{\overline{a}_c^{red}\,\overline{a}_X}{\overline{a}_c^{ox}\,\overline{a}_Y\,\overline{a}_{e^-}} \tag{105}$$

where n is the number of moles of electrons, F is the Faraday constant, $\overline{a}$ is the activity of the component, and $\overline{G}$ is the partial molar Gibbs free energy of the component. $\mathcal{E}^{\circ}_{1/2}$ is the standard electrode potential for the half-cell reaction. If the reaction has reached equilibrium, then $\overline{G}$ for each component is that of its standard state by definition, giving a standard electrode potential of

$$\mathcal{E}^{\circ}_{1/2} = -\frac{\Delta\overline{G}^{\circ}}{n\mathcal{F}} = -\frac{1}{n\mathcal{F}}\left(\overline{G}_{C}^{\circ\,red} - \overline{G}_{C}^{\circ\,ox} + \overline{G}^{\circ}_{Y} - \overline{G}^{\circ}_{X}\right) \tag{106}$$

Notice that by definition $\overline{G}^{\circ}_{e^-} = 0$ Recall also that this relation gives the equilibrium constant, K°, for the half-cell reaction

$$\mathcal{E}^{\circ}_{1/2} = -\frac{\Delta\overline{G}^{\circ}}{n\mathcal{F}} = \frac{RT}{n\mathcal{F}}\ln K^{\circ} \tag{107}$$

In order to obtain a complete redox reaction, the half-cell reaction must be combined with the half-reaction of a reductant, as shown earlier in Equation 67, giving

$$\begin{array}{c} \text{Clay}^{ox} + X + e^{-} = \text{Clay}^{red} + Y \\ \text{Reductant} = \text{Oxidant} + e^{-} \\ \hline \text{Clay}^{ox} + X + \text{Reductant} = \text{Clay}^{red} + Y + \text{Oxidant} \end{array} \tag{108}$$

If $\mathcal{E}^{\circ}_{1/2}$ is known for each half reaction, the whole-cell standard potential can be obtained by their algebraic sum, and the equilibrium constant can be calculated from Equation 65, *viz.*,

$$\mathcal{E}^{\circ} = \frac{RT}{n\mathcal{F}}\ln K_{eq} \tag{109}$$

Published values for $\mathcal{E}^{\circ}_{1/2}$ are readily available for a host of chemical compounds (Table 3), including a large number of those that contaminate soils and groundwaters [123,124]. Values are also available for many of the oxide, sulfate, sulfide, carbonate, and other minerals in soil environments (Table 3, [109,125,126]); but values for the half-cell reactions of clay minerals are virtually non-existent. Once these values are obtained, however, more complete estimates can be made regarding possible redox reactions with various redox-sensitive contaminants. Solid state electron transfer processes described above undoubtedly contribute to the overall redox potential of the clay.

The forgoing discussion of reduction processes in dioctahedral smectites was centered on results obtained using sodium dithionite in a citrate-bicarbonate (CBD) medium (pH ≈ 8). These arguments may not apply if unbuffered dithionite or other reducing agents and conditions are used. This is illustrated by results of Rozenson and Heller-Kallai [117,127] and Russell *et al.* [128] who reported considerable irreversible alterations in the nontronite structure using unbuffered dithionite or sodium disulfide. Russell *et al.* [128] showed rather marked differences in the Mössbauer spectra between an original and a reduced-reoxidized nontronite from California (designated KOE), using pH 6 dithionite as the reductant. These changes were attributed to Fe in a range of environments, and, since KOE contains high tetrahedral Fe, this would be expected if the reduction reaction produced irreversible changes in the mineral such as removal of tetrahedral Fe.

Rozenson and Heller-Kallai [117] reported that dithionite reduction produced a black-colored product that turned rust color upon reoxidation. Stucki and Roth [99] observed a transient black color only when reduction occurred in unbuffered dithionite medium. Rozenson and Heller-Kallai [117] reported that about 20% of the Fe in the original sample was liberated into solution by reduction-reoxidation-rereduction. In contrast, Stucki *et al.* [129] and Lear and Stucki [118] observed that less than 2% of the Fe was lost during a single reduction in CBD, and the Mössbauer spectrum of reduced-reoxidized samples closely matched the spectra of unaltered samples. Note that the CBD used by Stucki and co-workers was a much lower concentration of dithionite than used by Mehra and Jackson [130] to remove Fe oxide coatings. Neither Russell *et al.* [128] nor Rozenson and Heller-Kallai [117] used citrate-bicarbonate in combination with the dithionite. Also, Ericsson *et al.* [131] reported that CBD treatment of two soils and a Wyoming bentonite had almost no effect on the clay mineral structure except to alter the Fe^{2+}/Fe^{3+} ratio.

Reduction with sodium sulfide dissolves some of the Fe from the clay, which precipitates as ferrous sulfide and drives the reduction reaction to a rather considerable extent to the right [127]. The principal difference between sulfide and dithionite is that the former dissolves Fe from the structure to form an insoluble product, whereas the dissolution product of the latter is soluble.

Rozenson and Heller-Kallai [117] also found that dithionite and hydrazine behave differently toward the clay surfaces. Since dithionite is a negatively charged ion, a repulsive force must exist between the reducing anion and the basal surfaces. They suggested, therefore, that reduction by dithionite must take place from the pyramidal edges, and solid state electron diffusion occurs laterally through the octahedral sheet. This would explain why they observed comparatively low levels of reduction (Fe^{2+}:Fe^{3+} = 0.3 to 0.78) with dithionite. However, this hypothesis is questionable because Komadel *et al.* [119] and Stucki *et al.* [129] observed virtually complete reduction of ferruginous smectite and Upton montmorillonite without much difficulty using CB buffered dithionite.

Russell *et al.* [128] observed that the rate or extent of unbuffered dithionite reduction varied directly with the amount of tetrahedral Fe present in the clay, which demonstrates that increased tetrahedral charge has a positive effect on dithionite reduction. A side effect, however, is that some tetrahedral Fe is dissolved in the process. Tetrahedral Fe^{3+} could facilitate electron transfer through the clay layer, regardless of whether the reducing agent were at the pyramidal edges or the basal surfaces. The fact that layer charge also increased with increasing tetrahedral Fe^{3+} indicates that either reducing action is primarily from the pyramidal edges or coulombic repulsion at basal surfaces is unimportant. Dissolution of structural Fe could require a more intimate association between reducing agent and basal O than would be possible if dithionite were only at the edges of the layers.

Hydrazine, on the other hand, is a base and thus able to approach the basal surfaces rather easily. Rozenson and Heller-Kallai [117] suggested that proton-electron pairs penetrate the basal surface along the c-axis to the octahedral Fe. They argue that montmorillonite is more completely reduced than nontronite because the higher tetrahedral charge in nontronite serves as a potential barrier to the electron from hydrazine. Results for beidellite are consistent with this trend.

Gan *et al.* [132] recently took a different approach, using ESR (electron spin resonance) spectroscopy and three different reducing agents. The standard electrode reduction potential, $\mathcal{E}^\circ$, of sodium dithionite ($Na_2S_2O_4$) is approximately -1.12 V [133]. But hydrazine (N_2H_4, or its conjugate acid $N_2H_5^+$ if hydrated), which generally reduces only about 10% of the structural Fe in these same clays [95,99,116,117,127], has an $\mathcal{E}^\circ$ of -0.94 V [134]. Obviously, the difference between these two reducing agents must be due to something other than their standard electrode potentials.

Figure 6 reveals the reducing abilities of three different reducing agents in ferruginous smectite (SWa-1) at room temperature (25°C). Notice that dithionite is peculiar and demonstrates a much higher ability to reduce structural Fe^{3+} than any of the other compounds.

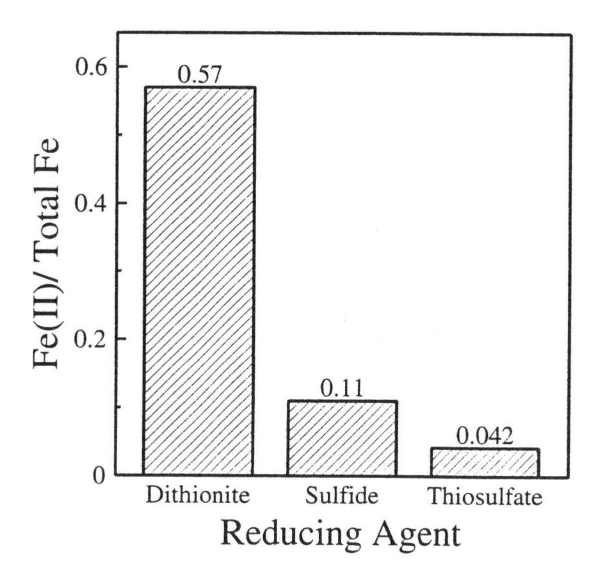

FIGURE 6. Reduction of Structural Fe in Na Smectite SWa-1 by Freshly Prepared 0.01 M Solutions of $Na_2S_2O_4$, Na_2S, or $Na_2S_2O_3$ for 24 hr at 25°C.

The ESR spectra of these same reducing agents revealed no signal from hydrazine or thiosulfate or sulfide, and a strong signal from dithionite (Figure 7). These results clearly indicate the presence of unpaired electrons in dithionite by the resonance signal centered at about g = 2.0091 for the solid phase and shifted to g = 2.0113 for the solution phase. The signals were most intense in the freshly prepared dithionite solutions, then decreased in intensity over time, disappearing completely after about 57 hrs in the 1.0 M solution, and after less than 4 hrs in the 0.01 M solution.

When dithionite was added to the clay suspension to make the final Na concentration 0.01 M, a paramagnetic signal was observed (Figure 8A). With time, the signal became even stronger than that of the pure dithionite solution (Figure 8B), and after 9 hrs of contact between clay and dithionite, the unpaired electron signal disappeared.

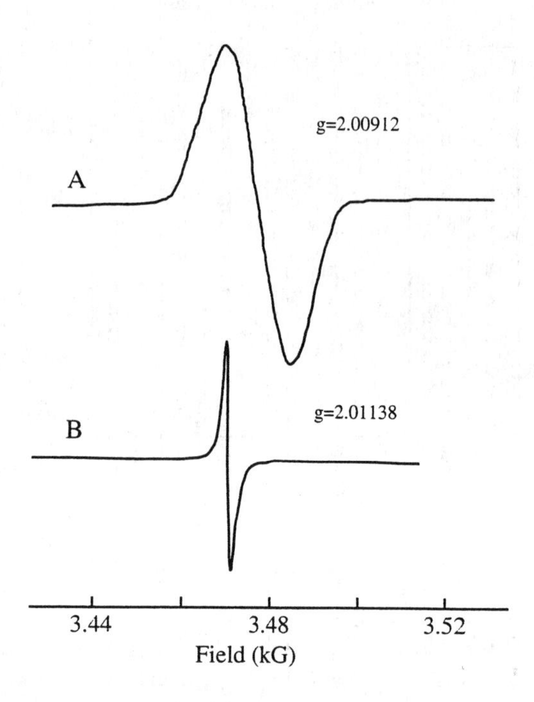

FIGURE 7. ESR Spectra of Solid (A) and 1.0 M Aqueous Solution (B) of $Na_2S_2O_4$.

These results indicate that a free radical (probably sulphoxylate, $SO_2^{-}\cdot$) from the dithionite solution is not only preserved when added to the clay but actually is enhanced. The clay apparently initially consumes some electrons from dithionite, but as the reaction proceeds, more unpaired electrons are produced. These unpaired electrons could be within the crystal structure of the clay or a combination of these sites and dithionite radical at the clay surface. This supports the hypothesis that the effectiveness of sodium dithionite in reducing the clay is due to the existence of a free radical.

Microorganisms apparently also reduce structural Fe in clay minerals [135-137]. The mechanisms for microbial reduction have yet to be identified, and many questions arise as to the precise role of microorganisms in Fe redox reactions. For instance, does the reduction occur because of a membrane-bound process requiring intimate contact between clay mineral and organism, or is it due to an extra-cellular or exudate compound from the organism? Is it an aerobic or anaerobic process? What are the metabolic sequences responsible for reduction?

Which organisms are most efficient? How does microbial reduction affect physical chemical properties of the minerals? Studies to answer these questions are currently underway (e.g., Gates *et al.*, [138]).

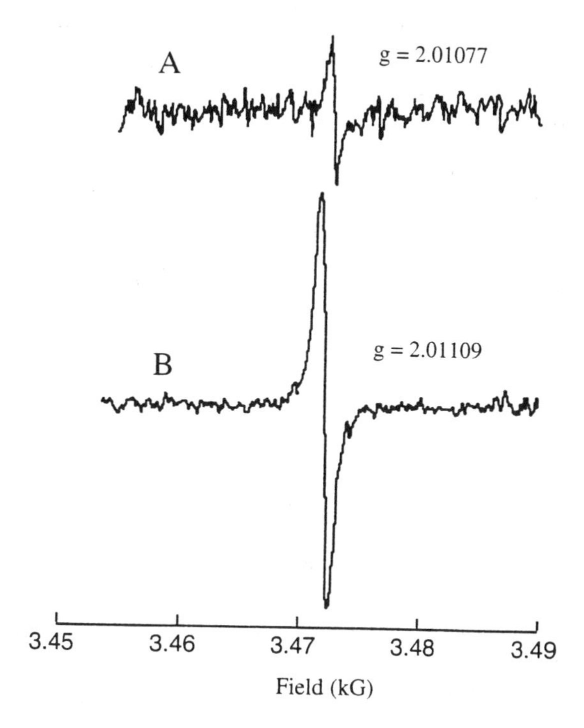

FIGURE 8. ***ESR Spectra of 0.01 M $Na_2S_2O_4$. in Smectite SWa-1 Suspension: A) Fresh, B) After 4 hrs.***

In summary, the reduction mechanism(s) for Fe^{3+} smectites depend on the nature of the reducing agent and medium in which reduction takes place. Laboratory methods may or may not replicate actual redox processes that occur in nature, but should provide insights as to the potential range of reactions that can occur. Standardization of techniques, natural variations among samples, and handling of air-sensitive clays all are factors that must be considered before valid comparisons between studies can be made.

7. SUMMARY

The chemical nature of mineral surfaces is widely recognized as one of the most significant factors contributing to the fate and behavior of contaminants, but these properties are generally considered to be unchanging during the time of contaminant exposure and reaction. The aforementioned studies clearly demonstrate, however, that mineral surface chemistry is greatly modified by changes in the oxidation state of Fe in the mineral crystal structure. The soil serves many functions in the environment. It acts as a geologic filter scavenging undesirable solutes from a percolating solution, or it may release previously sorbed substances back into the surrounding media depending on climatic and other conditions. The surfaces of soil minerals provide unique chemical environments that often facilitate the transformation of some chemical species to very different forms, some of which may be less desirable and more mobile, or vice-versa. Changing the surface properties of the minerals through oxidation or reduction of structural Fe will likely have a great impact on this chemical environment.

The redox activity of mineral surfaces also will affect the oxidation state, and consequently the speciation and chemical behavior, of redox-sensitive metal ions in the surrounding solution. One example of such a possible redox couple with a metal ion is Cr, which may exist as Cr(VI) anion in solution (CrO_4^{2-}, $Cr_2O_7^{2-}$). A reduced Fe clay will react with the Cr(VI) species, reducing it to Cr^{3+}, which will either be cationic or precipitated as the oxide. By this process, Cr will be less mobile and less hazardous in the soil environment. Geochemical models for predicting the fate and behavior of metals in the vadose zone, therefore, must account for redox transformations of the metal ion at mineral surfaces.

The exposure of soil minerals to redox-sensitive sorbates, solvents, humic materials, varying redox environments, and microorganisms may invoke surface chemical changes *in situ* over short time periods and thereby produce vital differences in the speciation and reactivity of all components in the soil-water system.

REFERENCES

1. Grim, R.E., "Clay Mineralogy," McGraw Hill, (New York, 1953).

2. Grim, R.E., "Clay Mineralogy, 2nd Edition," McGraw Hill, (New York, 1968).

3. Brindley, G.W. and G. Brown, "Crystal Structures of Clay Minerals and Their X-Ray Identification. Monograph No. 5," *Mineralogical Society*, (London 1980).

4. Ross, C.S. and S.B. Hendricks, "Minerals of the Montmorillonite Group. Their Origin and Relation to Soils and Clays," *U.S. Geol. Surv., Profess. Paper* 205b:23-79 (1945).

5. Kerr, P.F., "Preliminary Reports, Reference Clay Minerals," Amer. Petroleum Institute Research Project 49, Columbia University, (New York, 1951).

6. Brindley, G.W. and D.M.C. MacEwan, "Structural Aspects of the Mineralogy of Clays and Related Silicates," in. *Ceramics: A Symposium. The British Ceramic Society*, pp. 15-59. (London, 1953).

7. Weaver, C.E. and L.D. Pollard, "The Chemistry of Clay Minerals," (Elsevier, Amsterdam 1973).

8. Martin, R.T., S.W. Bailey, D.D. Eberl, D.S. Fanning, S. Guggenheim, H. Kodama, D.R. Pevear, J. Srodon and F.J. Wicks, "Report of the Clay Minerals Society Nomenclature Committee: Revised Classification of Clay Materials," *Clays Clay Miner.* 39:333-335 (1991).

9. Eslinger, E., P. Highsmith, D. Albers and B. De Mayo, "Role of Iron Reduction in the Conversion of Smectite to Illite in Bentonite in the Disturbed Belt, Montana," *Clays Clay Miner.* 27:327-338. (1979).

10. Malathi, N., S.P. Puri and I.P. Saraswat, "Mossbauer Studies of Iron in Illite and Montmorillonite. II: Thermal Treatment," *J. Phys. Soc. Jap.* 31:117-122 (1971).

11. Burst, J.F., "Mineral Heterogenity in Glauconite Pellets," *Am. Mineral.* 45:481-489 (1958).

12. McRae, S.G., "Glauconite," *Earth Sci. Rev.* 8:397-440 (1972).

13. Thompson, G.R. and J. Hower, "The Mineralogy of Glauconite," *Clays Clay Miner.* 23:289-300 (1975).

14. McConchie, D.M., J.B. Ward, V.H. McCann and D.W. Lewis, "A Mossbauer Investigation of Glauconite and Its Geological Significance," *Clays Clay Miner.* 27:339-348 (1979).

15. Rozenson, I. and L. Heller-Kallai, "Mossbauer Spectra of Glauconites Reexamined," *Clays Clay Miner.* 26:173-175 (1978).

16. Tardy, Y. and O. Touret, "Hydration Energies of Smectites: A Model for Glauconite, Illite, and Corrensite Formation," in *Proc. Int. Clay Conf., Denver, 1985,* L.G. Schulze, H. van Olphen, and F.A. Mumpton, Eds. The Clay Minerals Society, pp. 46-52. (Bloomington, IN 1987).

17. Buckley, H.A., J.C. Bevan, K.M. Brown, L.R. Johnson and V.C. Farmer, "Glauconite and Celadonite: Two Separate Mineral Species," *Mineral. Mag.* 42:373-382 (1978).

18. Ross, G.J., "Experimental Alteration of Chlorites into Vermiculites by Chemical Oxidation," *Nature* 255:133-134 (1975).

19. Ross, G.J. and H. Kodama, "Experimental Alteration of a Chlorite into a Regularly Interstratified Chlorite-vermiculite by Chemical Oxidation", *Clays Clay Miner.* 24:183-190 (1976).

20. Ballet, O. and J.M.D. Coey, "Magnetic Properties of Sheet Silicates; 2:1 Layer Minerals," *Phys. Chem. Miner.* 8:218-29 (1982).

21. Olivier, P.D., J.C. Vedrine and H. Pezerat, "Application de La Resonance Paramagnetique Electronique a La Localization du Fe^{3+} das Les Smectites," *Bull. Groupe Fr. Argiles* 27:153-165 (1975).

22. Goodman, B.A. and D.C. Bain, "Mossbauer Spectra of Chlorites and Their Decomposition Products," *Dev. Sedimentol.* 27:65-74 (1979).

23. Borggaard, O.K., H.B. Lindgreen and S. Morup, "Oxidation and Reduction of Structural Iron in Chlorite at 480 Degree Celsius," *Clays Clay Miner.* 30:353-363 (1982).

24. Scott, A.D. and J. Amonette, "The Role of Iron in Mica Weathering," in *Iron in Soils and Clay Minerals* J.W. Stucki, B.A. Goodman, and U. Schwertmann Eds. pp. 537-623. D. Reidel, (Dordrecht, 1988).

25. Bradley, W.F., "Structure of Attapulgite," *Am. Mineral.* 25:405-410 (1940).

26. Rozenson, I. and L. Heller-Kallai, "Mossbauer Studies of Palygorskite and Some Aspects of Palygorskite Mineralogy," *Clays Clay Miner.* 29:226-232 (1981).

27. Stucki, J.W., "Structural Iron in Smectites," in *Iron in Soils and Clay Minerals* J.W. Stucki, B.A. Goodman, and U. Schwertmann Eds. pp. 625-675, D. Reidel, (Dordrecht, 1988).

28. Jepson, W.B., "Kaolins: Their Properties and Uses," *Phil. Trans. R. Soc. Lond. A* 311:411-432 (1984).

29. Chukhrov, F.V., B.B. Zvyagin, V.A. Drits, A.I. Gorshkov, L.P. Ermilova, E.A. Goilo and E.S. Rudnitskaya, "The ferric Analog of Pyrophyllite and Related Phases," in *Proc. Int. Clay Conf. 1979*. M.M. Mortland and V.C. Farmer Eds. pp. 55-64. Oxford, (Elsevier, Amsterdam 1978).

30. Coey, J.M.D., F.V. Chukhrov and B.B. Zvyagin, "Cation Distribution, Mossbauer Spectra, and Magnetic Properties of Ferripyrophyllite," *Clays Clay Miner.* 32:198-204 (1984.)

31. Gruner, J.W., "The Composition and Structure of Minnesotaite, a Common Iron Silicate in Iron Formations," *Am. Mineral* 29:363-372 (1944).

32. Bailey, S.W., "Structures of Layer Silicates," in *Crystal Structures of Clay Minerals and their X-ray Identification. Monograph No. 5* G.W. Brindley and G. Brown Eds. pp. 1-123. The Mineralogical Society, (London, 1980).

33. Guggenheim, S. and S.W. Bailey, "The Superlattice of Minnesotaite," *Can. Mineral* 20:579-584 (1982).

34. Kager, P.C.A. and I.S. Oen, "Iron-rich Talc-opal-minnesotaite Spherulites and Crystallochemical Relations of Talc and Minnesotaite," *Mineral. Mag.* 47:229-231 (1983).

35. Ballet, O., J.M.D. Coey, P. Mangin and M.G. Townsend, "Ferrous Talc -- A Planar Antiferromagnet," *Solid State Comm.* 55:787-790 (1985).

36. McBride, M.B. and M.M. Mortland, "Copper(II) Interactions with Montmorillonite: Evidence from Physical Methods," *Soil Sci. Soc. Am. Proc.* 38:408-415 (1974).

37. Protod'yakonova, Z.M. and M.R. Enikeev, "Phyllosilicates: Montmorillonite group. Saponite $(Ca,Na)_x [Mg_3Al_x Si_{4-x}]O_{10} (OH)_2{\cdot}4H_2O)$, zinc saponite (sauconite) $(Zn,Mg,Al,Fe)_3[Al_xSi_{4-x}]O_{10} (OH)_2{\cdot}4H_2O)$, and copper saponite $((Ga,Na)_x[(Mg,Cu)_3Al_xSi_{4-x}]O_{10} (OH)_2{\cdot}4H_2O)$," *Miner. Uzb.* [*From Chem. Abstr. 91:126188*] 3:334-339 (1976).

38. Maksimovi'c, Z. and G.W. Brindley, "Hydrothermal Alteration of a Serpentinite Near Takovo, Yugoslavia, to Chromium-bearing Illite/Smectite, Kaolinite, Tosudite, and Halloysite," *Clays Clay Miner.* 28:295-302 (1980).

39. Brindley, G.W., "Order-disorder in Clay Mineral Structures," in *Crystal Structures of Clay Minerals and their X-ray Identification. Monograph No. 5* G.W. Brindley and G. Brown Eds. pp. 125-195. Mineralogical Society, (London, 1980).

40. Bailey, S.W., "Report of AIPEA Nomenclature Committee," *Supplement to AIPEA Newsletter No. 22,* (February, 1986).

41. Ross, C.S., "Sauconite: A Clay Mineral of the Montmorillonite Group," *Am. Mineral.* 31:411-424 (1946).

42. Maksimovi'c, Z., "-Kerolite-pimelite Series from Goles Mountain, Yugoslavia," in *Proc. Int. Clay Conf., Jerusalem, 1966, Vol. 1.* L. Heller and A. Weiss Eds. pp. 97-105. Israel Prog. Sci. Transl., (Jerusalem, 1966).

43. Brindley, G.W. and J.V. De Souza, "Nickel-containing Montmorillonites and Chlorites from Bazil, with Remarks on Schuchardite," *Mineral. Mag.* 40:141-152 (1975).

44. Brindley, G.W., D.L. Bish and H.-M. Wan, "Compositions, Structures, and Properties of Nickel-containing Minerals in the Kerolite-pimelite Series," *Am. Mineral.* 64:615-625 (1979).

45. Decarreau, A., F. Colin, A. Herbillon, A. Manceau, D. Nahon, H. Paquet, D. Trauth-Badaud and J.J. Trescases, "Domain Segregation in Ni-Fe-Mg-smectites," *Clays Clay Miner.* 35:1-10 (1987).

46. McBride, M.B., T.J. Pinnavaia and M.M. Mortland, "Perturbation of Structural Fe^{+3} in Smectites by Exchange Ions," *Clays Clay Miner.* 23:103-107 (1975).

47. McBride, M.B., M.M. Mortland and T.J. Pinnavaia, "Exchange Ion Positions in Smectite: Effect on Electron Spin Resonance of Structural Ion," *Clays Clay Miner.* 23:162-164 (1975).

48. McBride, M.B., "Reactivity of Adsorbed and Structural Iron in Hectorite as Indicated by Oxidation of Benzidine," *Clays Clay Miner.* 27:224-230 (1979).

49. Osthaus, B.B., "Chemical Determination of Tetrahedral Ions in Nontronite and Montmorillonite," *Clays Clay Miner.* 2:404-417 (1953).

50. Quakernaat, J., "A New Occurrence of a Macrocrystalliner Form of Saponite," *Clay Miner.* 8:491:493 (1970).

51. Goodman, B.A., J.D. Russell, A.R. Fraser and F.W.D. Woodhams, "A Mossbauer and I.R. Spectroscopic Study of the Structure of Nontronite," *Clays Clay Miner.* 24:53-59 (1976).

52. Eggleton, R.A., "Nontronite: Chemistry and X-ray Diffraction," *Clays Clay Miner.* 12:181-94 (1977).

53. Besson, G., A.S. Booking, L.G. Dainyak, M. Rautureau, S.I. Tsipursky, C. Tchoubar and V.A. Drits, "Use of Diffraction and Mossbauer Methods for the Structural and Crystallochemical Characterization of Nontronites," *J. Appl. Cryst.* 16:374-383 (1983).

54. Bonnin, D., G. Calas, H. Suquet and H. Pezerat, "Site Occupancy of Fe^{3+} in Garfield Nontronite: A Spectroscopic Study," *Phys. Chem. Miner.* 12:55-64 (1985).

55. Cardile, C.M. and J.H. Johnston, "Structural Studies of Nontronites with Different Iron Contents by ^{57}Fe Mossbauer Spectroscopy," *Clays Clay Miner.* 33:295-300 (1985).

56. Johnston, J.H. and C.M. Cardile, "Iron Sites in Nontronite and the Effect of Interlayer Cations from Mossbauer Spectra," *Clays Clay Miner.* 33:21-30 (1985).

57. Luca, V. and C.M. Cardile, "Improved Detection of Tetrahedral Fe^{3+} in Nontronite SWa-1 by Mossbauer Spectroscopy," *Clay Miner.* 24:555-559 (1989).

58. Luca, V. and D.J. MacLachlan, "Site Occupancy in Nontronite Studied by Acid Dissolution and Mössbauer Spectroscopy," *Clays Clay Miner.* 40:1-7 (1992).

59. Mering, J., and A. Oberlin, "Electron-optical Study of Smectites," in. *Clays and Clay Minerals, Proc. 15th Natl. Conf., Pittsburgh, Pennsylvania, 1966.* S.W. Bailey Ed. pp. 3-25. Pergamon Press, (New York, 1967).

60. Petruk, W., D.M. Farrell, E.E. Laufer, R.J. Tremblay and P.G. Manning, "Nontronite and Ferruginous Opal from the Peace River Iron Deposit in Alberta, Canada," *Can. Mineral.* 15:14-21 (1977).

61. Goodman, B.A., "An Investigation by Mossbauer and EPR Spectroscopy of the Possible Presence of Iron-rich Impurity Phases in Some Montmorillonites". *Clay Miner.* 13:351-356 (1978).

62. Goodman, B.A., "The Mossbauer Spectra of Nontronites: Consideration of an Alternative Assignment," *Clays Clay Miner.* 26:176-177 (1978).

63. Besson, G., C. de la Calle, M. Rautureau, C. Tchoubar, S.I. Tsipurski and V.A. Dritz, "X-ray and Electron Diffraction Study of the Structure of the Garfield Nontronite," in *Proc. Int. Clay Conf. 1982.,* H. van Olphen and F. Veniale Eds. pp. 29-40. Bologna and Pavia, (Elsevier, Amsterdam 1981).

64. Tsipursky, S.I. and V.A. Drits, "The Distribution of Octahedral Cations in the 2:1 Layers of Dioctahedral Smectites Studied by Oblique-texture Electron Diffraction," *Clay Miner.* 19:177-193 (1984).

65. Murad, E., "Mossbauer Spectra of Nontronites: Structural Implications and Characterization of Associated Iron Oxides," *Z. Pflanzenern hr. Bodenk.* 150:279-285 (1987).

66. Lear, P.R. and J.W. Stucki, "Intervalence Electron Transfer and Magnetic Exchange Interactions in Reduced Nontronite," *Clays Clay Miner.* 35:373-378 (1987).

67. Lear, P.R. and J.W. Stucki, "Magnetic Ordering and Site Occupancy of Iron in Nontronite," *Clay Miner.* 25:3-13 (1990).

68. Low, P.F., "Force Fields and Chemical Equilibrium in Heterogeneous Systems with Special Reference to Soils," *Soil Sci.* 71:409-418 (1951).

69. Gibbs, J.W., "The Collected Works of J. Willard Gibbs," Vol. 1. Yale University Press, (New Haven, CN 1948).

70. Lewis, G.N., "Valence and the Structure of Atoms and Molecules," in *The Chemical Company Catalog,* pp. 141-142. (NY, 1923).

71. Lewis, G.N. and M. Randall, (Revised by K.S. Pitzer and L. Brewer), *Thermodynamics, Second Edition.*, pp. 158-161. McGraw-Hill, (New York, 1961).

72. Lindsay, W.L. and M. Sadiq, "Use of pe + pH to Predict and Interpret Metal Solubility Relationships in Soils," *Sci. Total Environ.* 28:169-178 (1983).

73. Licht, T.S. and A.J. de Bethune, "Recent Developments Concerning the Signs of Electrode Potentials," *J. Chem. Ed.* 34:433-440 (1957).

74. Ayers, G.H., "Quantitative Chemical Analysis, 2nd Edition." p. 710. Harper and Row, (New York, 1968).

75. Barrow, G.M., "Physical Chemistry, 3rd Edition". p.78. McGraw Hill, (New York, 1973).

76. Gratzel, M., "Heterogeneous Photochemical Electron Transfer," pp. 21-25. CRC Press, (Boca Raton, FL 1989).

77. Kahn, S.U.M. and J.O'M. Bockris, "Electronic States in Solution and Charge Transfer," *J. Phys. Chem.* 87:2599-2603 (1983).

78. Theis, T.L. and P.C. Singer, "The Stabilization of Ferrous Iron by Organic Compounds in Natural Waters," in *Trace Metals and Metal Organic Interactions in Natural Waters.* P.C. Singer Ed. pp. 303-320. Ann Arbor Science, (Ann Arbor, MI 1973).

79. Pearson, R.G., "Hard and Soft Acids and Bases," *J. Am. Chem. Soc.* 85:3533-3539 (1963).

80. Pearson, R.G., "Acids and Bases," *Science* 151:172-177 (1966).

81. Pearson, R.G., "Hard and Soft Acids and Bases," *Chem. Brit.* 3:103-107 (1967).

82. Pearson, R.G., "Application of the Principle of Hard and Soft Acids and Bases to Organic Chemistry," *J. Am. Chem. Soc.* 89:1827-1836 (1967).

83. Pearson, R.G., "Hard and Soft Acids and Bases, Part II, Underlying Theories," *J. Chem. Edu.* 45:643-648 (1968).

84. Pearson, R.G., "Hard and Soft Acids and Bases, Part I, Fundamental Principles," *J. Chem. Edu.* 45:581-587 (1968).

85. Pearson, R.G., "Hard and Soft Acids and Bases," Dowden, Hutchinson, and Ross, Eds. (Stroudsburg, PA 1973).

86. Arhland, S., J. Chatt and N.R. Davies, "The Relative Affinities of Ligand Atoms for Acceptor Molecules and Ions," *Quarterly Rev., Chem. Soc.* XII: 265-276 (1958).

87. Klopman, G., "Chemical Reactivity and the Concept of Charge- and Frontier-controlled Reactions," *J. Am. Chem. Soc.* 90:223-234 (1968).

88. Klopman, G., "Reactivity and Reaction Paths," Wiley Interscience, (New York, 1974).

89. Misono, M., E. Ochiai, Y. Saito and Y. Toneda, "A New Dual Parameter Scale for the Strength of Lewis Acids and Bases with the Evaluation of Their Softness," *J. Inorg. Nucl. Chem.* 29:2685-2691 (1967).

90. Yu, Tian-ren, "Physical Chemistry of Paddy Soils," Springer-Verlag, (Berlin, 1985).

91. Patrick, W.H., "The Role of Inorganic Redox Systems in Controlling Reduction in Paddy Soils," in *Proc. Symp. on Paddy Soil.* Science Press, pp. 107-117. (Beijing, 1981).

92. Moore, P.A. Jr. and W.H. Patrick, Jr., "Manganese Availability and Uptake by Rice in Acid Sulfate Soils," *Soil Sci. Soc. Am. J.* 53:104-109 (1989).

93. Moore, P.A. Jr. and W.H. Patrick, Jr., "Iron Availability and Uptake by Rice in Acid Sulfate Soils," *Soil Sci. Soc. Am. J.* 53:471-476 (1989).

94. Moore, P.A. Jr. and W.H. Patrick, Jr., "Calcium and Magnesium Availability and Uptake by Rice in Acid Sulfate Soils," *Soil Sci. Soc. Am. J.* 53:816-822 (1989).

95. Stucki, J.W. and P.R. Lear, "Variable Oxidation States of Iron in the Crystal Structure of Smectite Clay Minerals," in *Structures And Active Sites Of Minerals*. L.M. Coyne, D. Blake, and S. McKeever Eds. Chapter 17. American Chemical Society, (Washington, D.C. 1989).

96. Walker, G.F., "The Decomposition of Biotite in the Soil," *Mineral. Mag.* 28:693-703 (1949).

97. Robert, M. and G. Pedro, "Etude des Relations Entre les Phenomenes D'oxydation et L'aptitude a L'ouverture Dans Les Micas Trioctahedriques," *Proc. Int. Clay Conf. 1969* pp.455-473 (1969).

98. Stucki, J.W., P.F. Low, C.B. Roth and D.C. Golden, "Effects of Oxidation State of Octahedral Iron on Clay Swelling," *Clays Clay Miner.* 32:357-362 (1984).

99. Stucki, J.W. and C.B. Roth, "Oxidation-reduction Mechanism for Structural Iron in Nontronite," *Soil Sci. Soc. Am. J.* 41:808-814 (1977).

100. Lear, P.R. and J.W. Stucki, "Effects of Iron Oxidation State on the Specific Surface Area of Nontronite," Clays Clay Miner. 37:547-552 (1989).

101. Jepson, W.B., "Structural Iron in Kaolinites and Associated Ancillary Minerals," in *Iron in Soils and Clay Minerals*. J.W. Stucki, B.A. Goodman, and U. Schwertmann Eds. pp. 467-536. D. Reidel, (Dordrecht, 1988).

102. Babcock, K.L., "Theory of Chemical Properties of Soil Colloidal Systems at Equilibrium," *Hilgardia* 34:417-542 (1963).

103. Low, P.F., "Nature and Properties of Water in Montmorillonite-Water Systems," *Soil Sci. Soc. Am. J.* 43:651-658 (1979).

104. Low, P.F., "The Swelling of Clay: II. Montmorillonite," *Soil Sci. Soc. Am. J.* 44:667-676 (1980).

105. Ravina, I. and P.F. Low, "Relation Between Swelling, Water Properties, and b-dimension in Montmorillonite-water Systems," *Clays Clay Miner.* 20:109-123 (1972).

106. Ravina, I. and P.F. Low, "Change of b-Dimension with Swelling of Montmorillonite," *Clays Clay Miner.* 25:196-200 (1977).

107. Oliphant, J.L. and P.F. Low, "The Relative Partial Specific Enthalpy of Water in Montmorillonite-water Systems and Its Relation to the Swelling of These Systems," *J. Colloid Interface Sci.* 89:366-373 (1982).

108. Zhang, Z.Z. and P.F. Low, "Relation Between the Heat of Immersion and the Initial Water Content of Li-, Na-, and K-montmorillonite," *J. Colloid Interface Sci.* 133:461-472 (1989).

109. Sposito, G., "The Thermodynamics of Soil Solutions," pp. 223. Oxford University Press, (New York, 1981).

110. Foster, M.D., "Geochemical Studies of Clay Minerals: II. Relation Between Ionic Substitution and Swelling in Montmorillonites," *Am. Mineral.* 38:994-1006 (1953).

111. Wu, J., P.F. Low and C.B. Roth, "Effects of Octahedral-iron Reduction and Swelling Pressure on Interlayer Distances in Na-nontronite," *Clays Clay Miner.* 37:211-218 (1989).

112. Chen, S.Z., P.F. Low, and C.B. Roth, "Relation Between Potassium Fixation and the Oxidation State of Octahedral Iron," *Soil Sci. Soc. Am. J.* 51:82-86 (1987).

113. Khaled, E.M., and J.W. Stucki, "Effects of Iron Oxidation State on Cation Fixation in Smectites," *Soil Sci. Soc. Am. J.* 55:550-554 (1991).

114. Stucki, J.W. and D. Tessier, "Effects of Iron Oxidation State on the Texture and Structural Order of Na-nontronite," *Clays Clay Miner.* 39:137-143 (1991).

115. Addison, W.E., and J.H. Sharp, "Redox Behavior of Iron in Hydroxylated Silicates," *Clays Clay Miner.* 11:95-104 (1963).

116. Stucki, J.W., C.B. Roth and W.E. Baitinger, "Analysis of Iron-bearing Clay Minerals by Electron Spectroscopy for Chemical Analysis (ESCA)," *Clays Clay Miner.* 24:289-292 (1976).

117. Rozenson, I. and L. Heller-Kallai, "Reduction and Oxidation of Fe^{3+} in Dioctahedral Smectite--1: Reduction with Hydrazine and Dithionite," *Clays Clay Miner.* 24:271-282 (1976).

118. Lear, P.R. and J.W. Stucki, "The Role of Structural Hydrogen in the Reduction and Reoxidation of Iron in Nontronite," *Clays Clay Miner.* 33:539-545. (1985).

119. Komadel, P., P.R. Lear and J.W. Stucki, "Reduction and Reoxidation of Iron in Nontronites: Rate of Reaction and Extent of Reduction," *Clays Clay Miner.* 38:203-208 (1990).

120. Turner, F.T. and W.H. Patrick, "Chemical Changes in Waterlogged Soils as a Result of Oxygen Depletion," *Trans. Int. Congr. Soil Sci. 9th* 4:53-65 (1968).

121. Ponnamperuma, F.N., "The Chemistry of Submerged Soils," *Adv. Agronomy* 24:29-96 (1972).

122. Liu, Z., "Oxidation-reduction Potential," in *Physical Chemistry of Paddy Soils.* Tian-ren Yu, Ed. pp. 1-26. Springer-Verlag, (Berlin, 1985).

123. Rai, D. and J.M. Zachara, "Chemical Attenuation Rates, Coefficients, and Constants in Leachate Migration. Volume 1: A critical review," *Report to Electric Power Research Institute,* February, 1984. Batelle, Pacific Northwest Laboratories, (Richland, WA 1984).

124. Rai, D., J.M. Zachara, R.A. Schmidt and A.P. Schwab, "Chemical Attenuation Rates, Coefficients, and Constants in Leachate Migration. Volume 2: An Annotated Bibliography," *Report to Electric Power Research Institute,* February, 1984. Batelle, Pacific Northwest Laboratories, (Richland, WA 1984).

125. Lindsay, W.L., *Chemical Equilibria in Soils.* Wiley-Interscience, (New York 1979).

126. Lindsay, W.L., "Solubility and Redox Equilibria of Iron Compounds in Soils," in *Iron in Soils and Clay Minerals.* J.W. Stucki, B.A. Goodman, and U. Schwertmann Eds. pp. 37-62. D. Reidel, (Dordrecht, 1988).

127. Rozenson, I. and L. Heller-Kallai, "Reduction and Oxidation of Fe^{3+} in Dioctahedral Smectite--2: Reduction with Sodium Sulphide Solution," *Clays Clay Miner.* 24:283-288 (1976).

128. Russell, J.D., B.A. Goodman and A.R. Fraser, "Infrared and Mossbauer Studies of Reduced Nontronites," *Clays Clay Miner.* 27:63-71 (1979).

129. Stucki, J.W., D.C. Golden and C.B. Roth, "Effects of Reduction and Reoxidation of Structural Iron on the Surface Charge and Dissolution of Dioctahedral Smectites," *Clays Clay Miner.* 32:350-356 (1984).

130. Mehra, O.P. and M.L. Jackson, "Iron Oxide Removal from Soils and Clays by a Dithionite-citrate System Buffered with Sodium Bicarbonate," in *Clays and Clay Minerals, Proc. 7th Natl. Conf., Washington, D.C., 1958.* A. Swineford Ed. pp. 317-327. Pergamon Press, (New York, 1960).

131. Ericsson, T., J. Linares and E. Lotse, "A Mossbauer Study of the Effect of Dithionite/citrate/bicarbonate Treatment on a Vermiculite, Smectite and a Soil," *Clay Miner.* 19:85-91 (1984).

132. Gan, H., J.W. Stucki, and G.W. Bailey, "Free Radical Reduction of Structural Iron in Ferruginous Smectite," *Clays Clay Miner.* (1992).

133. Lide, D.R., Ed. *CRC Handbook of Chemistry and Physics, 72nd Edition.* CRC Press, (Boca Raton 1992).

134. Douglas, B., D.H. McDaniel and J.J. Alexander, *Concepts and Models of Inorganic Chemistry.* John Wiley and Sons, (New York, 1983).

135. Komadel, P., J.W. Stucki and H.T. Wilkinson, "Reduction of Structural Iron in Smectites by Microorganisms," in *Proc. Sixth Meeting of the European Clay Groups, Seville, 1987.* E. Galán, J.L. Pérez-Rodriguez, and J. Cornejo Eds. pp. 322-324. Sociedad Española de Arcillas, (Sevilla, 1987).

136. Stucki, J.W., P. Komadel and H.T. Wilkinson, "Microbial Reduction of Structural Iron(III) in Smectites," *Soil Sci. Soc. Am. J.* 51:1663-1665 (1987).

137. Wu, J., P.F. Low and C.B. Roth, "Biological Reduction of Structural Iron in Sodium-nontronite," *Soil Sci. Soc. Am. J.* 52:295-296 (1988).

138. Gates, W.P., H.T. Wilkinson and J.W. Stucki, "Effects of Microbial Reduction on Clay Swelling," *Agronomy Abstracts 1991*:365 (1991).

139. Vanysek, P., "Electrochemical Series," in *CRC Handbook of Chemistry and Physics, 71st ed.* Lide, D.R. Ed. pp. 8-16 to 8-23. CRC Press, (Boca Raton, FL 1992).

140. Williams, D.R., *The Metals of Life: The Solution Chemistry of Metal Ions in Biological Systems.* Van Nostrand Reinhold Co., (New York, 1971).

141. Misono, M. and Y. Saito, "Evaluation of Softness from the Stability Constants of Metal-ion Complexes," *Bull. Chem. Soc. Japan* 43:3680-3684 (1970).

6

DETECTION OF ANIONIC SITES ON BACTERIAL WALLS, THEIR ABILITY TO BIND TOXIC HEAVY METALS AND FORM SEDIMENTABLE FLOCS AND THEIR CONTRIBUTION TO MINERALIZATION IN NATURAL FRESHWATER ENVIRONMENTS

Terry J. Beveridge, Susanne Schultze-Lam and Joel B. Thompson
Department of Microbiology
University of Guelph
Guelph, Ontario, N1G 2W1, Canada

1. INTRODUCTION

Bacteria are ubiquitous throughout the natural environment and possess surfaces that interact strongly with metal ions. Since they are small (approximately 1.0 - 1.5 μm^3) the cells have an extremely high surface area to volume ratio, which makes them especially efficient at sorbing toxic heavy metals from solution and immobilizing them in the natural environment. The sorption is usually due to carboxyl or phosphoryl groups within the bacterial surface fabric which nucleate the formation of fine-grained minerals whose composition depends on the available cations and anions within the external aqueous solution. These begin as amorphous precipitates and, through time, grow and become crystalline. The neutralization of the anionic chemical sites by the metals also induces flocculation, leading to the formation of small particles that fall through the water column to the sediment, collecting metals as they go. Studies with bacteria-clay composites show that the organic fraction frequently dominates metal sorption and that metals cannot be as easily remobilized from this fraction as from the inorganic fraction. Sometimes, as in our Fayetteville Green Lake study, a bacterium's (in this case the cyanobacterium *Synechococcus*) metabolism can produce a

0-87371-697-3/95/$0.00+$.50

microenvironment around the cell which helps determine the mineral type that will precipitate from solution. In this case photosynthetic alkalization converts conditions necessary for gypsum mineralization to those that are more conducive to the formation of calcite.

This article reviews much of our previous work on metal sorption and mineralization on bacterial surfaces and is a condensation of a review article entitled "Metal Ion Immobilization by Bacterial Surfaces in Freshwater Environments" published in 1993 in a special issue of the *Water Pollution Research Journal of Canada* devoted to the dynamics and biotoxicity of metals in the freshwater environment [1]. To familiarize yourself with the chemistry and structure of bacterial surfaces we refer you to the following references: [2-7].

2. METAL BINDING AND MINERAL FORMATION

In nature, bacteria are surrounded by aqueous solutions containing many different organo- and metallo-ions. This is true even for those bacteria that live in seemingly dry conditions under the mineral varnishes of soil aggregates and rocks in both hot and cold desert regions. For at least part of the year (during which the cells are active) they are surrounded by a thin film of water. Bacteria require water not only to maintain cellular integrity (through turgor pressure) and proper metabolic functioning, but also to carry nutrients to the cell. They rely entirely on diffusion from their immediate local environment to obtain necessary compounds and to disperse metabolic waste products (see [4] for details). It is the surface layers that have first exposure to the diffusible components of the external milieu. Knowing this, it would seem unreasonable to assume that bacteria would not have metal ions and other environmentally-derived compounds intimately associated with their surfaces.

The bacterial cell surface can be visualized as a bristling forest of molecules that protrudes into the environment and provides a large surface area for interaction with the metal ions in which the cell is constantly being bathed. In fact, of all cellular life forms, bacteria have the largest surface area to volume ratio for these sorts of interactions [4]. Although in most natural environments these metal ions are present in low concentrations, bacterial cells show a remarkable ability to concentrate metal ions from aqueous solutions.

2.1 Cell walls

Regardless of whether bacterial walls are of the gram-positive or gram-negative variety [2,3,7,8], they tend to have a net negative charge at circumneutral pH. Those of *Escherichia coli* have a remarkable capacity to accumulate metal which is attributable to this charge character [9]. Their negative charge is contributed mainly by phosphoryl groups present in the core oligosaccharide and the N-acetylglucosaminyl residues in the lipid A moiety of the lipopolysaccharide molecule [10,11]. Although there are carboxyl groups present in the 2-keto-3-deoxyoctonate residues of the lipopolysaccharide molecule, studies with *E. coli* K12, a mutant lacking the O-antigen portion of its lipopolysaccharide, have shown that only one of three carboxyl residues is available for metal binding, the others being cross-linked to amino groups within the molecule [11]. The peptidoglycan layer of gram-negative bacteria is also capable of binding metals. Its capacity on a percent dry weight basis is similar to that of gram-positive bacteria, but on a per cell basis it is less since there is less of this polymer present [12].

Gram-positive surfaces have an even higher capacity to bind metal (Table 1)[13]. The prime site of metal binding in the wall of gram-positive bacteria seems to be the carboxyl residues of peptidoglycan and teichuronic acids and the phosphoryl groups of teichoic acids. The contribution of each to the overall metal binding capacity of the cell depends on how much of each polymer is present. For example, *B. subtilis* cells grown in media with sufficient magnesium and phosphate have walls consisting of 54% teichoic acid and 45% peptidoglycan [14-16]. Extraction of the teichoic acid shows that the majority of the metal associated with the cell was bound to the peptidoglycan portion of the wall [16,17]. On the other hand, *B. licheniformis* cells which have walls consisting of 26% teichuronic acid, 52% teichoic acid and only 22% peptidoglycan lose most of the wall-associated metal when the two acid polymers are removed. This indicates that in this case the peptidoglycan layer is not the major metal binding entity in the cell wall [18].

So far the mechanistic details of metal binding to bacterial walls are known only for *E. coli* K12 and *B. subtilis* 168. However, the fact that the heavy metal stains routinely used for electron microscopy bind to the surface of most bacteria and that the bacteria observed by electron microscopy include a massive diversity of ultrastructural and chemical types attests to the fact that the ability to bind metals is not an isolated peculiarity confined to these two organisms. It is a widespread reflection of the general anionic nature of bacterial cell walls.

Table 1. Metal Binding By Bacterial Walls [a]

Metal	*B. subtilis*			*B. licheniformis*		*E. coli*[b]	
	Native[b]	Neutralized COO^- [b,c]	No TA[d,f]	Native[b]	No TA or TUA[d,f]	Murein	Outer Membrane
Na^+	2.697	0	1.497	0.910	0.080	0.290	0.081
K^+	1.944	0	0.782	0.560	0	0.058	0.025
Mg^{2+}	8.226	0.520	7.683	0.400	0.024	0.035	0.019
Ca^{2+}	0.399	0.380	0.012	0.590	0.096	0.038	0.020
Mn^{2+}	0.801	0.732	0.656	0.662	0.004	0.052	0.012
Fe^{3+}	3.581	2.260	1.720	0.760	0.172	0.100	0.233
Ni^{2+}	0.107	0.024	0.021	0.520	0	0.019	0.019
Cu^{2+}	2.990	0.993	2.488	0.490	0	ND[e]	ND
Au^{3+}	0.363	0.214	0.265	0.031	0.012	ND	ND

[a]These values were obtained by suspending 0.1 to 5.0 mg (dry weight) of walls in metal salt under saturating conditions. For details see Beveridge *et al.* [18], Beveridge and Murray [14-16] and Beveridge and Koval [9].

[b]Micromoles of metal per milligram (dry weight) of walls.

[c]Neutralized by the addition of glycine ethyl ester to carbodimide-activated carboxylate groups.

[d]The dry weight of these walls has been adjusted downward to reflect the loss of mass due to the extraction.

[e]ND, Not determined.

[f]TA = teichoic acid; TUA = teichuronic acid.

Reprinted from Beveridge [13] with permission from the American Society for Microbiology.

In a broader sense, this ability to bind metals for electron microscopy includes all biological tissue and is one of the tenets of biological electron microscopy.

2.2 Capsules

So far we have considered only the immediate surface of the bacterium. However the ability of bacterial capsules to bind metals has been known for some time. Capsular material is a major component of biofilms and since biofilms are present in almost every natural environment, it follows that capsular material is also a major component of the total organic fraction of these natural systems.

Most polysaccharidic capsules are composed of repeating units of 2 to 6 sugars. The predominant organic species present are uronic acids which may make up to 25% of the capsule polymer [19], yet the bulk of the capsule is made up of water (about 99% by weight). As a result this highly hydrated gel-like structure must have a highly diffusive nature.

The capsule acts as a sort of "buffer zone" between the cell and its environment and is the first structure encountered by metal ions when they are in the vicinity of the cell. Due to their chemical nature capsules are ideal cation scavengers, resembling cation exchange resins, and can accumulate large quantities of metal ions [20]. *Zoogloea ramigera*, a bacterium that forms an extensive capsule, was found to contain 25% by weight of metals after growing in sewage sludge [19]. Other studies indicate that up to 3 mmol Cu/g dry weight of polymer can be accumulated by these cells [21].

The major functional groups of capsules responsible for metal binding are carboxyl and hydroxyl groups. Of the two chemical groups carboxyl groups are the most active (e.g., the gamma-glutamyl groups of the *B. licheniformis* capsule [22]). For these charged polysaccharides, metals are usually bound by cross-bridging between anionic groups. This neutralization of charge by the metal ions often results in coprecipitation of the metal-polymer composite resulting in floc formation [19]. Thus metal binding to the extracellular polymers of planktonic bacteria can provide an effective means of transporting metals from the aqueous phase to the sediment in natural aqueous environments.

Although most capsular polymers are charged, uncharged polymers can also be found. In this case weak electrostatic interactions between metals and hydroxyl groups become important. The affinity of uncharged polysaccharides for the metal ions generally decreases with increasing radius of the hydrated metal ion. In general, an acid-base reaction is involved which results in the liberation of protons [23]. Such reactions appear to be significant in the corrosion of metal surfaces by bacteria bound to them.

Due to their exceptional ability to accumulate metals, capsules have been advocated as an additional design strategy by which bacteria protect themselves against toxic metal concentrations. There is a fine line between the concentrations at which metals are essential or toxic. Since metal concentrations often fluctuate widely in the natural environment, it seems reasonable that bacteria have a mechanism by which the concentration of metal that actually reaches the cell surface can be controlled [19]. Bitton and Freihofer [24] have shown that unencapsulated mutants of *Klebsiella aerogenes* were killed by metal concentrations to which their encapsulated counterparts were resistant.

The presence of metals can even enhance capsule formation. An increase in the concentration of Cr^{3+} led to increased polymer production by a coryneform bacterium isolated from Cr polluted water [25]. Capsule composition can also be influenced by the metals present. The removal of Ca^{2+} from the growth medium of *Azotobacter vinlandii* resulted in an increased proportion of manuronic acid in the exopolymers produced by this bacterium [26,27]. It is apparent that bacteria can change their capsules, both in quantity and quality, in response to changes in external metal type and concentration. This gives them greater flexibility in dealing with the types and concentrations of metals they are faced with.

3. MINERAL FORMATION

Beveridge and Murray [14] proposed a two-step mechanism for the deposition of metal ions in the bacterial cell wall. The first step involves a stoichiometric interaction between metal cations and active sites within the wall. This interaction provides a nucleation site for the deposition of more metal from solution. The metal aggregate grows within the wall until it is physically constrained by the intermolecular spaces in the wall fabric itself. As a result, metals deposited in the wall are not easily redissolved by water or replaced by protons or other metallo-ions. Therefore the matrix of the bacterial wall provides a special environment for the nucleation and growth of metal aggregates. Since bacterial walls are among the most resilient biological structures known, this results in an effective immobilization of metal ions and may be an important means of transporting metals from a dilute aqueous solution (such as lake water [28]) to the sediment where they may undergo geochemical processing which transforms them into minerals (i.e., metal oxides and hydroxides, carbonates, sulfates, etc. which can be either amorphous or crystalline precipitates).

The idea that bacteria could act as nucleation sites and templates for the formation of minerals provides some explanation for the existence of microfossils [29]. Beveridge *et al.* [30] undertook low temperature diagenesis simulations of geological processes to see what types of minerals would form on metal-loaded *B. subtilis* walls exposed to artificial sediments containing either calcite, quartz or a 1:1 mixture of the two in the presence of either sulfur or magnetite as redox buffers. They found that when sulfur was present sulphitic minerals formed. However, in the absence of sulfur the dominant mineral types were phosphates with the phosphate being provided by the bacterial walls. The mineral composition seemed to depend on the type of metal sorbed to the walls and the mobile ions present in the simulated sediment.

The actual formation of highly mineralized bacteria resembling intact microfossils was successfully simulated by Ferris *et al.* [31] in 1988. Whole cells were artificially aged for up to 150 days at 70°C in the presence of silica. However only those cells that had been pretreated with iron remained structurally intact and recognizable as bacteria. It was concluded that the binding of heavy metal ions such as iron by bacterial cells inactivated cell wall hydrolyzing enzymes and were an important factor in preserving cellular shape and contributing to the fossilization of microorganisms.

3.1 *In situ* observations

When we venture out into the natural environment and start looking at samples from many different sites possessing thin coatings of microbes or "biofilms" (Figure 1; see also [32,33]), we invariably find evidence of microbial metal binding and mineralization. The easiest and quickest way to verify this is to examine unstained samples by transmission electron microscopy. This reveals aggregates of (usually) amorphous minerals often arranged to conform to the outline of a recognizable microorganism (Figure 2). The identity of the minerals can be established by compositional methods such as energy dispersive x-ray spectroscopy (EDS) and crystallographic techniques such as selected area electron diffraction (SAED). The morphology of the microorganism can be examined by subsequent staining of the sample material with heavy metals such as uranium and lead (Figure 3). This provides confirmation that mineralization is taking place on or within the microbial cells which are present, the majority of which are bacteria.

FIGURE 1. Scanning Electron Micrograph of a Biofilm Grown in situ on Granite Rock in the Speed River near Guelph, Ontario. Note the diversity of microbial types present; the organisms are predominantly bacteria. Bar = 3 μm.

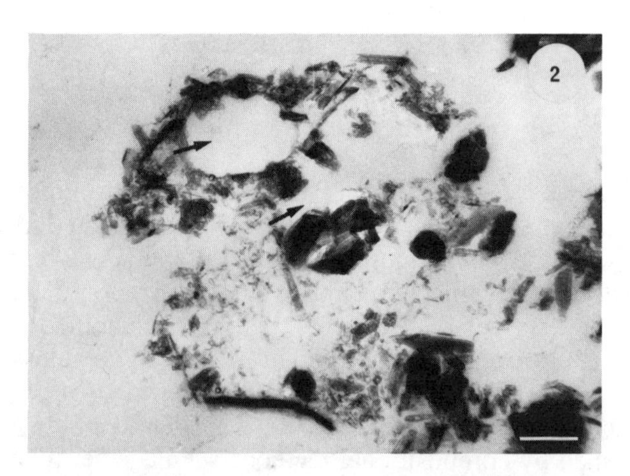

FIGURE 2. Thin Section Showing Mineral Aggregates Roughly Outlining the Shapes of Bacteria (arrows) in an in situ Sample from Gros Morne National Park, Newfoundland. Electron contrast was provided solely by the minerals present in the sample. Bar = 200 nm.

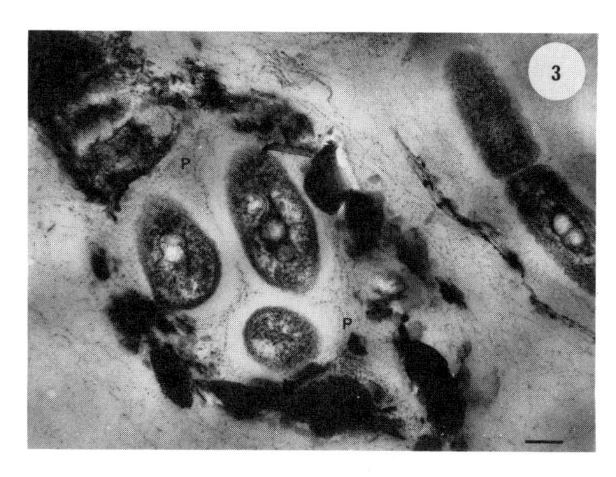

FIGURE 3. ***Thin Section Profile Similar to that Shown in Figure 6 but Which has been Stained with Uranyl Acetate and Lead Citrate to Show the Organic Components Present. Note the three bacterial cells which are occluded within mineral precipitates that appear to have formed on the polymers (p) surrounding the cells. Bar = 200 nm.***

Some bacteria, such as *Leptothrix* and *Sphaerotilus*, both ensheathed in amorphous tubes composed of loosely arranged polymers, are able to build up immense concentrations of manganese and iron oxides, respectively (Figure 4 [34]). In fact, for *Leptothrix discophora*, the sheath possesses enzymatic oxidizing activity [35,36].

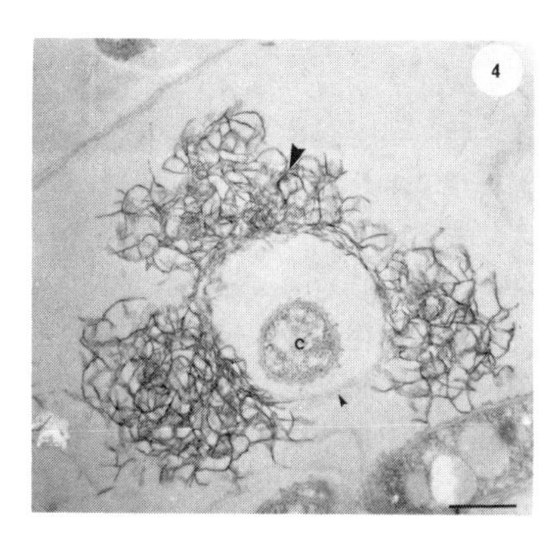

FIGURE 4. ***Thin Section Showing a Cell of Leptothrix discophora (c), a Filamentous Bacterium that Grows Enclosed within an Amorphous Sheath (small arrow). In this micrograph fibrous MnO_2 precipitates (large arrow) are seen encrusting the sheath. Bar = 500 nm.***

These can become significant agents for metal immobilization and transport in the environments they inhabit. Extensive manganese oxide deposition has been found to occur in microbial mats in hot springs [37] as well as a wide range of other environments including lakes [38] and the ocean floor [39]. Nucleation of iron-silica crystallites was found to occur in the sediment of an acidic hot spring in Yellowstone National Park [40].

In situ examination of microbial biofilms emphasizes their importance in metal binding and mineralization in the natural environment. For example, Ferris *et al.* [41] studied metal-contaminated lake water and revealed biofilm metal adsorption (per cm^2) up to 3 orders of magnitude greater than was present per mL of water. This corresponded to the observation that, throughout the 17-week study period, there were 10 to 100 times more bacterial cells present in the biofilms (per cm^2 of surface) than per mL of the bulk fluid phase. It is probable that the extracellular polymers which knitted the cells together in the biofilm also contributed to the greater metal sorption. Indeed, electron microscopy and EDS revealed abundant iron-rich precipitates, tentatively identified as ferrihydrite, throughout its polymeric framework.

Perhaps the most striking and widespread examples of the ability of biofilms to effect mineralization are those seen in microbialitic structures such as stromatolites and thrombolites. These are laminated or open porous carbonaceous structures, respectively, which have been formed by the growth and subsequent mineralization of oxygenic photosynthetic phototrophs; predominantly cyanobacteria. This special case of microbial mineralization is explored further in the following section.

3.2 Fayetteville Green Lake

Fayetteville Green Lake is a small lake situated within Green Lakes State Park near Fayetteville, New York. It is a fairly deep (approximately 55 m at the deepest point) meromictic lake with an unusual chemistry that has attracted researchers for well over a century. A thorough review of the fascinating history of research of this lake along with an investigation of its microbially-driven geochemistry is provided by Thompson *et al.* [42]. Reviews of the lake chemistry and physical characteristics are given by Brunskill and Ludlam [43] and Torgersen *et al.* [44].

The unique chemistry of this lake, particularly the high carbonate (3.57 - 7.43 mmol/L; [44]), sulphate (11.66 - 15.09 mmol/L; [43]) and very low ammonium and iron concentrations [43] ensure that eukaryotic phototrophs are at a disadvantage compared to prokaryotic phototrophs such as cyanobacteria and anaerobic purple and green sulfur bacteria. Prokaryotes are extremely efficient scavengers of nutrients in oligotrophic environments [45] and can successfully out-compete eukaryotic phototrophs in environments like Green Lake, thereby becoming the dominant phytoplankton species. In the anoxic zone, which begins at the permanent chemocline in this lake, green and purple sulfur bacteria thrive on the high levels of sulphide emanating from the lower regions of the lake. Thus Green Lake seems to have the unusual characteristic of having a prokaryote-dominated microbial community.

The lake also contains extensive carbonate bioherms (modern stromatolites/thrombolites; Figure 5) of which the origin was long a mystery until the discovery of a small (< 1 μm cell diameter) unicellular cyanobacterium. This appears to be the only significant phytoplankton species in the lake and is invariably found in association with bioherm material when viewed by light and electron microscopy. In 1963 W.H. Bradley noted small bacterial cells occluded within some calcite grains from the bottom sediment of Fayetteville Green Lake. Subsequently, Thompson *et al.* [42] demonstrated that the occluded cells were in fact small cyanobacterial cells of the genus *Synechococcus;* (Figure 6).

FIGURE 5. ***A Submerged Portion of the Extensive Carbonaceous Bioherms Present in Fayetteville Green Lake which is Covered by Aquatic Mosses (arrow). Note the irregular texture of the bioherm material. This results from unicellular cyanobacterial cells being entombed within the calcareous material they precipitate. Bar = 0.5 m. (Reprinted from Thompson et al. [42] with permission from the Society for Sedimentary Geology).***

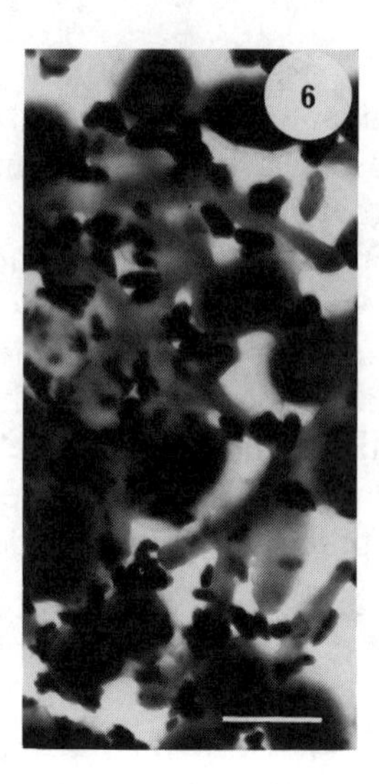

FIGURE 6. Unstained Whole Mount Transmission Electron Micrograph of an Experimental Synechococcus Culture Isolated from Fayetteville Green Lake Showing the Precipitation of Calcite within the Colony and on the Surface of the Synechococcus Cells. Bar =1 μm. (Reprinted from Thompson et al. [42] with permission from the Society for Sedimentary Geology).

Synechococcus thrives in the oxygenated zone of the lake. Its growth is seasonal and peaks during the warmer months of July and August. Interestingly, it is during these months that calcite mineralization is greatest, yet the exact reason for calcite precipitation was at first unclear. At the circumneutral pH of the lake precipitation should be in the gypsum solid field; calcite precipitation requires a more alkaline pH. We attribute calcite mineralization to the ability of *Synechococcus* to initiate carbonate deposition within its immediate vicinity by the photosynthetically-driven alkalization of the microenvironment surrounding the cell. In this process, bicarbonate is taken into the cell where it is "fixed" by the Calvin-Benson cycle for energy and reducing power during photosynthesis. As a result of this activity hydroxyl ions are released by the cell and produce an alkaline microenvironment immediately peripheral to the cell surface [46]. This alkalization moves calcium precipitation from gypsum into the calcite field and provides for the marl sediments and calcareous bioherms [47].

Laboratory simulations using lake water and isolated *Synechococcus* cells confirmed alkalization of the water by these bacteria and revealed their ability to promote epicellular calcite mineralization [47]. In this study the mechanism for calcium sequestering on the cell surface was also addressed; the surface of this *Synechococcus* appears to have an affinity for calcium ions. When the extracellular pH becomes high enough, calcite forms as both an original calcium carbonate precipitate and as a result of chemical exchange between carbonate and the sulphate present in the gypsum (Figure 7).

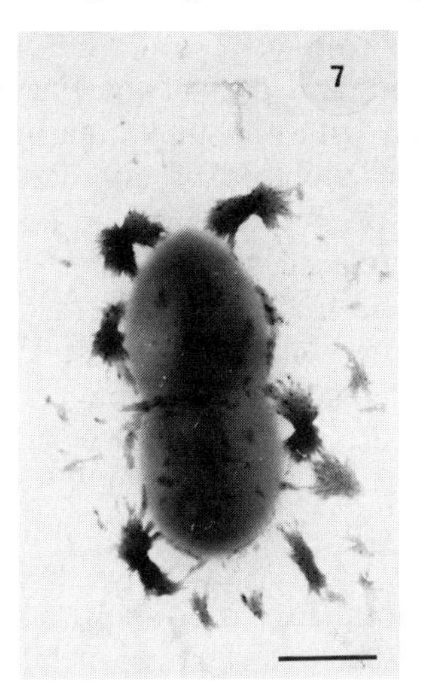

FIGURE 7. Transmission Electron Micrograph of a Dividing Synechococcus Cell from Experimental Culture Showing Gypsum Precipitation on the Cell Surface (unstained sample). Bar = 500 nm. (Reprinted from Thompson and Ferris [47] with permission from the Geological Society of America).

Recent structural studies of this *Synechococcus* reveal a gram-negative wall ultrastructure. In addition, the bacterium is surrounded by a proteinaceous paracrystalline hexagonal surface array or S-layer [48]. Since some S-layers, such as those of *Aquaspirillum* spp. require and sequester Ca for assembly [15], it is possible that the *Synechococcus* S-layer also concentrates this metal.

Fayetteville Green Lake provides an excellent example of the ability of microorganisms to effect overt physical changes in their environment as a result of microbial mineralization processes. The appearance and growth of the carbonate bioherms and the substantial thickness of marl sediment at the lake bottom provide evidence that microbial life has played (and is still playing) a major part in the formation of geologically significant structures [42,47]. The discovery of the mechanism of bioherm formation in the lake has far-reaching implications for the origin of similar structures found in the fossil record and in other parts of the modern world.

4. METAL TRANSPORT AND THE IMMOBILIZATION OF TOXIC HEAVY METALS

Metals are an integral part of the Earth's crust, mostly present through complexation with phosphate, carbonate, silicate, hydroxide and sulphide ions as insoluble precipitates and minerals. Natural geochemical dissolution, weathering and microbial leaching [49] are responsible for their release in a soluble form and account for their natural concentrations in freshwater systems. Perhaps a greater load of metal ions enters the environment as a result of human activity such as mining, heavy industry and metal refining. Heavy metal pollution has become a major concern in recent years and it is important for us to understand what happens to the metals in the natural environment. This will hopefully lead us to ways of containing and possibly recycling these metals for future use. It is here that we expect to find that bacteria play an important role in impeding the transport of metals in natural systems. The fact that bacteria can bind metals is perhaps not surprising but the tremendous quantity they can accumulate, especially from dilute solution, is remarkable. The major factors responsible for this ability are their small size and the physicochemical attributes of their surfaces. Their large surface area to volume ratio promotes the efficiency of the immobilization process. The fact that bacteria are ubiquitous and present in enormous numbers makes them a force to be reckoned with when it comes to affecting the transport of metals in natural systems.

In the water column of a temperate climate lake, we envision planktonic microorganisms and their remains forming a season-variable light rain of microscopic particles which cleanse the waters of dilute, soluble, toxic heavy metals, forming aggregates which, over time, increase in size through flocculation. These gently settle to the sediments and account for the high concentrations of heavy metals found at these depths.

Biofilms, on the other hand, form heterogenous but stable outer layers on hard surfaces such as rocks that can approach millimeters in thickness. They are avid concentrators of toxic metals and become, themselves, mineralized over time. Indeed, they can account for large proportions of the varnishes and other mineralized films we see on such surfaces. Certainly, as in the Fayetteville Green Lake system, cyanobacterial communities and their alkalization can account for tremendous mineralization and can effectively control the "flavor" of the mineral type.

Although we have been stressing the importance of bacteria as the major presence responsible for metal binding in natural systems, it is prudent to point out that there are other components present in the environment that are capable of presenting a significant impediment to the mobility of metal ions. Other biological particulates are essentially made out of the same "stuff" as bacteria and should also interact with toxic metals. Some of the most active components are soluble organics such as humic and fulvic acids, but in particular, small inorganic minerals, such as clays, can play a prominent role in metal immobilization.

The small size (<2 μm diameter) and overall electronegative charge density of clays should make them very similar to bacteria in terms of metal binding ability. Walker *et al.* [50] were interested in finding out how bacteria and clays interact with each other and how this interaction may affect the metal binding capacity of each component. Does this interaction enhance or impede the metal binding capacity of the system? This interest arose from the recognition that the laboratory simulations with bacterial surfaces were an oversimplification of the conditions actually present in the natural environment. Now that the ability of bacteria to bind metals and the mechanisms by which the binding occurs are fairly well understood [13,34], and the associations between clays and metals are well recognized [51,52], it was important to understand how these two components interact to immobilize metals.

Walker *et al.* [50] used *B. subtilis* walls and *E. coli* envelopes as microbial particulates. These were mixed with each of two types of clay, kaolinite and smectite (montmorillonite). The metal binding ability of each separate component as well as of composites was determined as well as the manner in which the two components aggregated. Bacterial-clay composites formed best when mixed in a 1:1 proportion (on a dry weight basis) and the clays showed a preference for edge-on orientation with the organic fraction (Figure 8). This indicated that binding occurred between the negatively charged walls and envelopes and the positive edges of the clays. The addition of multivalent metal ions increased the incidence of planar surface orientations indicating that the metal ions were acting as bridges between the two particulate components.

The comparative metal binding abilities of the components tested in terms of metal bound per dry weight of clay, walls, envelopes, or composites was determined [50]. The results indicated that gram-positive walls bind a greater quantity of metal than gram-negative envelopes and that both bind more metal than the clays. The sum

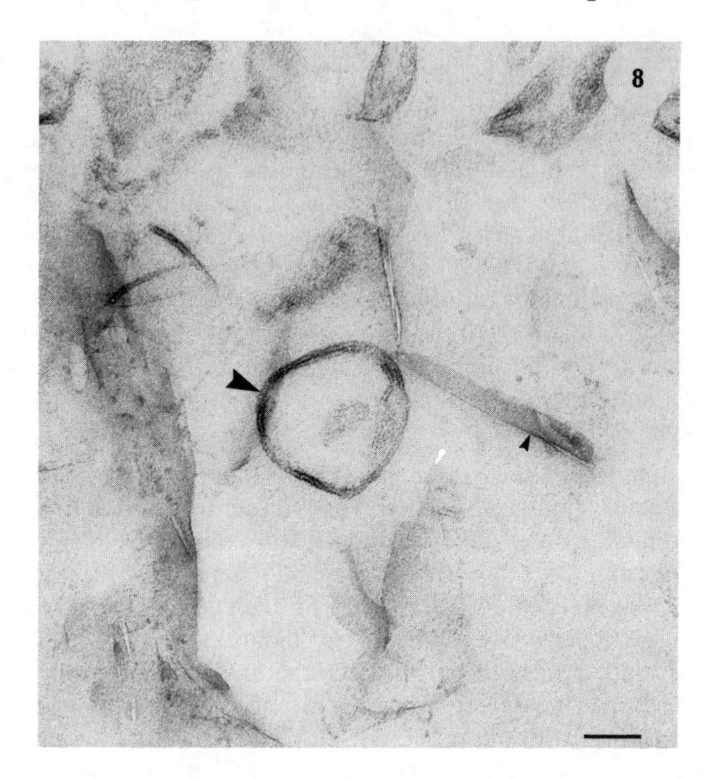

FIGURE 8. ***Electron Micrograph of a Thin-sectioned E. coli Envelope (large arrow) Sorbed to Smectite (small arrow) and Stained with Uranyl Acetate and Lead Citrate. Bar = 100 nm.(Reprinted from Walker et al. [50] with permission from the American Society for Microbiology).***

of the metal binding capacity of each of the individual components exceeded that of the composites. This was thought to be due to available metal binding sites on the surfaces of the particulates being eliminated during the binding between the organic component and the clays. On a dry weight basis the envelope-clay and wall-clay mixtures bound 20% to 90% less metal than equal amounts of the individual components. The organic portion of the composite accounted for most of the binding capacity (Table 2 [50]).

These experiments were followed up by Flemming *et al.* [53] who wished to determine what conditions were necessary in order to remobilize the metals bound to the bacteria-clay systems. This could give some indication of the nature and strength of the forces binding the metals to the walls, envelopes, and clays.

Table 2. Metal Proportions Associated With The Cellular Constituents In The Clay-Wall And Envelope-Clay Composites.

Metal	% of metal associated with cellular constituents in			
	Wall + smectite	Wall + kaolinite	Envelope + kaolinite	Envelope + smectite
Ag	91	100	100	80
Cu	73	99	97	47
Ni	79	99	98	52
Cd	100	99	97	100
Pb	82	99	99	68
Zn	94	99	93	72
Cr	92	98	93	72

Reprinted from Walter *et al.* [50] with permission from the American Society for Microbiology.

Several concentrations of nitric acid, calcium nitrate, EDTA, fulvic acid (a major component of soil organic matter) and lysozyme (an enzyme that degrades the peptidoglycan in bacterial cell walls) were tested. The remobilization of the sorbed metals depended on the physical properties of the organic and clay surfaces and on the character and concentration of the leaching agents. Although each leaching agent was effective for mobilization of a certain metal (in this study Ag^{+}, Cu^{2+}, and Cr^{3+} were used), the greatest mobility was achieved at acidic pH or with elevated Ca^{2+} levels (160 ppm). Under all conditions, less metal was resolubilized from the organic fraction than from the clays. However, although the results were highly reproducible, it was difficult to recognize a clear pattern for the remobilization of each metal. This points out how complicated metal interactions in the natural environment actually are and adds a note of caution to the conclusions drawn from laboratory simulations. Yet, it is clear that bacteria not only compete better than clays for soluble metal, but that they also are less willing to give the metal up when challenged by competing chemicals such as EDTA. We believe that bacteria and their polymeric debris are among the major barriers to the mobility of toxic metals within natural ecological systems.

5. CONCLUSION

Bacteria are ubiquitous and possess innate characteristics, such as their small size and the physicochemical nature of their surface, which make them ideal scavengers of metal ions and nucleators of mineral formation. Although there are other components of natural systems that are capable of binding metals, so far, bacteria seem to be the most potent. As such, the presence of these small microorganisms likely represents one of the major cleansing mechanisms available to the natural environment when threatened with heavy metal contamination. Further studies, both in the laboratory and especially *in situ*, are needed if we are to understand the role of bacteria in metal immobilization and transport in natural systems and, perhaps, exploit this aspect of microbiology to deal with the problems of metal pollution that we have inherited from mining and industry.

Acknowledgments: All research reported from the authors' laboratory in this review has been supported by an on-going operating grant from the Natural Sciences and Engineering Council of Canada to T.J. Beveridge. S. Schultze-Lam was also supported by this grant and J.B. Thompson was supported by a grant from the Ontario Geological Survey to T.J. Beveridge.

REFERENCES

1. Schultze-Lam, S., J.B. Thompson and T.J. Beveridge, "Metal Ion Immobilization by Bacterial Surfaces in Freshwater Environments," *Water Poll. Res. J. Can.* 28:51-81 (1993).

2. Beveridge, T.J., "Ultrastructure, Chemistry, and Function of the Bacterial Wall," *Int. Rev. Cytol.* 72:229-317 (1981).

3. Beveridge, T.J., "Wall Ultrastructure: How Little We Know" in *Antibiotic Inhibition of Bacterial Cell Surface Assembly and Function,* P. Actor, L. Daneo-Moore, M.L. Higgins, M.R.J. Salton and G.D. Shockman, Eds. pp. 3-20. American Society for Microbiology, (Washington, D.C. 1988).

4. Beveridge, T.J., "The Bacterial Surface: General Considerations Towards Design and Function," *Can. J. Microbiol.* 34:363-372 (1988).

5. Beveridge, T.J., "The Structure of Bacteria," in *Bacteria in Nature Vol. 3,* J.S. Poindexter and E.R. Leadbetter, Eds. pp. 1-65. Plenum, (New York 1989).

6. Beveridge, T.J., "Role of Cellular Design in Bacterial Metal Accumulation and Mineralization," *Annual Review of Microbiology* 43:147-171 (1989).

7. Beveridge, T.J. and L.L. Graham, "Surface Layers of Bacteria," *Microbiol. Rev.* 55:684-705 (1991).

8. Beveridge, T.J. and J.A. Davies, "Cellular Responses of *Escherichia coli* and *Bacillus subtilis* to the Gram Stain," *J. Bacteriol.* 156:846-858 (1983).

9. Beveridge, T.J. and S. Koval, "Binding of Metal Ions to Cell Envelopes of *Escherichia Coli* K-12," *Appl. Env. Microbiol.* 42:325-335 (1981).

10. Nikaido, H. and M. Vaara, "Outer Membrane," in *Escherichia coli and Salmonella typhimurium: Cellular and Molecular Biology*, Volume One, F.C. Neidhardt, J.L. Ingraham, K.B. Law, B. Megasanik, M. Schaechter and H.E. Umberger, Eds. pp. 7-23. American Society for Microbiology, (Washington, D.C. 1987) .

11. Ferris, F.G. and T.J. Beveridge, "Site Specificity of Metallic Ion Binding" in *Escherichia coli* Lipopolysaccharide," *Can. J. Microbiol.* 32:52-55 (1986).

12. Hoyle, B.D. and T.J. Beveridge, "Metal Binding by the Peptidoglycan Sacculus of *Escherichia coli* K12," *Can. J. Microbiol.* 30:204-211 (1984).

13. Beveridge, T.J., "Mechanisms of the Binding of Metallic Ions to Bacterial Walls and the Possible Impact on Microbial Ecology," in *Current Perspectives in Microbial Ecology*, M.J. Klug and C.A. Reddy, Eds. pp. 601-607. American Society for Microbiology (Washington, D.C. 1984).

14. Beveridge, T.J. and R.G.E. Murray, "Uptake and Retention of Metals by Cell Walls of *Bacillus subtilis*," *J. Bacteriol.* 127:1502-1518 (1976).

15. Beveridge, T.J. and R.G.E. Murray, "Dependence of the Superficial Layers of *Spirillum putridiconchylium* on Ca^{2+}," *Can. J. Microbiol.* 22:1233-1244 (1976).

16. Beveridge, T.J. and R.G.E. Murray, "Sites of Metal Deposition in the Cell Wall of *Bacillus subtilis*," *J. Bacteriol.* 141:876-887 (1980).

17. Doyle, R.J., T.H. Matthews and U.N. Streips, "Chemical Basis for Selectivity of Metal Ions by the *Bacillus subtilis* Cell Wall," *J. Bacteriol.* 143:471-480 (1980).

18. Beveridge, T.J., C.W. Forsberg and R.J. Doyle, "Major Sites of Metal Binding in *Bacillus licheniformis* Walls," *J. Bacteriol.* 150:1438-1448 (1982).

19. Geesey, G.G. and L. Jang, "Interactions Between Metal Ions and Capsular Polymers," in *Metal Ions and Bacteria*, T.J. Beveridge and R.J. Doyle, Eds. pp. 325-358. John Wiley and Sons, (New York, 1989).

20. Geesey, G.G., L. Jang, J.G. Jolley, M.R. Hankins, T. Iawoka and P.R. Griffiths, "Binding of Metal Ions by Extracellular Polymers of Biofilm Bacteria," *Water Sci. Technol.* 20:161-165 (1988).

21. Norberg, A.B. and H. Persson. "Accumulation of Heavy-metal Ions by *Zoogloea ramigera*," *Biotechnol. Bioeng.* 26:239-246 (1984).

22. McLean, R.J.C., D. Beauchemin, L. Clapham and T.J. Beveridge, "Metal-Binding Characteristics of the Gamma-Glutamyl Capsular Polymer of *Bacillus licheniformis* ATCC 9945," *Appl. Env. Microbiol.* 56:3671-3677 (1990).

23. Mittelman, M.W. and G.G. Geesey, "Copper-binding Characteristics of Exopolymers from a Freshwater Sediment Bacterium," *Appl. Env. Microbiol.* 49:846-851 (1985).

24. Bitton, G. and V. Freihofer, "Influence of Extracellular Polysaccharide on the Toxicity of Copper and Cadmium Toward *Klebsiella aerogenes*," *Microb. Ecol.* 4:119-125 (1978).

25. Bremer, P.J. and M.W. Loutit, "The Effect of Cr(III) on the Form and Degradability of a Polysaccharide Produced by a Bacterium Isolated From a Marine Sediment," *Mar. Env. Res.* 20:249-259 (1986).

26. Couperwhite, I. and M.F. McCallum, "The Influence of EDTA on the Composition of Alginate Synthesized by *Azotobacter vinelandii*," *Arch. Microbiol.* 97:73-80 (1974).

27. Annison, G. and I. Couperwhite, "Consequences of the Association of Calcium with Alginate During Batch Culture of *Azotobacter vinelandii*," *Appl. Microbiol. Biotechnol.* 19:321-325 (1984).

28. Mayers, I.T. and T.J. Beveridge, "The Sorption of Metals to *Bacillus subtilis* Walls From Dilute Solutions and Simulated Hamilton Harbour (Lake Ontario) Water," *Can. J. Microbiol.* 35:764-770 (1989).

29. Barghoorn, E.S. and S.A. Tyler, "Microorganisms From the Gunflint Chert," *Science* 147:563-577 (1965).

30. Beveridge,T.J., J.D. Meloche, W.S. Fyfe and R.G.E. Murray, "Diagenesis of Metals Chemically Complexed to Bacteria: Laboratory Formation of Metal Phosphates, Sulfides, and Organic Condensates in Artificial Sediments," *Appl. Env. Microbiol.* 45:1094-1108 (1983).

31. Ferris, F.G., W.S. Fyfe and T.J. Beveridge, "Metallic Ion Binding by *Bacillus subtilis*: Implications for the Fossilization of Microorganisms," *Geology* 16:149-152 (1988).

32. Costerton, J.W., K.J. Cheng, G.G. Geesey, T.I. Ladd, J.C. Nickel, M. Dasgupta and T.J. Marrie, "Bacterial Biofilms in Nature and Disease," *Ann. Rev. Microbiol.* 41:435-464 (1987).

33. Fletcher, M., "The Attachment of Bacteria to Surfaces in Aquatic Environments," in *Adhesion of Microorganisms to Surfaces*, D.C. Ellwood, J. Melling and P. Rutter, Eds. pp. 87-108. Academic Press (London, 1979).

34. Beveridge, T.J., "Metal Ions and Bacteria," in *Metal Ions and Bacteria*, T.J. Beveridge and R.J. Doyle, Eds. pp 1-30. John Wiley and Sons, (New York, 1989).

35. Boogerd, F.C. and J.P.M. deVrind, "Manganese Oxidation by *Leptothrix discophora,"* *J. Bacteriol.* 169:489-494 (1987).

36. Adams, L.F. and W.C. Ghiorse, "Characterization of Extracellular Mn^{2+} Oxidizing Activity and Isolation of a Mn^{2+}-oxidizing Protein From *Leptothrix discophora*," *J. Bacteriol.* 169:1279-1285 (1987).

37. Ferris, F.G., W.S. Fyfe and T.J. Beveridge, "Manganese Oxide Deposition in a Hot Spring Microbial Mat," *Geomicrobiol. J.* 5:33-41 (1987).

38. Chapnick, S.D., W.S. Moore and K.H. Nealson, "Microbially Mediated Manganese Oxidation in a Freshwater Lake," *Limnol. Oceanogr.* 26:1004-1014 (1982).

39. Ghiorse, W.C., "Biology of Iron- and Manganese-depositing Bacteria," *Ann. Rev. Microbiol.* 38:515-550 (1984).

40. Ferris, F.G., T.J. Beveridge and W.S. Fyfe, "Iron-silica Crystallite Nucleation by Bacteria in a Geothermal Sediment," *Nature* 320:609-611 (1986).

41. Ferris, F.G., S. Schultze, T.C. Witten, W. E.S. Fyfe and T.J. Beveridge, "Metal Interactions with Microbial Biofilms in Acidic and Neutral pH Environments," *Appl. Env. Microbiol.* 55:1249-1257 (1989).

42. Thompson, J.B., F.G. Ferris and D.A. Smith, "Geomicrobiology and Sedimentology of the Mixolimnion and Chemocline in Fayetteville Green Lake, New York," *Palaios* 5:52-75 (1990).

43. Brunskill, G.J. and S.D. Ludlam, "Fayetteville Green Lake, New York I: Physical and Chemical Limnology," *Limnol. Oceanogr.* 14:817-829 (1969).

44. Torgersen, T., D.E. Hammond, W.B. Clarke and T.-H. Peng, "Fayetteville Green Lake, New York: ^{3}H- ^{3}He Water Mass Ages and Secondary Chemical Structure," *Limnol. Oceanogr* 26:110-122 (1981).

45. Fogg, G.E., "Picoplankton," in *Perspectives in Microbial Ecology*, F. Megusar and M. Gantar, Eds. pp. 96-101. Slovene Society for Microbiology, (Ljubljana, 1986).

46. Miller, A.G. and B. Colman, "Evidence for HCO_3 Transport by the Blue-green Alga (Cyanobacterium) *Coccochloris peniocystis,*" *Plant Phys.* 65:397-402 (1980).

47. Thompson, J.B. and F.G. Ferris, "Cyanobacterial Precipitation of Gypsum, Calcite and Magnesite From Natural Alkaline Lake Water," *Geology* 18:995-998 (1990).

48. Sleytr, U.B. and P. Messner, "Crystalline Surfaces Layers in Prokaryotes," *J. Bacteriol.* 170:2891-2897 (1988).

49. Ehrlich, H.L. and C.L. Brierley, "Bioleaching and Biobeneficiation," in *Microbial Mineral Recovery*, H.L. Ehrlich and C.L. Brierley, Eds. pp. 1-182. McGraw-Hill, (New York, 1990).

50. Walker, S.G., C.A. Flemming, F.G. Ferris, T.J. Beveridge and G.W. Bailey, "Physicochemical Interaction of *Escherichia coli* Cell Envelopes and *Bacillus subtilis* Cell Walls with Two Clays and Ability of the Composites to Immobilize Metals From Solution," *Appl. Env. Microbiol.* 55:2976-2984 (1989).

51. Carroll, D., "Role of Clay Minerals in the Transport of Iron," *Geochim. Cosmochim. Acta* 14:1-27 (1958).

52. Pickering, W.F., "Copper Retention by Soil/Sediment Components," in *Copper in the Environment: Part I, Ecological Cycling,* J.O. Nriagu, Ed. pp. 217-253. John Wiley and Sons, (Toronto, 1979).

53. Flemming, C.A., F.G. Ferris, T.J. Beveridge and G.W. Bailey. "Remobilization of Toxic Heavy Metals Adsorbed to Bacterial Wall-Clay Composites," *Appl. Env. Microbiol.* 56:3191- 3203 (1990).

7

TRACE METAL CHEMICAL REACTIONS IN GROUNDWATER: PARAMETERIZING COUPLED CHEMISTRY TRANSPORT MODELS

Thomas L. Theis and Ramesh Iyer
Department of Civil and Environmental Engineering
Clarkson University
Potsdam, NY 13699

1. INTRODUCTION

As our understanding of the behavior of trace elements in soils has advanced, and as the need to apply this knowledge to the movement of these substances in the subsurface environment has become more pressing, a number of coupled chemistry transport models have been developed and codified during the past decade or so [1-7]. These models sometimes differ in the method of formulating and solving the problem, and in the degree of sophistication in the chemical and transport aspects, however, they all recognize the importance of trace element partitioning to local solid phases as a factor in defining their subsurface flux. As an example, consider the model schematic shown in Figure 1 for the FIESTA code [8]. In this case, the model consists of three submodels: hydrologic, transport, and soil chemistry. The latter, additionally, involves coupled solution phase and surface phase speciation. Since groundwater contaminant transport can be viewed as a two-part problem, this model first solves the groundwater flow problem, resulting in an average velocity, which is input to the transport problem.

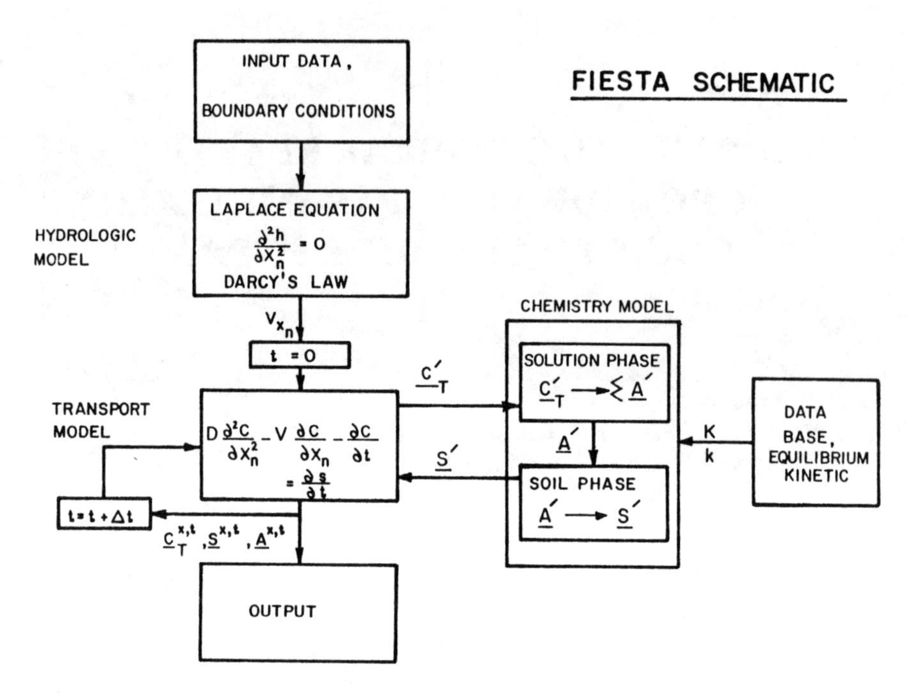

FIGURE 1. Computational Schematic of the Fiesta Model for Multicomponent Solute Transport [8].

The transport and chemistry submodels, although formulated separately, are coupled via the source/sink term of the time and space dependent mass balance equations, one for each chemical component. These equations are, in general, coupled to each other because of the multicomponent nature of most trace element speciation and partitioning reactions.

Of special concern in this paper is the manner in which the chemical partitioning coefficients that are input to these models are determined experimentally and the confidence with which they can be used for the given partition model. In general, it is necessary to determine these coefficients under experimental conditions removed from the actual subsurface environment, often at the laboratory scale. There can be substantial uncertainties both in the representativeness of results obtained in relation to the environment and the way in which the data are best analyzed for purposes of parameterizing coupled transport models.

2. SOLUTE PARTITIONING

There are many conceptual models of solute partitioning including instantaneous linear and nonlinear equilibrium, non-equilibrium with or without transformation, and multiple binding site models. Each of these has advantages and disadvantages, and one is usually chosen for a given application based upon the best available data and sometimes our intuitive perception of the important factors in a system. As we shall see, one of the most crucial aspects of the parameterization exercise is that once an apparently robust model is chosen, it is the degree to which the experimental data subsequently gathered (or, if already on hand, applied) represents and stresses the model in its critical or most sensitive areas.

For purposes of illustration in this paper, we will examine the partitioning of lead ion (Pb^{2+}) onto a granular porous iron oxide consisting of the hydrous oxide, goethite (FeOOH). This reaction has been studied in considerable detail, and can be represented by the reaction stoichiometry [9].

$$\equiv FeOOH + Pb^{2+} \leq \ \equiv FeOOPb^{2+} + H^{+} \tag{1}$$

Application of mass action and surface balance equations results in an isotherm of the form

$$\{Pb\}_{ads} = \frac{K_{app} S_T \left[Pb^{2+}\right]}{\left[H^{+}\right] + K_{app}\left[Pb^{2+}\right]} \tag{2}$$

where $\{Pb\}_{ads}$ is the concentration of adsorbed lead ion, S_T is the total available binding site concentration, and K_{app} is the apparent sorption constant, i.e., inclusive of electrostatic interactions. Such influences will not always be negligible and often the electrostatic factor is broken out as a separate term, however, lead adsorption to goethite has a strong chemical reaction component and changes in the electrostatic factor are small for small pH changes. The constant derived from Equation (2) should be valid for most computational purposes.

Equation (2) can be transformed, using the total soluble lead mass balance, to

$$\{Pb\}_{ads} = \frac{K_{app}S_T\left([Pb]_T - \{Pb\}_{ads}\right)}{\left[H^+\right] + K_{app}\left([Pb]_T - \{Pb\}_{ads}\right)} \quad (3)$$

This approach has been suggested as a way of avoiding negatively correlated errors during regression-based estimation; the negative correlation arising because adsorbed concentrations are usually measured, in practice, indirectly as the difference between total initial solute and solute in solution, which is measured directly [10].

3. EXPERIMENTAL APPROACHES

The granular iron oxide used in this research was generated in conjunction with personnel at the Alcoa Technical Center, New Kensington, Pennsylvania. The process consisted of the cementation of iron oxide particles using a proprietary binding material. The specific oxide used was goethite (α-FeOOH) obtained commercially from Pfizer, Inc., as a powder consisting of 1-2 μm particles [11]. The adsorbent was supplied as 1 mm extrudates which were then ground and sieved to a uniform size of 0.5-1.0 mm granular particles. The granular sorbent was subjected to a BET-N_2 surface area analysis, yielding a value of 20.0 m^2/g, and bulk and skeletal density analyses, giving values of 1.51 and 3.86 g/cm^3, respectively. The goethite, both in granular and powder forms, was also analyzed by mercury pore intrusion. This suggested that the manufacturing process tended to eliminate the very largest interparticle pores and created smaller relatively uniform pores between 70-100 nm, roughly one-tenth the size of the parent goethite particles.

Two types of sorption experiments were carried out as part of this research. Initially, pH-adsorption edges for lead removal were generated using the "bottle-point" method, in which pH is systematically varied using HNO_3 or KOH in agitated batch reactors containing known amounts of granular oxide and total lead ion (added as $Pb(NO_3)_2$). The reactors were equilibrated for a period of four hours after which the solids were separated by centrifugation and filtration, and the sample was acidified prior to storage and subsequent analysis. Quantification of soluble lead was by graphite furnace atomic absorption. Data are reported as percentage of lead removal as a function of pH for various total lead concentrations.

A second type of experimentation involved the use of small column reactors. This approach is inherently similar to the subsurface soil water system, and permits the dynamic evaluation of important operational parameters. The column reactors used were constructed of acrylic and contained a small but variable bed volume (generally 0.2 to 3.0 cm^3) into which the oxide was placed. Influent aqueous solution consisted of deionized distilled water to which was added an appropriate pH buffer, KNO_3 for ionic strength adjustment, and, for adsorption experiments, lead nitrate. The solution was purged with high purity nitrogen to eliminate carbon dioxide and passed through the column at a constant flow rate using a peristaltic pump. The pH was monitored in both the influent and effluent, the latter being routed to a fraction collector. Samples were analyzed for lead ion as before. Further details on the use and operation of these small column reactors can be found elsewhere [12].

Column experimentation consisted of both the adsorption and desorption of lead, the direction of the reaction being controlled by pH: adsorption at approximately pH 7.2, desorption at pH 3.2 with these values having been suggested from batch experiments. The pH buffers used were N,N,-bis(2-hydroxyethyl)-2-aminoethane sulfonic acid (BES) for pH 7.2, and acetic acid for pH 3.2. Both of these buffers are relatively non-nucleophilic and had been previously tested to ensure no interference in metal adsorption reactions at the buffer concentrations used (10^{-3} M) [13].

Data from both batch and column experiments were analyzed using appropriate balance equations (as indicated below). Coefficients were derived by employing the Levenberg-Marquardt nonlinear least square optimization technique [14,15].

4. RESULTS AND DISCUSSION

4.1 Batch experiments

Figure 2 shows results of four batch adsorption experiments in the form of pH adsorption edges. The data indicate the adsorption envelope lies in the pH range of four to seven with a slight decline above pH 7, assumed to be due to the formation of non-adsorbing complexes of lead with carbonate ion.

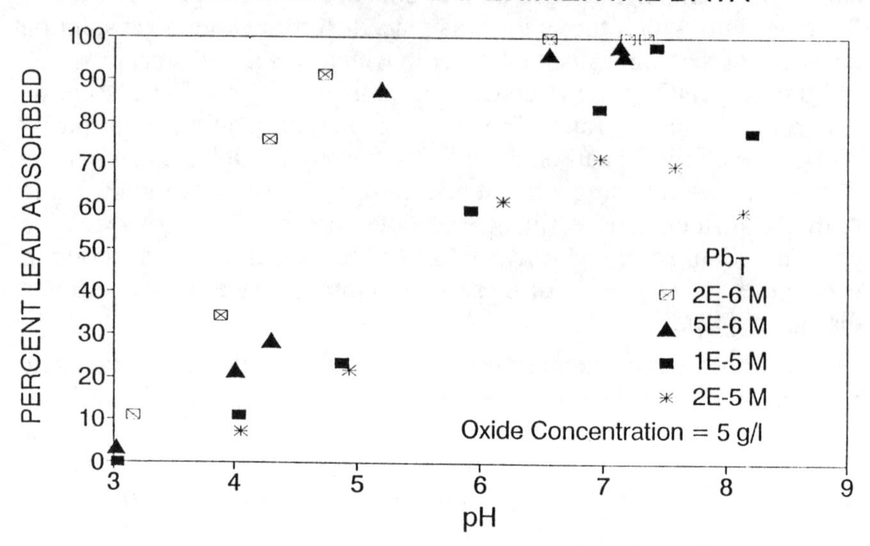

FIGURE 2. Adsorption of Lead Ion onto Granular Goethite Particles as a Function of pH in Batch Reactors. Equilibration Period Equals Four Hours.

Analysis of the data of Figure 2 was accomplished through the use of nonlinear regression using Equation (3), which resulted in the estimation of the two parameters, S_T and K_{app}. As an initial exercise, and in order to illustrate the importance of data quality on the results obtained, regression using a single data set $Pb_T = 2x10^{-6}$ M) is shown in Figure 3. This data set consists of seven observations and appears to yield an acceptable fit as measured by the correlation coefficient. Inspection of both individual and joint 95 percent confidence intervals for this data set, however, presents a very different picture, as shown in Figure 4. Although estimates for K_{app} and S_T are calculated, their confidence intervals, normalized so that they can be more easily accommodated on the graph, are so large as to render the conclusion essentially meaningless. In addition, the tilt of the joint interval ellipse, at minus 45 degrees, indicates a high degree of inverse correlation between the two parameters, suggesting difficulty in distinguishing the individual contributions of the two parameters to Equation (3).

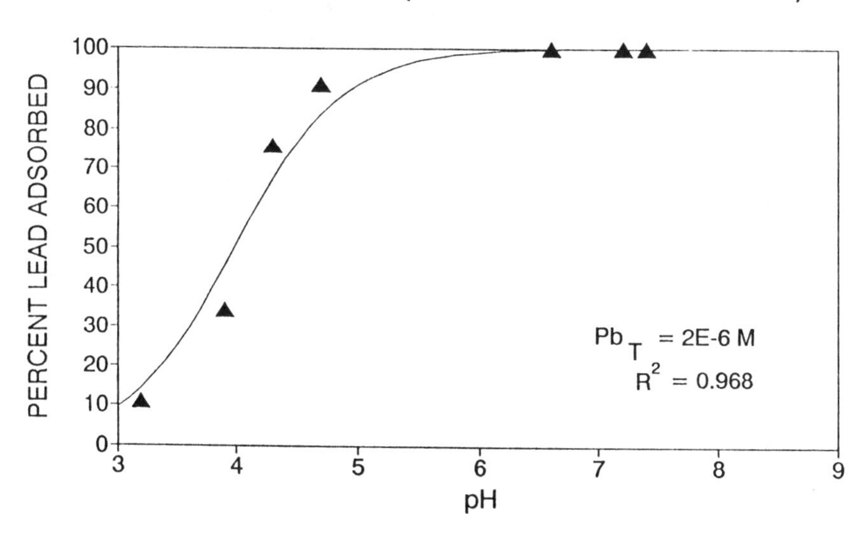

FIGURE 3. ***Results of the Nonlinear Regression for a Single Batch Reactor Adsorption Data Set using Equation (3).***

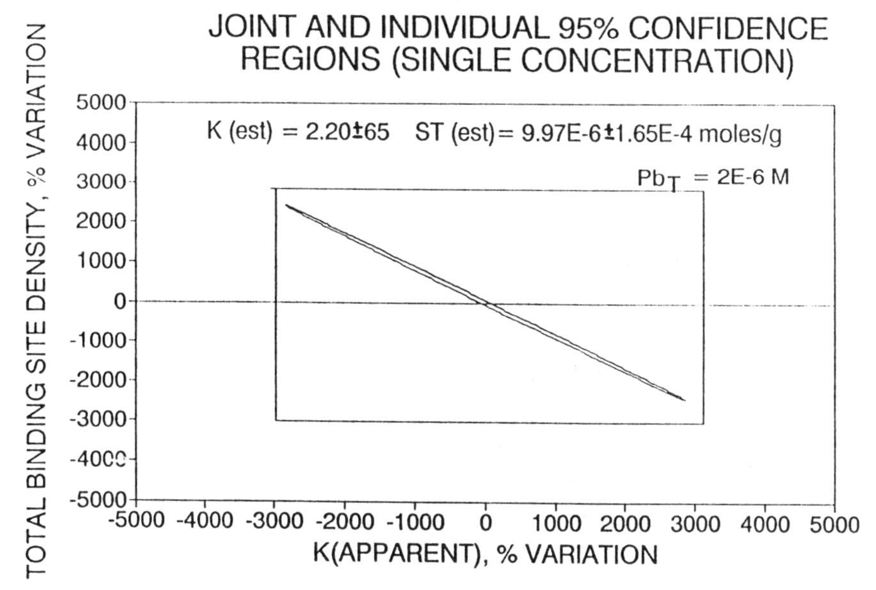

FIGURE 4. ***Individual and Joint Ninety-five Percent Confidence Intervals for the Adsorption Parameters K_{app} and S_T using the Data Set of Figure 3.***

It is instructive to inquire into the reason for such meaningless parameter values obtained from such an apparently good fit of the model to the data. Although the data set shown in Figure 3 contains seven observations, only the first four (left to right) exhibit significant change in adsorbed lead ion as a function of pH. The final three data points illustrate the same feature, virtually all lead has been removed from solution. "One hundred percent" adsorption means that measurements two orders of magnitude below Pb_T are being taken. Apart from the increased analytical errors associated with measurements near the limit of detection, these observations add no new information for use in the parameterization exercise. Thus, the problem is simply one of too few high quality observations; that is, observations which result from having stressed the model (in this case Equation 3) in regions of model sensitivity.

Some improvement in the estimates can be obtained by using the other data sets shown in Figure 2; however, each of these exhibits a leveling off of adsorbed lead as pH increases such that the final results are still less than desirable. The best estimates obtainable from the batch data shown involve the simultaneous use of all 28 observations in Figure 2. The results of this regression are presented in Figure 5 (a and b), wherein the individual data sets and their respective regressions are divided for clarity. Here the "fit" of the regression model must be viewed collectively since the least squares criterion applies to all data, not just a single set. Although the correlation coefficient for Figure 5 is less than before, the confidence intervals, shown in Figure 6, are substantially better with the variation in K_{app} of about ± 80 percent and in binding site density of about ± 20 percent. In addition, joint and individual confidence regions agree closely, there being only a slight degree of inverse correlation between the parameters. This indicates each parameter has an independent effect on the prediction of the model, as the supporting theory for the site-binding approach used suggests they should.

4.2 Column experiments

The character of the data obtained from the short column reactor is illustrated for the solutes, arsenate, cadmium, and lead, in Figure 7, although only the lead data are analyzed in this paper. This type of reactor is capable of yielding a relatively large number of pore or bed volumes in a short amount of time, thus solutes tend to break through rapidly. Analysis of these data proceeds in a manner similar to batch reactor data in the sense that the appropriate model is used via the nonlinear optimizer. This type of reactor requires that a column balance

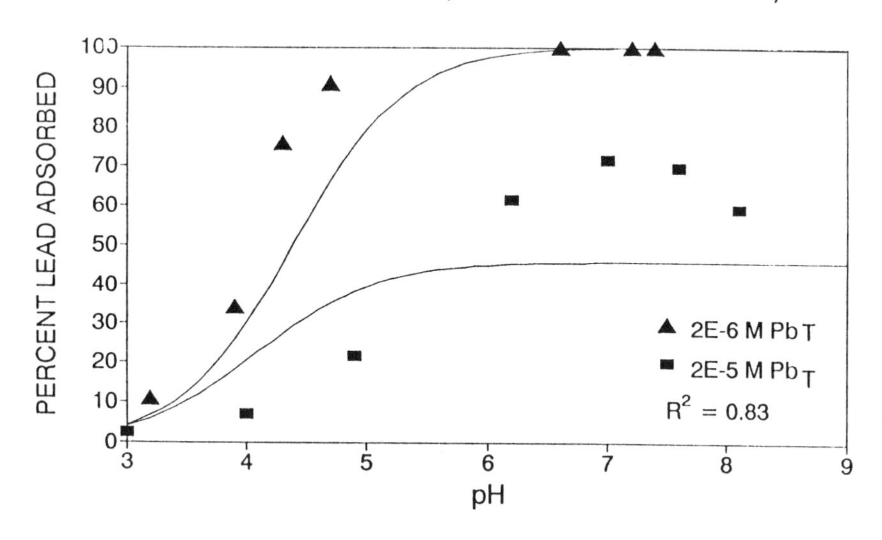

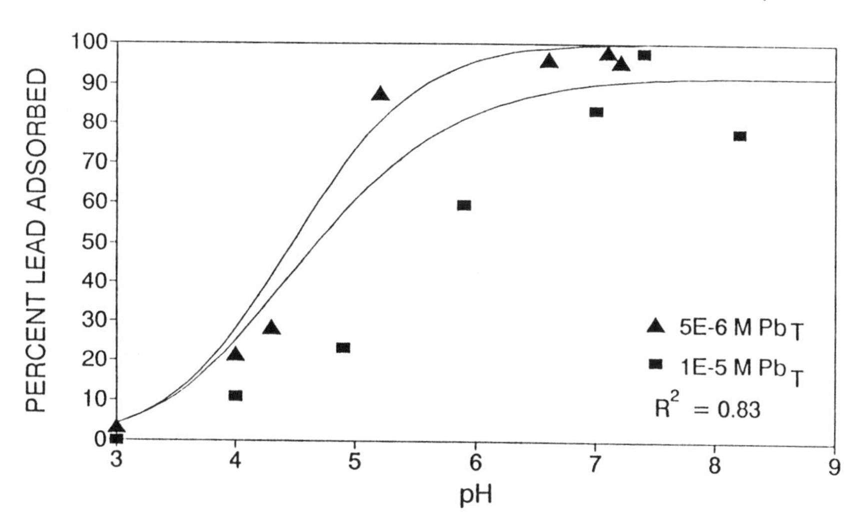

FIGURE 5. Results of the Nonlinear Regression for the Complete 5a and 5b Batch Reactor. Adsorption Data Set of Figure 2 using Equation (3).

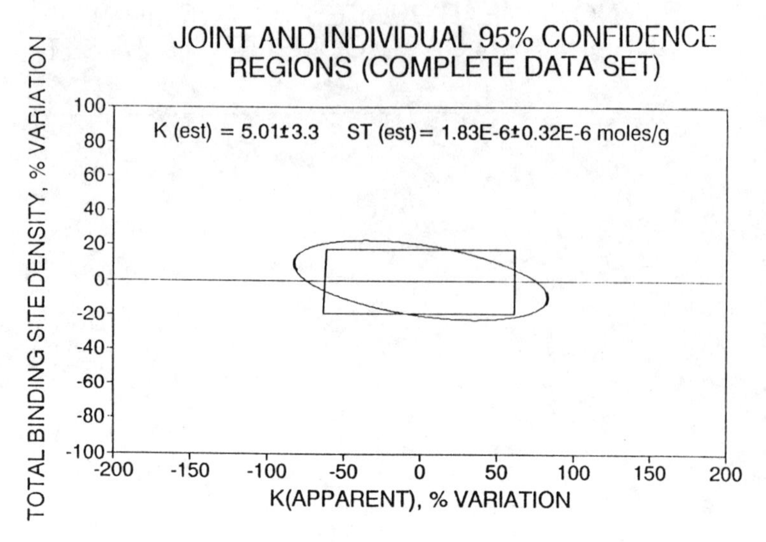

FIGURE 6. Individual and Joint Ninety-five Percent Confidence Intervals for the Adsorption Parameters K_{app} and S_T using the Complete Data Set of Figure 2.

be written and the dynamic nature of the data set (concentration versus time) plus the porosity of the granular oxide suggest the need for a rate-dependent sorption model.

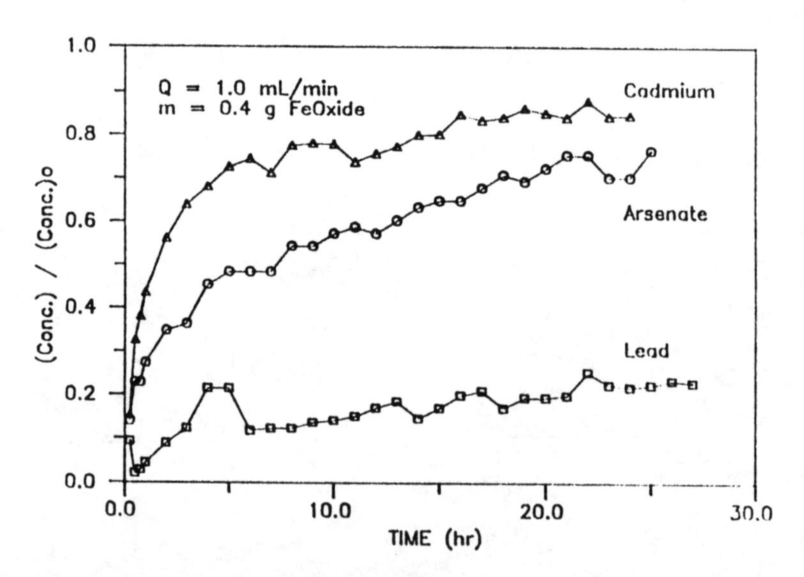

FIGURE 7. Typical Solute Breakthrough Data for Short-column Reactor used in this Research.

Because of its small size, the column is taken as a plug flow reactor; the non-steady balance on soluble lead ion in the axial direction is given as :

$$\frac{\partial[Pb]}{\partial t} = -u\frac{\partial[Pb]}{\partial x} - \left[\frac{1-\varepsilon}{\varepsilon}\right]ak_c\left([Pb]-[Pb]_f\right) \tag{4}$$

where u is fluid velocity, ε is bed porosity, a is the area/volume ratio of the granular particles, k_c is the film mass transfer coefficient, and [Pb] and $[Pb]_f$ are the bulk and film concentrations of lead in the aqueous phase. Once lead has crossed the mass transfer zone, it is perceived as adsorbing to immediately accessible sites on the particle surface, or diffusing through interior pores, for which the balance equations are :

$$V_f\frac{\partial[Pb]_f}{\partial t} + V_{s1}\frac{\partial\{Pb\}_{a1}}{\partial t} = ak_c V_S\left([Pb]-[Pb]_f\right) - V_{s2}\frac{\partial\{\overline{Pb}\}_{a2}}{\partial t} \tag{5}$$

$$\frac{\partial\{Pb\}_{a2}}{\partial t} = D_s\frac{\partial}{\partial t}\left[r^2\frac{\partial\{Pb\}_{a2}}{\partial r}\right] \tag{6}$$

In these equations V_f, V_{sl}, V_{s2}, and V_T are film, accessible adsorption zone, intrapore adsorption zone, and total column volumes, respectively; $\{Pb\}_{a1}$, $\{Pb\}_{a2}$, and $\{\overline{Pb}\}_{a2}$ are the surface, instantaneous intrapore, and average intrapore adsorbed lead concentrations; and D_S is the intrapore surface diffusion coefficient. Note that Equation (5) is written in the radial dimension of the particle. In general, in the neutral pH range, the concentration of lead in the pore fluid is negligible due to its strong tendency to adsorb, thus diffusion along pore surfaces is the dominant mechanism by which lead migrates to interior pores of the oxide. The relationship between the average particulate lead and the instantaneous intraparticle lead concentrations is found by averaging the latter within each particle, i.e.,

$$\{\overline{Pb}\}_{a2} = \frac{3}{R^3}\int_0^R \{Pb\}_{a2}\, r^2 dr \tag{7}$$

where R is the particle radius. A final equation needed for the description of the adsorption is the lead partitioning relation given by Equation (3). As indicated previously, desorption of lead from the oxide was achieved by changing the influent pH, from 7.2 to 3.2. Although the change was instantaneous as the fluid was input to the column, the change in pH across the axial direction was somewhat slower as higher pH solution was displaced, and slower yet as H^+ ions diffused into the pores of the oxide grains. The axial gradient was assumed to be linear; however, the pore was modeled as a dual diffusion problem, H^+ diffusing into pores and Pb^{2+}, desorbed in accordance with Equation (2), diffusing outward to bulk solution. Under these conditions, lead can no longer be assumed to diffuse only via the surface pathway, thus it is necessary to permit transport in the pore fluid, the proportion of lead in each sector dictated by Equation (2). During desorption Equations (6) and (7) are replaced by:

$$V_{s2}\frac{\partial\{Pb\}_{a2}}{\partial t} + V_p\frac{\partial[Pb]_p}{\partial t} = D_s\frac{\partial}{\partial r}\left[r^2\frac{\partial\{Pb\}_{a2}}{\partial t}\right] + D_p\frac{\partial}{\partial r}\left[r^2\frac{\partial[Pb]_p}{\partial r}\right] \tag{8}$$

$$\{\overline{Pb}\}_{a2} = \frac{3}{R^3}\int_0^R\left(V_{s2}\{Pb\}_{a2} + V_p[Pb]_p\right)r^2dr \tag{9}$$

where $[Pb]_p$ is the pore concentration of lead, V_p is the intrapore volume, and D_p the pore (liquid) diffusion coefficient.

Equation sets (4)-(7) and (4), (5), (8) and (9) appear to expand greatly the number of parameters which must be supplied; however, the values for several of these were obtained from independent measurements. The fluid velocity and total volume were measured directly. The bed porosity was computed from the bulk density measurement, and the particle area/volume ratio was found through BET-N_2 adsorption. The film volume (V_f) was estimated from a previous study in which flow rate was systematically varied [12]. Porosimetry provided a distinction between easily accessible and pore binding site volumes (V_{s1} and V_{s2}) and total intrapore volume (V_p). Values for all physical data are given in Table 1.

Table 1. Physical Parameters Used During Column Data Analysis

PARAMETER	VALUE
V_T	0.264 cm^3
V_{s1}	0.030 cm^3
V_{s2}	0.074 cm^3
V_P	0.101 cm^3
V_f	0.016 cm^3
a	7.69×10^5 cm^2/cm^3
ε	0.61
R	0.1 cm
u	0.013 cm/s

The equation sets have five remaining parameters which must be estimated: S_T and K_{app} as before, D_s and D_p, the surface and pore diffusion coefficients, and k_c, the mass transfer coefficient. For estimating these parameters there are two essentially independent data sets, one for adsorption of lead at pH 7.2 and one for desorption as the pH drops to 3.2. In performing the nonlinear regression on the data for adsorption, it was found that the results were insensitive to the sorption constant. This is understandable in view of the strong sorptive tendency of lead at the pH of the experiment and the rapid adsorption kinetics of lead on goethite (9). Since the adsorption equations do not contain an aqueous pore diffusion mechanism, this leaves three parameters, S_T, D_s and k_c to be estimated from the adsorption data. Values for D_s and k_c were then input to the desorption equations for estimation of S_T, D_p and K_{app}. Of course it is not necessary to repeat the estimation procedure for S_T, but in so doing a check for consistency is incorporated into the procedure.

Figure 8 shows the results for the regression analysis and parameter estimation for the lead adsorption data. The interval expressed for each parameter represents the 95 percent confidence region. This interval is less than ten percent for the mass transfer coefficient (estimated as ak_c), and approaches 80 percent for both S_T and D_s. Such a variation is not necessarily unacceptable, falling within a factor of two; however, it does suggest that the binding site density and surface diffusion coefficient are correlated. This was checked by performing a two parameter estimation, holding ak_c constant at

7.98x10^{-3} sec^{-1}, for S_T and D_s, the result of which is given in Figure 9. These estimates for each parameter are similar to those obtained previously and confidence intervals have narrowed somewhat; however, the joint interval indicates a substantial degree of inverse correlation.

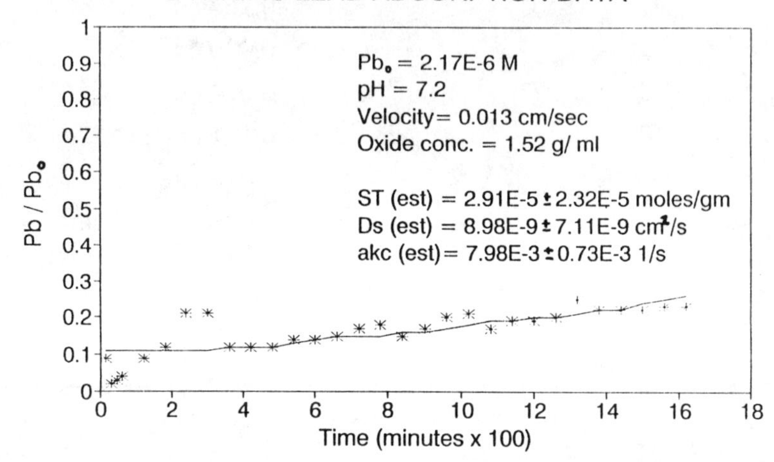

FIGURE 8. Results of the Nonlinear Regression for Lead Adsorption in the Column Reactor using Equations 4 and 7.

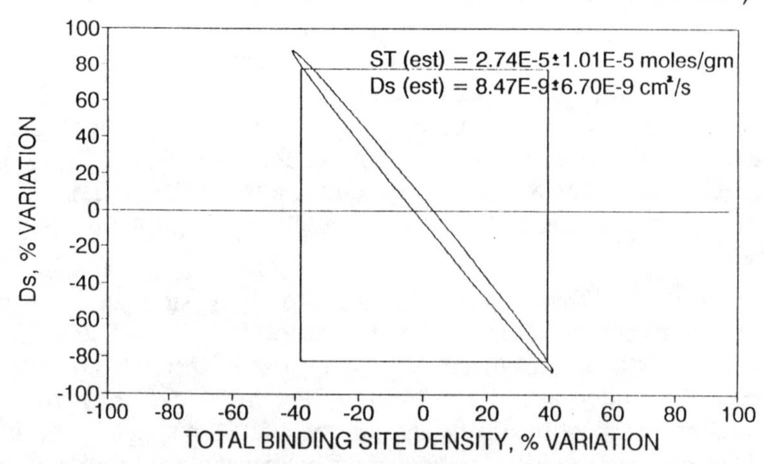

FIGURE 9. Individual and Joint Ninety-five Percent Confidence Intervals for Nonlinear Regression on Column Adsorption Data (ak_c = 7.98 x 10^{-3} sec^{-1}).

Figure 10 shows the lead desorption data, the model regression result, and the parameter estimates for S_T, D_p, and K_{app}. In this case the confidence intervals are about 16 percent, 48 percent, and 40 percent, respectively. Additional sets of two-parameter estimations (holding the third constant) did not reveal any strong correlations among parameters, indicating that independent estimations within a factor of 1.5 were obtained from this data set. The estimates obtained for the surface and pore diffusion coefficients are in close agreement with reported values for the diffusivities of ionic solutes in exchange resins and bulk aqueous solution, respectively [16, 17].

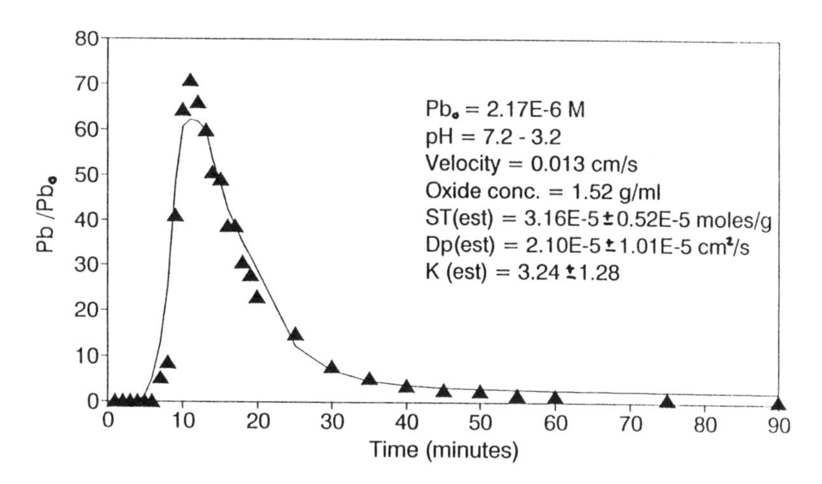

FIGURE 10. Result of the Nonlinear Regression for Lead Desorption in the Column Reactor using Equations 4, 5, 8 and 9.

In comparing the two experimental methods used, involving batch and column reactors, several aspects should be noted. There is a clear trade-off between complexity of the approach and type and quantity of information obtained. The short-column reactors are somewhat more difficult to set up and run than completely mixed batch reactors, and data analysis is substantially more complicated. The plug flow configuration is a superior method for determining diffusional parameters, which are gradient driven, since a large concentration difference between bulk solution and the particle surface is continuously maintained. It is conceptually possible to determine these parameters in batch systems by, for example, taking samples as a function of time; however, the gradient generally declines as the experiment progresses unless total solute mass is added to the reactor over time. In a closed batch system, calculations suggest that several weeks are required in order to achieve the same degree of sorption saturation that the column reaches in one to two days.

If diffusional parameters are not required for a given application, the batch approach might be the system of choice for determining isotherm sorption parameters were it not for a significant degree of underestimation of the maximum sorption binding site density, S_T. In this research, the best value obtained from batch experiments was about 17 times lower than the column value. This is a direct consequence of the time allowed for batch adsorption, four hours in this case, which is sufficient for the aqueous phase to reach apparent steady state, but is not adequate for a sufficient number of binding sites throughout the particle to be filled. Interestingly the values obtained between the two approaches for the binding constant, K_{app}, agree within statistical accuracy (5.01±3.3 and 3.24±0.28). In view of the large difference in the concentration of solids for these reactors, the results obtained do not support the so called "solids effect" noted by other researchers [18].

5. CONCLUSIONS

It is difficult to overemphasize the importance of establishing meaningful error bounds on the parameter estimations obtained for solute sorption, regardless of the experimental configuration used. When multiple parameters are determined, as is the case in most groundwater contaminant transport applications, it is the joint confidence interval which is generally the most useful, since it indicates the degree of correlation among parameters and can be used to narrow the confidence intervals of the parameters if an independent measurement of one parameter is available. Reliance on single "goodness-of-fit" statistics, such as the correlation coefficient, is ill-advised in most multiple parameter estimations, particularly those with a minimum set of observations. In this respect, continuous flow column reactors, once in operation, are capable of producing relatively large amounts of data in comparison with most batch reactors. The minimum number of observations required to product acceptable parameter estimations, regardless of the experimental configuration, depends on a number of factors including the extent to which each observation stresses the model being used (that is, avoids regions of minimum sensitivity), and the number of parameters to be estimated. In this research, the complete batch data set (Figure 2) provided enough information for independent estimates of the isotherm parameters K_{app} and S_T within approximate factors of 1.7 and 1.2, respectively (notwithstanding the underestimation of S_T, as indicated above). Single data sets, at a given total lead concentration, gave much poorer results. column adsorption and desorption data sets yielded estimates for five parameters, K_{app}, S_T, D_S, D_P, and ak_c within factors of 1.4, 1.2, 1.8, 1.5, and 1.1, respectively, although there was a considerable degree of correlation between D_s and S_T. This suggests that four parameters may be the limit for independent estimations from these data. Greater distinction between these coefficients would require additional experimentation designed to differentiate their individual effects on the model system.

Acknowledgments: This research was supported, in part, by the New York State Center for Hazardous Waste Management and the Industrial Waste Elimination Research Center of the U.S. Environmental Protection Agency. The authors wish to acknowledge the contribution of Sonia Ellis who assisted in the performance of laboratory experimentation and analysis. Optimization computations were conducted, in part, using the Cornell National Supercomputer Facility, a resource of the Cornell Theory Center, which receives major funding from the National Science Foundation and the IBM Corporation, with additional support from New York State and members of its Corporate Research Institute.

REFERENCES

1. Cederberg, G.A., R.L. Street and J.O. Leckie, "A Groundwater Mass Transport and Equilibrium Chemistry Model for Multicomponent Systems," *Water Resour. Res.*, 21:1095-1104 (1985).

2. Electric Power Research Institute, "FASTCHEM Package: User's Guide to EICM, The Coupled Geohydrochemical/Transport Code," Report No. EA-5870-CCM, Vol. 5 (1989).

3. Jennings, A.A., D.J. Kirkner and T.L. Theis, "Multicomponent Equilibrium Chemistry in Groundwater Quality Models," *Water Resour. Res.*, 18:1089-1096 (1982).

4. Kirkner, D.J., T.L. Thesis and A.A. Jennings, "Multicomponent Solute Transport with Sorption and Soluble Complexation," *Advances in Water Resources*, 7:120-125 (1984).

5. Lichtner, P.C., "Continuum Model for Simultaneous Chemical Reactions and Mass Transport in Hydrothermal Systems," *Geochim. Cosmochim. Acta*, 49:779-800 (1985).

6. Miller, C.W. and L.V. Benson, "Simulation of Solute Transport in a Chemically Reactive Heterogeneous System: Model Development and Application," *Water Resour. Res.*, 19:381-391 (1983).

7. Yeh, G.T. and V.S. Tripathi, "A Critical Evaluation of Recent Developments in Hydrogeochemical Transport Models of Reactive Multichemical Components," *Water Resour. Res.*, 25:93-108 (1989).

8. Theis, T.L., "Reactions and Transport of Trace Metals in Groundwater," in *Metal Speciation: Theory, Analysis and Application*, J.R. Kramer and H.E. Allen, Eds. pp. 81-98. Lewis Publishers, (Chelsea, MI 1988).

9. Hayes, K.F., and J.O. Leckie, "Mechanism of Lead Ion Adsorption at the Goethite/Water Inferface," in *Geochemical Processes at Mineral Surfaces*, ACS Symposium Series 323, J.A. Davis and K.F. Hayes, Eds. pp. 114-141 American Chemical Society, (Washington, DC 1986).

10. Kinniburgh, D.G., "General Purpose Adsorption Isotherms," *Environ. Sci. and Tech.*, 20:895-904 (1986).

11. Pfizer Color Oxide Technical Data. Pfizer, Inc., Minerals, Pigments and Metals Division, (Clifton, NJ 1981).

12. Theis, T.L., R. Iyer, and L.W. Kaul, "Kinetic Studies of Cadmium and Ferricyanide Adsorption on Goethite," *Environ. Sci. Tech.* 22:1013-1017 (1988).

13. Kiphart, K., "The Kinetics of Sorption Reactions at the Goethite-Aqueous Interface," Ph.D. Dissertation, University of Notre Dame (Notre Dame, IN 1983).

14. Draper, N.R. and H. Smith, *Applied Regression Analysis*, John Wiley and Sons, (New York, 1981).

15. Seinfield, J.H., and L. Lapidus, *Mathematical Methods in Chemical Engineering, Vol. 3*, Process Modeling, Estimation and Identification, Prentice-Hall, Inc. (Englewood Cliffs, NJ 1974).

16. Hasanian, M.A., A.L. Hines, and D.O. Cooney, "Adsorption Kinetics for Systems That Exhibit Nonlinear Equilibrium Isotherms," in *Adsorption and Ion Exchange*, Y.H. Ma, D.O. Cooney, and A.L. Hines, Eds. AIChE Symposium Series, 79:60-66 (1983).

17. Price, D.M. and A. Varma, "Effective Diffusivity Measurement Through an Adsorbing Porous Solid," in *Diffusion and Convection in Porous Catalysts,* I.A. Webster and W.C. Strieder, Eds. AIChE Symposium Series 266, 84:88-96 (1988).

18. DiToro, D.J., J.D. Mahony, P.R. Kirchgraber, A.L. O'Byrne, L.R. Pasquale and D.C. Piccirilli, "Effects of Nonreversibility, Particle Concentration, and Ionic Strength on Heavy Metal Sorption," *Environ. Sci. Technol.*, 20:55-61 (1986).

8

ASSESSMENT OF METAL MOBILITY IN SOIL - METHODOLOGICAL PROBLEMS

Walter W. Wenzel and Winfried E.H. Blum
Institute of Soil Science
University of Agriculture
A-1180 Vienna, Austria

1. INTRODUCTION

Metal mobility in soils is generally determined by chemical extraction techniques using water, salt solutions or complexing agents on disturbed, air-dried soil samples [1]. In such standard procedures possible influences as the season of sampling, of sample preparation techniques, such as homogenizing and sieving, as well as sample storage are not currently considered.

On the other hand, in several countries such extraction procedures are used for the definition of threshold values [2-4] as well as for soil monitoring for environmental purposes, e.g. aimed at long term changes in soils [5]. For example, cation exchange capacity (CEC) and base saturation data are used as important criteria in soil classification [6].

This contribution aims at identifying some critical sources of error caused by soil sampling procedures, sample preparation and extraction techniques.

2. MATERIALS AND METHODS

2.1 Experiments

Table 1 shows some possible sources of error resulting from sampling procedures, sample preparation and extraction techniques, and the respective experiments undertaken to identify them. Some

0-87371-697-3/95/$0.00+$.50

characteristics of the soils used in these experiments are shown in Table 2.

Table 1. Analytical Procedures, Sources of Error and Experimental Layout.

Analytical Procedure	Source of error	Experiment
Sampling	Sampling density	Experiment 1
	Season of sampling	Experiment 2
Sample preparation	Air-drying	Experiment 3
	Homogenizing and sieving	Experiment 4
	Sample storage	Experiment 5
Extraction techniques	Extraction pH	Experiment 6

Table 2. Characteristics of Soils Used in This Work.

Soil (FAO)	Horizon (FAO)	Texture (FAO)	pH (H_2O)	Organic Matter %
Experiment 1				
Eutric Cambisol (grassland)	Ah	sandy loam	5.7	4.7
Experiment 2				
Dystric Cambisol (forest)	Ah	sandy loam	5.4	9.2
Experiment 3				
Dystric Cambisols (forest)	Ah	sandy loam	4.0-5.5	3.5-10
Experiment 4				
Chromic Luvisols (forest)	CvBtck	loamy sand	4.5	0.1
Experiment 5				
Dystric Cambisols (forest)	Ah	sandy loam	4.0-5.5	3.5-10
Experiment 6				
Cambisols	Ah, Ap	sand -	3-8	0.1-15
Luvisols	E, Bt	clay loam		
Chernozems	Bv, Cv			
Gleysols	Bg, Br			

Details of each experiment are given in the following discussion.

Experiment 1: Effects of sampling density on the accuracy of plot means of stable (aqua regia) and mobile (EDTA) metal fractions in an Eutric Cambisol under grassland were determined.

Eight subsamples were taken from the Ah-horizon (0-10 cm) within a plot of 100 m^2, air-dried, passed through a 2 mm-sieve and analyzed separately for Fe, Mn, Zn and Cu dissolved by aqua regia and 0.05 M Na_2-EDTA [5,7]. Arithmetic mean values, standard deviations and coefficients of variation were calculated as well as the number of subsamples required for limits of accuracy of 5%, 10% and 20%, respectively [8].

Experiment 2: Effect of the season of sampling on pH and exchangeable aluminum in an Ah-horizon of a Dystric Cambisol under forest was analyzed.

Within a plot of about 7 m^2 (diameter 3 m) twenty subsamples of an Ah-horizon (0-5 cm) were taken in June, July, August, October, November, December 1990 and in January, February 1991. Field-moist samples were mixed together and passed through a 2mm-sieve. The pH(H_2O) at a soil:solution ratio of 1:2.5 and the exchangeable aluminum (by 0.1 M $BaCl_2$) at a soil:solution ratio 1:20 [5] were analyzed.

Experiment 3: Effects of air-drying on pH, soluble C and metal mobility in Ah-horizons of Dystric Cambisols under forest were investigated.

Samples of Ah-horizons (0-5 cm) were taken in June, July, August, October, November and December 1990 from 13 plots with diameters of 3 m along a transect with litter translocation in a beech-oak stand. The distances between the plots were 20 m. Twenty field-moist subsamples from each plot were mixed together, homogenized and passed through a 2mm-sieve. One part of each sample was stored at 4°C, the other part was air-dried and stored at room temperature (20°C - 25°C). Samples were analyzed for exchangeable Fe and Al (by 0.1 M $BaCl_2$) at a soil to solution ratio of 1:20, pH(H_2O) as well as the organic carbon soluble using 1:5 water extracts [5].

Experiment 4: Effects of sieving and homogenization intensity on pH and exchangeable Al and Ca of a BtCvck-horizon for a Chromic Luvisol under forest.

The pH of non-carbonatic fine earth of subsamples of a BtCvck-horizon [6] showing carbonate nodules (< 5 cm) and weathered rock fragments was measured in the field. In the laboratory 3 subsamples were prepared for analysis by three different intensities of homogenization and passed through a 2 mm-sieve. For each of the subsamples pH(H_2O) 1:2.5, and exchangeable Al and Ca (by 0.1 M $BaCl_2$) at a soil solution ratio of 1:20 were analyzed [5].

Experiment 5: Effects of one month of storage on the pH of field-moist and air-dried Ah-horizons (0-5 cm) of Dystric Cambisols under forest were investigated.

Ah-horizons from the plots of the transect of Experiment 3 were sampled in March 1991. The pH was analyzed immediately after sampling (March 25,1991) and one month later (April 24,1991).

Experiment 6: The pH(H_2O) of different soil samples was compared with the final pH of 0.1 M $BaCl_2$, 0.05 M Na_2-EDTA and water extracts.

The < 2mm-fraction of air-dried soil samples with a wide range of properties (Table 2) were shaken over-head during two hours with H_2O (1:5 soil to solution ratio), 0.1M $BaCl_2$ (1:2.5 soil to solution ratio) and 0.05 M Na_2-EDTA (1:10 soil to solution ratio). The pH of the filtrates was measured. These pH-values were compared with those obtained by the Austrian standard procedure (measurement of pH in H_2O 1:2.5 after 2 hours without shaking, see [5]).

3. RESULTS

The results of Experiment 1 are presented in Table 3. Experiment 1 shows that representative soil sampling must be based on the spatial variability of soil properties, often requiring an unreasonable number of subsamples within one sampling plot. In general, mobile element fractions (Na_2-EDTA) vary more than total contents aqua regia, (see also [9]).

Moreover, mobile fractions show considerable seasonal changes. Experiment 2 demonstrates seasonal changes of pH in a sandy Ah-horizon with a high content of organic matter, causing significant changes in aluminum mobility (Figure 1). This verifies that the season of sampling is very important, especially for exchangeable Al fractions. By contrast, mobile fractions are often regarded as relatively stable, and used as essential criteria in soil classification, as for instance CEC and base saturation [6].

In general, routine analysis is based on batching of disturbed soil samples assuming equilibrium, although it is evident that due to aggregation and tortuosity of pores, mobile phases are often in multi-phase nonequilibrium with solid soil components [10,11]. As a consequence, the results obtained by batch techniques differ considerably from those obtained by percolation of undisturbed soil columns. In addition, uncontrollable readsorption of metals at artificially created surfaces of disturbed aggregates must be considered [1].

Table 3. Experiment 1: Spatial Variability Fe, Mn, Zn and Cu Extractable by Aqua Regia and Na_2-EDTA Within the Ah-horizon (0-10 cm) of an Eutric Cambisol Within a Plot of 100 m^2 (N = 8).

		Mean Value	Standard Deviation	CV	No. of Subsamples Required for Limits of Accuracy of		
		mg kg^{-1}		%	5%	10%	20%
Fe	aqua regia	28871	1431	5	6	2	1
	Na_2-EDTA	442	111	25	100	27	9
Mn	aqua regia	425	76	18	51	15	6
	Na_2-EDTA	164	57	35	193	50	14
Zn	aqua regia	77.2	7.9	10	19	7	3
	Na_2-EDTA	4.5	1.3	29	193	35	11
Cu	aqua regia	18.9	2.8	15	37	11	5
	Na_2-EDTA	3.9	1.1	27	117	31	10

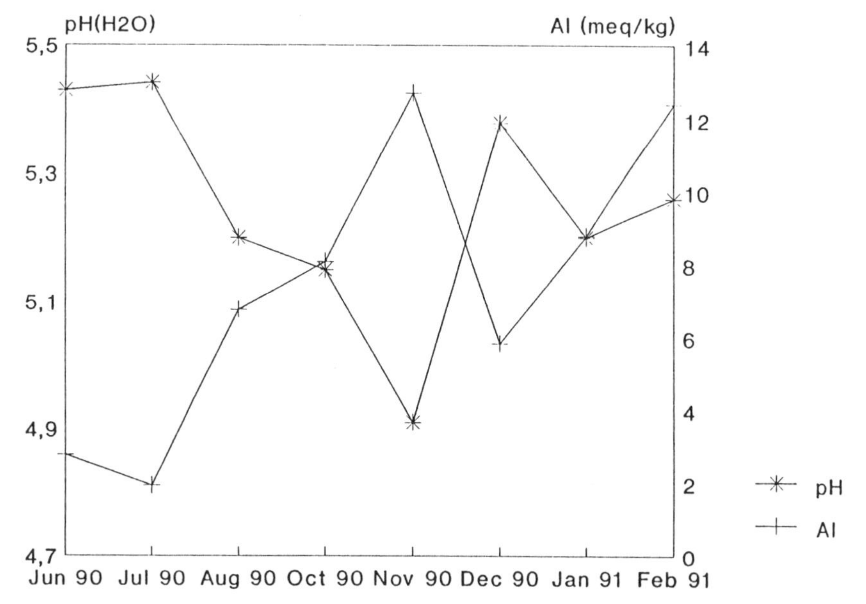

FIGURE 1. Seasonal Changes of pH and Exchangeable Al in an Ah-horizon (0-5 cm) of a Dystric Cambisol (Experiment 2).

Experiment 3 indicates that sample preparation is a further source of error when determining metal mobility. Air-drying of soil samples caused considerable mobilization of iron, probably due to the breakdown of organic matter during the process of air-drying and subsequent electron transfer, as well as a slight decrease in pH and substantial increase of soluble organic matter (Figure 2). However, no change in exchangeable Al was noticed.

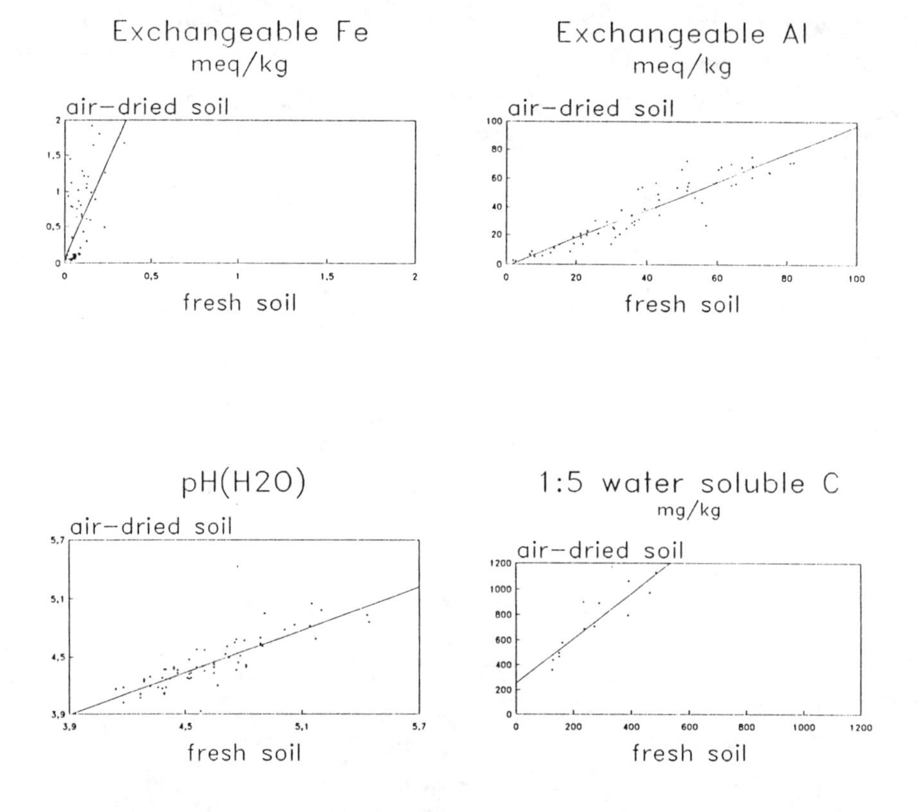

FIGURE 2. Effects of Air-drying on pH and Water-soluble C Influencing Metal Mobility in Ah-horizons (0-5 cm) of Dystric Cambisols (Experiment 3).

The effect of the homogenizing and sieving procedures on the pH and subsequent changes in metal mobility of soils with substantial amounts of weathered rock fragments and nodules of $CaCO_3$ is shown in Experiment 4 (Figure 3).

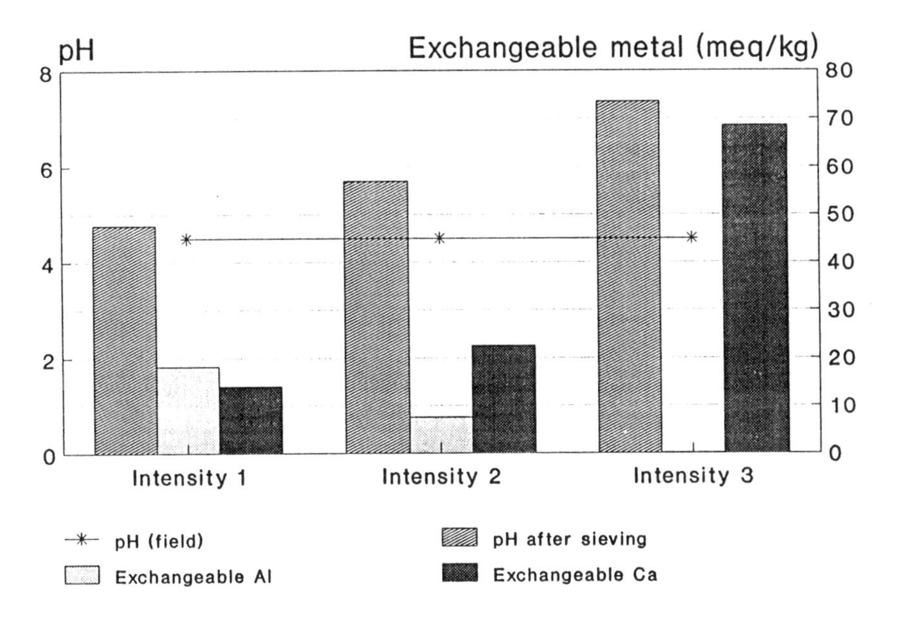

FIGURE 3. ***Effects of Homogenization and Sieving Intensity on pH and Exchangeable Metals in a BtCvck-horizon of a Chromic Luvisol (Experiment 4).***

Storing of soil samples can be a further source of error as exemplified by Experiment 5 (Figure 4). The changes in pH(H_2O) of Ah-horizons after one month are more pronounced in air-dried samples than in field-moist ones, although air-drying is thought to reduce further changes of soil characteristics.

Numerous extraction techniques are described in the literature, covering a wide range of extractants, concentrations used and differing enormously with respect to the duration of batching [1]. The chemical nature of extraction techniques for "mobile metal phases" varies from simple exchange reactions to more selective desorption processes, partly including more than one reaction mechanism, frequently resulting in a low selectivity of metal speciation [1]. Extraction pH, influenced by the chemical nature of the extractant and soil characteristics, is often not recorded, although being one of the most important factors of metal solubilization.

Experiment 6 indicates that during the extraction of soils with pH(H_2O) ranging from 4 to 7, 0.05 M Na_2-EDTA buffers the pH at 4.4-4.6, whereas in the presence of carbonates the final pH in the extracts is suddenly raised above 7. Because there is no correlation between extraction pH and soil pH in H_2O, Na_2-EDTA is not selective for exchangeable metals, which are regarded to be mobile (Figure 5).

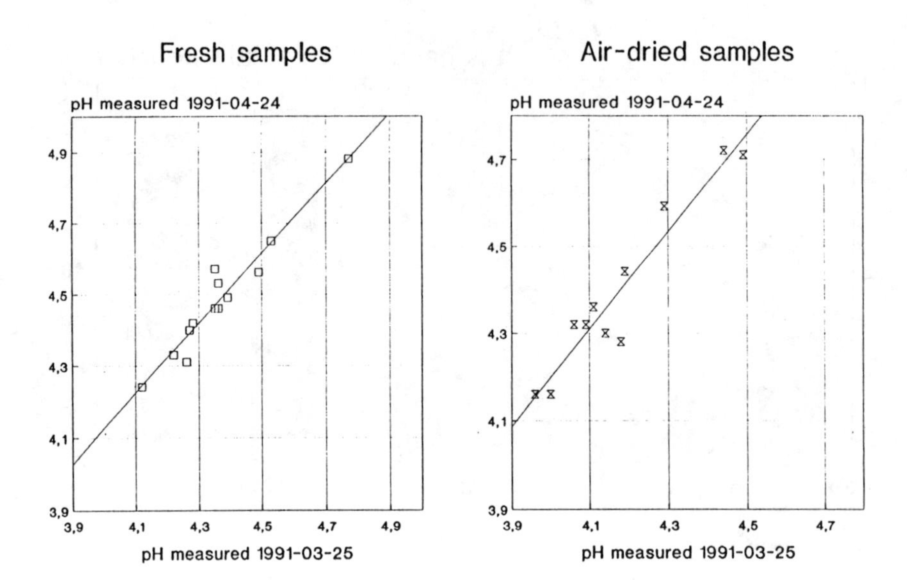

FIGURE 4. Effects of One Month of Storage on the pH of Field-moist and Air-dried Ah-horizons (0-5 cm) of Dystric Cambisols (Experiment 5).

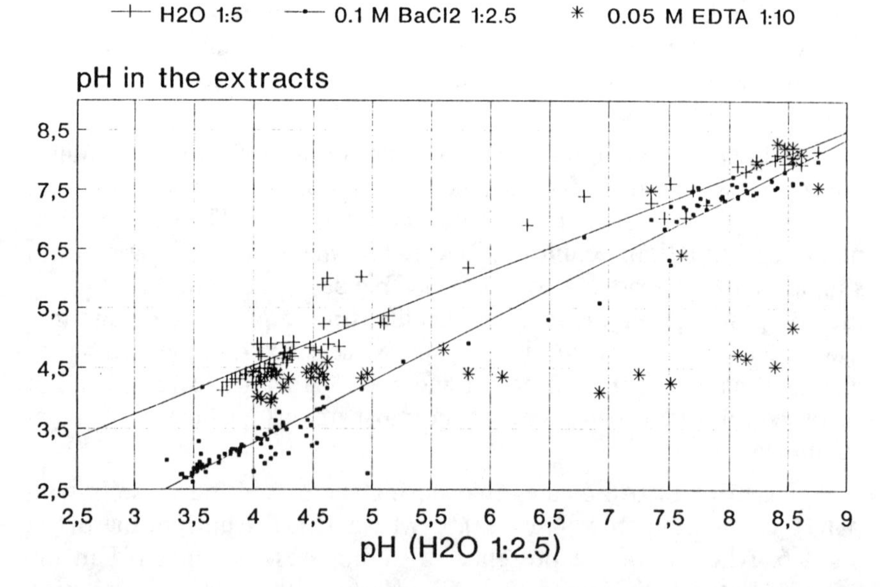

FIGURE 5. pH-values in Different Extracts Compared with pH(H_2O) (Experiment 6).

Moreover, the extraction pH of neutral salts like 0.1 M $BaCl_2$ is well correlated with pH(H_2O), but at a range of 1 unit lower. Raising the soil:solution ratio from 2.5 to 5 caused a marked increase in the final pH of water extracts. This should be considered when comparing extractions which differ in the soil:solution ratio.

4. CONCLUSIONS

The assessment of metal mobility by application of standard procedures implies many sources of possible errors, such as:

- non-representative soil sampling with respect to spatial and seasonal soil variability;
- sampling procedures which disturb the natural aggregation of soil;
- sample preparation procedures as sieving, homogenizing, drying and storing of soil samples; and
- extraction techniques which are non-selective or inadequate for the estimation of mobile metal phases due to the chemical nature of the extractants, the final extraction pH or the soil:solution ratio.

The experiments of this study, carried out on a limited number of soil samples with specific characteristics permit only some preliminary indications on those sources of error. More systematic work would be needed including a wide variety of soil characteristics. Moreover, comparative studies on standard extraction techniques used in different countries could provide a reliable basis for an international data base.

Acknowledgments: We thank Dr. A. Mentler, D.I. N. Farcas, Mr. A. Brandstetter and Ing. E.W. Brauner for analytical support.

REFERENCES

1. Beckett, P.H.T., "The Use of Extractants in Studies on Trace Metals in Soils, Sewage Sludges, and Sludge-treated Soils," *Adv. Soil Sci.* 9:143-176 (1989).

2. Delschen, T.U. and W. Werner, "Beitrag zur Ableitung “tolerierbarer” 0.1 M $CaCl_2$ löslicher Cadmium- und Zinkgehalte," *Landwirtschaftliche Forschung* 42:40-49 (1989).

3. Pruess, A., G. Turian, V. Schweikle and T. Noeltner, "Bestimmung pflanzenverfuegbarer Schwermetallgehalte in Boeden," *Mitt. Dtsch. Bodenkundl Gesellsch.* 61:123-125 (1989).

4. Bundesamt fuer Umweltschutz, Erläuterungen zur Verordnung vom 9. Juni 1986 ueber Schadstoffe im Boden (VSBo). Texte zum Umweltschutzgesetz, Dokumentationsdienst des Bundesamtes fuer Umweltschutz, p. 17 (Bern, 1987).

5. Blum, W.E.H., H. Spiegel and W.W. Wenzel, *Bodenzustandsinventur*, p. 95, Bundesministerium fuer Land- und Forstwirtschaft, (Vienna, 1989).

6. FAO, FAO-UNESCO Soil Map of the World, revised legend, *World Resources Report* 60, FAO, (Rome, 1988).

7. Blum, W.E.H., O.H. Danneberg, G. Glatzel, H. Grall, W. Kilian, F. Mutsch and D. Stoehr, *Waldbodenuntersuchung,* p. 76 Austrian Society of Soil Science, (Vienna, 1986).

8. McPherson, G., *Statistics in Scientific Investigation* p. 666 Springer, (New York, 1990).

9. Wilding, L.P. and L.R. Drees, "Spatial Variability and Pedology” in *Pedogenesis and Soil Taxonomy 1. Concepts and Interactions, Developments in Soil Science 11A,* L.P. Wilding, N.E. Smeck and G.F. Hall, Eds. pp. 83-116. Elsevier, (Wageningen, 1983).

10. Hildebrand, E.E., "A Procedure for Obtaining the Equilibrium Soil Pore Solution," *Z. Pflanzenernaehr Bodenk* 149:340-346 (1986).

11. Hantschel, R., M. Kaupenjohann, R. Horn and W. Zech, "Cation Concentrations of the Equilibrium Soil Solution (ESS) and Percolating Water (PW) - A Method Comparison," *Z. Pflanzenernaehr Bodenk* 149:136-139 (1986).

9

VIBRATIONAL AND NMR PROBE STUDIES OF SAz-1 MONTMORILLONITE

Cliff T. Johnston[1], William L. Earl[2] and C. Erickson[1]
[1] Department of Soil Science
University of Florida
Gainesville, FL 32611
[2] Chemical Science and Technology Division
Los Alamos National Laboratory
Los Alamos, NM 87545

1. INTRODUCTION

There is currently a lack of understanding about the molecular level mechanisms dominating subsurface transport and transformation of metals and organic contaminants. Before predictive models can be developed which accurately describe their geochemical behavior, information is needed to understand the fundamental physical-chemical mechanisms of surface interactions with naturally occurring substrates. The approach discussed here is to combine vibrational and magnetic spectroscopic techniques with macroscopic sorption measurements.

One of the primary problems with spectroscopic investigations of contaminants on surfaces is that the sensitivity of most spectroscopic techniques is insufficient to study the contaminants at environmentally relevant concentrations. In order to address this problem and to develop an unambiguous relationship between macroscopic sorption data and spectroscopic analysis, we have developed a method of collecting macroscopic sorption measurements and FTIR data simultaneously.

Although magnetic and vibrational spectroscopic methods have long been recognized as sensitive chemical techniques, interferences from paramagnetic impurities, water, and from the bulk adsorbent have restricted their application. Nonetheless, surface applications of these spectroscopic techniques can provide a large amount of mechanistic information.

In this paper we report on a study of the interactions of exchangeable metal cations with mineral surfaces using a combined spectroscopic/macroscopic approach. There are three primary advantages to the application of vibrational and magnetic spectroscopies to surface studies of sorbate-mineral complexes: (1) they can be applied *in situ*, providing nondestructive information about the sample. There is a need to obtain molecular information about the surface without disturbing or altering the interface. Many surface methods involve harsh preparation techniques such as desiccation. This results in a sample that is vastly different from those found in nature. One wonders whether the data obtained are real or are an artifact of the preparation: (2) there have been significant improvements in sensitivity and spectral discrimination due to advances in the past few years; and (3) these two experimental methods that are characterized by different time and energy scales. Structural information obtained by a particular method is intimately related to the time and energy scales associated with that approach. The time scale for NMR measurements is on the order of 10^{-3} to 10^{-7} seconds while vibrational spectroscopies probe in the 10^{-12} to 10^{-15} second range. Thus, a multiple-technique approach can generally provide a more coherent picture of the structure and dynamics of the system of interest.

When less than three molecular layers of water are present in the interlamellar region, the structure of water in the interlamellar region has been shown to be influenced predominantly by exchangeable cations [1-14]. Spectroscopic investigations of smectite-water interactions suggest that two distinct environments of sorbed water are present [7,9]: (i) water molecules coordinated to exchangeable metal cations, and (ii) physisorbed water molecules occupying interstitial pores, void spaces between exchangeable metal cations, and accumulated on external surfaces.

On the basis of NMR [12-14], complex permittivity [9], and dielectric [11,15] measurements, water molecules coordinated to metal cations are considered to be in less-mobile, restricted environments on the clay surface relative to those in bulk water. This molecular "picture" of coordinated water in the interlamellar region is consistent with vibrational studies of water sorbed on smectites. Russell and Farmer [16] found that the OH stretching and bending bands of water sorbed on montmorillonite and saponite were perturbed on going from higher to lower hydration states. In the case of the H-O-H bending band (υ_2 mode) in the 1610 to 1640 cm^{-1} region, the frequency decreased and its molar absorptivity increased upon dehydration. A similar result was reported by Poinsignon *et al.* [5] who observed that the molar absorptivities of the OH stretching and bending bands of sorbed water depended strongly on the hydration energy of the exchangeable metal cation, water content, and on the surface charge density of the clay surface.

There are a large number of NMR studies of water hydration in clays using one of the isotopes of hydrogen, the proton or the deuteron [17-19]. The difficulty with such studies is that the time scale probed by NMR is relatively long and it is impossible to distinguish between the protons of the water and those of surface hydroxyls. A potential variation is to use the cations exchanged in the clay as magnetic probes and to obtain information on the water structure and mobility from changes in the metal ion NMR spectra. Recently, there have been a few NMR studies of cation exchange but they have primarily concentrated on the structure and site of the cation itself [20-23]. In their work, Bank and co-workers make the observation that the Cs^+ ion must be quite mobile, sampling many sites in the clay interlayer. These cation probe studies have not concentrated on experiments as a function of clay hydration so do not directly obtain information about the nature of the water in the clays.

The objectives of the present study are to examine the use of water molecules and metal cations as molecular probes of smectite water interactions. Specifically, the ν_2 (H-O-H bending vibration) mode of water will be employed as a diagnostic vibrational band which is sensitive to changes in its local environment. An FTIR-gravimetric cell [24,25] will be used to examine the FTIR spectra of water on selected homoionic smectites as a function of water content and exchangeable cation. Although previous IR studies of water-smectite systems [5,16] have shown that water coordinated to exchangeable metal cations is distinct from physisorbed or bulk water, correlation of these data to the amount of water sorbed was either not measured or determined indirectly. Similar to the vibrational studies, Na will be used as a probe molecule of metal-water interactions on the surface of the smectite as a function of water content. The ^{23}Na NMR resonance is a sensitive function of the molecular mobility of the cation. Molecular mobility is, in turn, an indicator of the hydration of the cation and the extent to which the hydrated cation interacts with the clay surface.

2. EXPERIMENTAL

2.1 Vibrational spectroscopy

The samples studied were the Cheto montmorillonite (SAz-1) collected from Apache County, Arizona obtained from the Source Clays Repository of The Clay Minerals Society. A complete description of their physical properties are given by Van Olphen and Fripiat [26]. Prior to size fractionation, a homoionic Na-montmorillonite clay suspension was prepared by placing 10 g of the raw clay in 1.0 L of 0.5 M NaCl. The Na-montmorillonite suspension was then washed free of excess salts by repeated centrifugation with distilled-deionized water. The < 0.5 μm size fraction (e.s.d.) was collected by centrifugation. Stock clay suspensions of montmorillonite were prepared by adding approximately 0.7 L of 0.05 M solutions of the metal chloride to the salt-free clay suspension (volume =0.3 L) such that the total volume was 1.0 L; the final pH values of these suspensions were approximately 5. The washing and size fractionation steps were completed within a 24 h period to minimize degradation of the clay.

Self-supporting clay films were prepared for the FTIR study by washing 20 ml of the stock montmorillonite clay suspensions free of excess salts. The solids concentration of the suspension was diluted to 0.0025 g of clay/g of suspension. One ml aliquots of this suspension were deposited on a polyethylene sheet and allowed to dry. The dry clay films were then "peeled" off of the polyethylene sheet as self supporting clay films. These clay films were approximately 2.5 cm in diameter and weighed between 3.1 - 3.4 mg corresponding to cross sectional density of 0.65 mg/cm^2. The cation exchange capacity of the clay films was determined by placing a known amount of the dry clay film (3 to 5 mg) in 5 ml. of 0.25 M $MgCl_2$ and allowing the suspension to stand for 12 hours. A portion of the supernatant was removed for analysis of Na and Cu and the procedure was repeated twice. The cation exchange capacities of the clay films were 0.99 $mmol_c\ g^{-1}$ for Na-SAz-1 and 1.11 ± 0.02 $mmol_c\ g^{-1}$ for Cu-SAz-1.

For the coupled FTIR / gravimetric measurements, FTIR spectra were obtained using a Bomem DA3.10 FTIR spectrometer equipped with a MCT detector (D*=3.13 x109 cm $Hz^{0.5}$ and a low-frequency cutoff of 400 cm^{-1}) and a KBr beamsplitter. The Bomem DA3.10 spectrometer was controlled through a general-purpose-interface-bus (IEEE-488) interfaced to a DEC Vaxstation-II computer. The unapodized resolution for the FTIR spectra was 2.0 cm^{-1}.

A 15-cm pathlength gas cell, fitted with two ZnSe windows, Viton O-rings, and two stopcocks was modified to interface with a Cahn controlled-environment (Model D200-01) electrobalance (Figure 1). The modified gas cell allowed FTIR spectra and gravimetric data to be collected simultaneously; a complete description of the cell is given in Reference 25. The cell was connected to a gas /vacuum manifold which permitted the atmosphere in the cell to be controlled and the pressure to be monitored. A self-supporting homoionic montmorillonite clay film was placed on a 0.2 mm Pt hangdown wire in the Cahn electrobalance. The modified gas cell was mounted on an NRC optic table and was positioned using NRC x-y-z translational stages which permitted adjustment of the cell. Mounting the cell on the optic table mechanically decoupled the electrobalance assembly from vibrations originating from the FTIR spectrometer and other sources. The cell was positioned such that the clay film was centered on, and nearly perpendicular to, the incident modulated IR beam in the sample compartment.

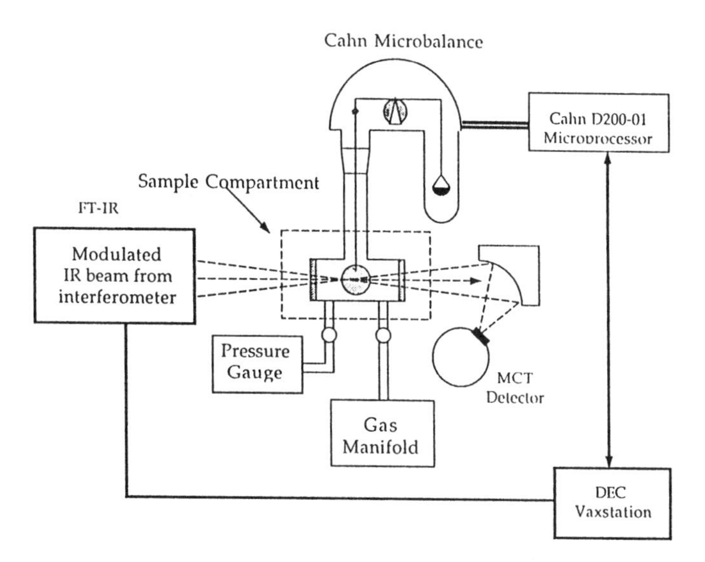

FIGURE 1. ***Schematic of the In-situ FTIR/gravimetric Cell Showing the Following: Gas Manifold and Pumping Station, Cahn D200-01 Electrobalance, 15 cm Path Length Gas Cell Fitted with Two 50 x 3 mm ZnSe Windows, Viton O-rings, Two Kontes Stopcocks, and a 164/45 Ground Glass Joint used to Interface the Hangdown Tub of the Cahn Balance with the Sample Compartment of the Bomem DA3.02 Spectrometer.***

Desorption isotherms were obtained by placing 1.5 ml of water in one of the hangdown tubes of the Cahn balance after the clay film had been mounted and tared on the balance. The water was then frozen by placing a dewar of liquid nitrogen around the hangdown tube. The Cahn balance and FTIR cell were then evacuated to 0.005 torr using a roughing pump. At this point, the isolation valve located between the Cahn balance/IR cell and the vacuum manifold was shut and the liquid nitrogen was removed. The water was allowed to slowly equilibrate and the pressure inside the Cahn balance/IR cell was determined using a Baratron gauge. The clay film was allowed to equilibrate for 24 hours in the presence of a small excess of liquid water to obtain the p/p_o value of 1.0. At this point the mass of the sample, pressure inside the cell, and the temperature were recorded along with the FTIR spectrum of the clay film. The mass of the clay film was recorded every 10 to 30 seconds using a Zenith 286 computer interfaced to the D200 balance. After the clay film had come to equilibrium, a small amount of vapor was removed from the cell by opening the isolation valve and the system was allowed to re-equilibrate.

2.2 NMR spectroscopy

The samples were prepared in a similar fashion to those for vibrational spectroscopy to the point of exchanging them with metals, i.e., size fractionated, Na exchanged SAz-1 was prepared. The clay was then dried in a 110°C oven overnight. It was split into three portions. One sample was stored over phosphorus pentoxide to be kept dry. A second sample was stored over pure water, yielding a p/p_o of 1.0. The third was stored over a saturated solution of NaBr which provides approximately 22% relative humidity or p/p_o of 0.22. These were then quickly packed into NMR rotors which were closed with o-ring caps. The o-rings prevent air from entering the rotors thus maintaining the relative hydration level of the clay.

NMR spectra were all obtained with a Varian Unity 400 spectrometer. The samples were rotated at 4.6 kHz at the magic angle with a Varian MAS probe except the dehydrated sample which was rotated at 6.7 kHz to clearly separate spinning sidebands. The spectra were simple Bloch decay spectra with no proton decoupling. The excitation pulse width was 1 μs, approximately $\pi/20$. This small pulse angle was used to prevent distortions in the spectrum due to the quadrupolar nature of ^{23}Na [27]. A spectral width of 100 kHz, recycle delay of 2 s, and 832 data points zero filled to 8 kpoints were used. Approximately 250 transients were acquired for the 22% and 100%

relative humidity samples and 450 transients were collected for the dehydrated spectrum because of the increased linewidth and resultant poorer signal to noise.

3. RESULTS

3.1 Desorption isotherms

Desorption isotherms of water from Na-, Cu-, K-, and Co-exchanged SAz-1 were obtained using the gravimetric/FTIR cell shown in Figure 1. Desorption isotherms obtained for water from self-supporting montmorillonite clay films using this cell are in reasonable agreement with reported desorption isotherms [1, 28, 29]. There is

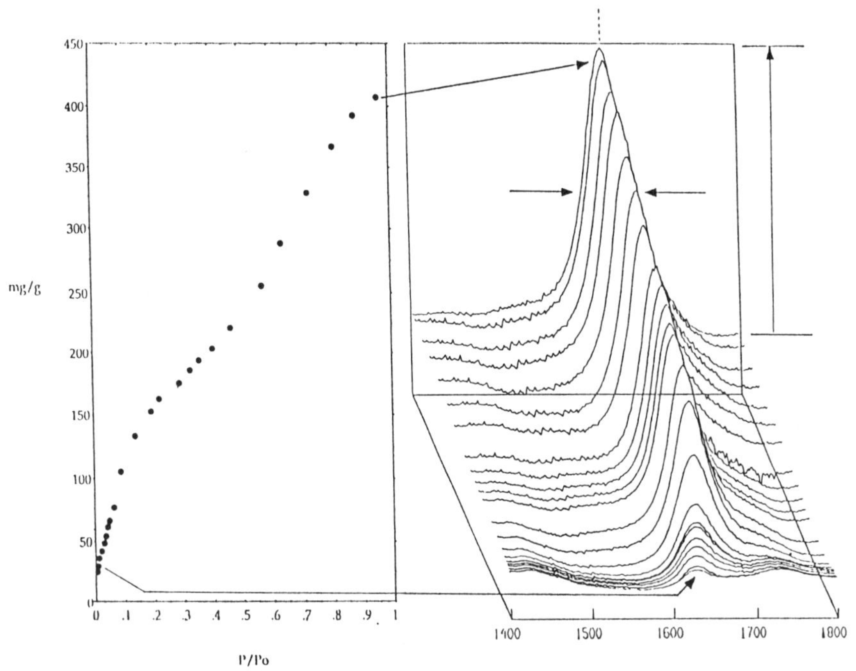

FIGURE 2. Comparison of the Desorption Isotherm of Water from a Self-supporting Clay Film of Na-SAz-1 Obtained at 24° C to the FTIR Spectra (right side) of Water Sorbed on the Na-SAz-1 Clay Film. Each Spectrum Shown on the Right Side Corresponds to One Data Point on the Desorption Isotherm.

relatively poor agreement, however, with our data and adsorption isotherms reported by Lipsicas *et al.* [30] due to hysteresis between adsorption and desorption cycles. As Mooney *et al.* [1] indicated, water adsorption isotherms are often non-reproducible due to slight variations in the amount of residual water present at the start of the adsorption run. The FTIR spectra shown on the right side of Figure 2 correspond to the desorption data plotted on the left side. Each FTIR spectrum shown on the right side of Figure 2 corresponds to one data point on the desorption isotherm.

The three dimensional plot of FTIR spectra shown in Figure 2 in the 1400 to 1800 cm^{-1} region is plotted with the z-axis representing the relative pressure of water. The band shown at 1640 cm^{-1} corresponds to the ν_2 mode of sorbed water on the Na-SAz-1 clay film. The intensity of the ν_2 band is directly proportional to the amount of water sorbed on the clay; upon lowering the water content, the intensity of the ν_2 band decreases. Simultaneous collection of the gravimetric and spectroscopic data provides a direct method of correlating the macroscopic sorption characteristics with the observed vibrational data.

3.2 FTIR spectra of sorbed water

The position of the ν_2 band of water sorbed on SAz-1 montmorillonite exchanged with Na^+, K^+, Cu^{+2}, and Co^{+2} are plotted in Figure 3 as a function of water content expressed as the number of water molecules per metal cation. At high water contents, the position of the ν_2 band was relatively stable. In the case of Na-SAz-1, the high water content limit on the position of the ν_2 band was 1640 cm^{-1} which compares to a value of 1635 for Cu-SAz-1. Upon reducing the water content, however, to less than 10 water molecules per Cu^{2+} ion or 6 water molecules per Na^+ ion, the position of the ν_2 band was observed to shift to lower frequency in agreement with previous IR studies of water on smectite [5, 16]. In a related study, Mortland and Raman [31] showed that the surface acidity of montmorillonite increased significantly upon lowering the water content. Thus, the shift in frequency of the ν_2 band may reflect the change in surface acidity upon dehydration of the clay surface.

The decrease in frequency of the ν_2 band upon lowering the water content is attributed to a decrease in hydrogen bonding [16]. The more labile water is removed from the interlamellar region upon dehydration leaving the water molecules which are coordinated to exchangeable metal cations in the interlamellar region. A decrease in the ν_2 band generally accompanies an increase in the ν_1 and ν_3 (O-H stretching bands). The position of the OH stretching bands could not be determined unambiguously in this study due to significant overlap and broadening of the OH stretching bands with each other and with the structural ν(O-H) band at 3630 cm^{-1}.

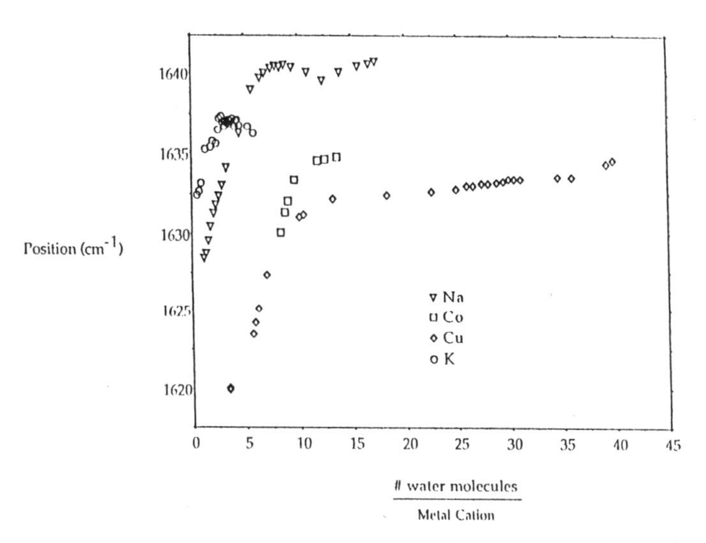

FIGURE 3. Position of the ν_2 Band of Water Sorbed on SAz-1 Montmorillonite Exchanged with Na^+, K^+, Co^{2+}, and Cu^{2+} Plotted as a Function of Water Content Expressed as the Number of Water Molecules Sorbed Per Exchangeable Metal Cation.

The surface concentration of water sorbed on the clay films was determined using the familiar expression for the Bouguer-Beer-Lambert law (Beer's law) expressed as:

$$A(\nu) = \varepsilon(\nu) c d = \ln\left(\frac{I_o(\nu)}{I(\nu)}\right) \tag{1}$$

Where A(ν) is the absorbance of the band of interest at frequency ν, e(ν) is the molar absorptivity at ν_2 (cm^2/mmole), c is the concentration of the solute expressed as (mmoles/cm^3), d is the thickness of the clay film in cm, $I_o(\nu)$ is the intensity of the incident beam, and I(ν) is the intensity of the transmitted beam. For solute sorption on thin films, it is more convenient to replace the quantity (c · d) with $\bar{c}$ which has units of (mmoles/cm^2) and corresponds to the number of adsorbers per unit area defined by

$$\bar{c} = c d \qquad \left(\text{mmoles/cm}^2\right) \tag{2}$$

i.e., (mmoles of solute)/(cm^2 of clay film)

$$\bar{c} = (\text{single solute}) = \frac{(m_{cs} - m_c)}{(a_c M_{solute})} \tag{3}$$

Where m_{cs} is the total mass of the clay film (combined mass of clay and sorbed solute) in mg, m_c is the dry mass of the clay film in mg, a_c is the area of the clay film in cm^2, and M_{solute} is the molecular weight of the solute (i.e., water in this case).

Using the measured peak height of the ν_2 band of water, the molar absorptivity of this vibrational mode was determined for Na- and Cu-exchanged montmorillonite using Equation (1) and is shown in Figure 4 as a function of water content. At high water contents, the observed molar absorptivity of it is the same as that of bulk water. Upon lowering the water content, however, the molar absorptivity value increases. In the case of Cu-SAz-1, a steady increase in the molar absorptivity occurs until the lowest water content of approximately 3 water molecules/cation is reached. In contrast to water sorbed on Cu-SAz-1, a smaller increase is observed for the Na-exchanged clay film. A maximum molar absorptivity of 35 cm $mmol^{-1}$ occurs at a (H_2O/Na^+) ratio of 6. This implies 6 waters coordinating the Na^+. Upon reducing the water content further, the molar absorptivity value decreases to a value slightly less than that of bulk water for Na-SAz-1.

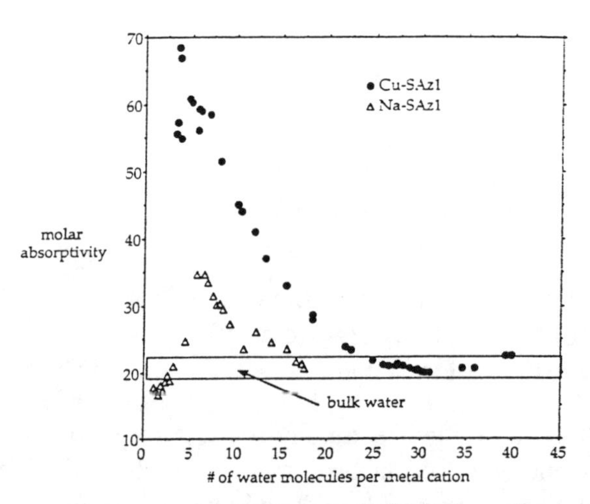

FIGURE 4. Change in the Molar Absorptivity of the ν_2 Band of Water Sorbed on SAz-1 Montmorillonite Exchanged with Na^+ and Cu^{2+} Plotted as a Function of Water Content Expressed as the Number of Water Molecules Sorbed per Exchangeable Metal Cation.

Although not measured directly, Russell and Farmer [16] and Poinsignon *et al.* [5] provided supporting evidence that the molar absorptivity of the ν_2 band of water increased at low water content. Poinsignon suggested that the observed increase was related to polarization of the intercalated water species by exchangeable cations. This hypothesis is consistent with the observed increase in surface acidity by Mortland and Raman [31] at low water content.

3.3 NMR spectra of sodium ions

The three samples for NMR were prepared at p/p_o's of 0, 0.22, and 1.0. These partial pressures of water were selected to correspond to a clay with zero, one, and multiple layers of water in the interlamellar space. At 0% relative humidity, with the sample stored over P_2O_5, all interlamellar water except the immediate water of hydration of the Na^+ ions is removed. Total dehydration of the sample would require very high temperature treatment under vacuum. The data in Figure 4 indicates that at p/p_o between about 0.2 and 0.4 there is only a single layer of water in the interlamellar space. At $p/p_o = 1.0$ there are approximately two layers of water in the interlamellar space. Figure 5 contains the ^{23}Na NMR spectra of the exchanged SAz-1 at the three hydration levels. The interpretation of these data are colored by the fact that ^{23}Na is a "quadrupolar" nucleus; i.e., one with a spin quantum number greater than 1/2. The linewidth in nuclei with spins greater than 1/2 is a very sensitive function of molecular motion [32]. The lower spectrum ($p/p_o = 1.0$) has a chemical shift very close to that of an aqueous solution of Na^+. The linewidth is very narrow indicating a high degree of mobility, i.e., the electric quadrupolar interaction is virtually totally averaged by molecular motion. The upper spectrum ($p/p_o = 0$) contains several very interesting features. First, it is significantly shifted from the resonance of Na^+ ion in solution. Secondly, it is clear that the resonance is composed of at least two overlapping peaks. And thirdly, there are small, but observable sidebands at the positions marked SSB on the figure. These sidebands are shifted by 6700 Hz (the angular rotation frequency of MAS in this particular spectrum) and are a result of MAS in the NMR of quadrupolar nuclei. The central spectrum in Figure 5, with approximately one layer of water in the interlamellar space has both a linewidth and resonance position between the saturated and dehydrated samples.

The spectrum of the water saturated sample is the easiest to interpret. The Na^+ ion is a fairly small ion and two layers of water open the interlamellar space sufficiently that this ion has virtually free mobility in the interlamellar region. This spectrum is entirely consistent with a picture of a solvated Na^+ ion with a freely exchanging coordination sphere of water. The aqueous complex is free to rotate and to move laterally between the clay layers. An alternative interpretation is that the waters in the first coordination sphere of the Na^+ ion are free to exchange rapidly producing the NMR averaging necessary to account for the narrow line with a shift virtually identical to that of an aqueous Na^+ solution.

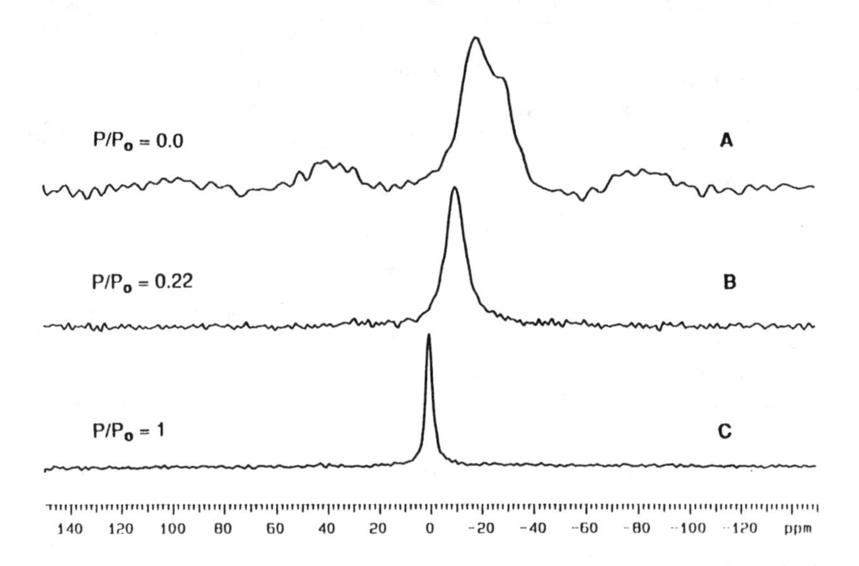

FIGURE 5. ***Three ^{23}Na NMR Spectra of Sodium Substituted SAz-1 Montmorillonite as a Function of Hydration. Spectrum A was taken of a dried sample held over P_2O_5. Spectrum B is of a sample held over saturated NaBr which results in an atmosphere with about 22% relative humidity. Spectrum C is of a sample held over pure water, i.e., 100% relative humidity.***

Although more complex, there is a straightforward interpretation of the upper spectrum in Figure 5. First, the quadrupole coupling constant (QCC) is probably relatively small. Small quadrupole coupling constants are the result of small intrinsic quadrupole moments or high molecular symmetry about the nucleus of interest. The quadrupole moment of ^{23}Na is intermediate in size. We expect the Na^+ ion to be solvated by 4 waters in a roughly tetrahedral geometry or perhaps 6 waters in an octahedron as suggested by our FTIR data, *vide supra.* We also expect rapid exchange of the waters of hydration. The high symmetry and rapid exchange would produce a small QCC. The magnitude of QCC is significant in interpreting the shift of the peaks relative to aqueous Na^+ ion. In ^{23}Na, NMR, the measured shift can come from both the well known chemical shift and from a second order quadrupole shift. If the QCC is small, it follows that the second order quadrupole shift is also small and we can assign the peaks seen in the upper spectrum to chemically shifted Na in the system. This interpretation is consistent with work by Laperche and co-workers [20] who conclude that the QCC for Na^+ adsorbed on vermiculite is very small. In either case (chemical or second order quadrupolar shift) the resonance position is indicative of an electronic interaction of the Na^+ ion with the surface of the clay. The fact that the resonance is clearly composed of two peaks is indicative of at least two chemically distinct sites for Na^+ on the clay. At present, our data are insufficient to assign those sites to well known structural components of montmorillonite.

Finally, how do we interpret the central spectrum (p/p_o about 0.22)? Since it is clear that there is abundant molecular motion in the hydrated sample and very little motion in the dry one, it is reasonable that with a single layer of water in the interlamellar space there should be some motion or chemical exchange. The single, rather broad peak, intermediate in shift between the other two spectra is diagnostic of "intermediate" chemical exchange. We believe that in this intermediate regime of hydration, there are some hydrated Na^+ ions bound to the clay surface and some freely moving in the interlamellar space. A given Na ion can exchange between those two sites at a rate of about 10^{-3}s, the reciprocal of the approximate shift between the fully hydrated and the dry samples.

4. CONCLUSIONS

There are strong changes in both the position and absorptivity of the vibrational spectroscopy of water bending modes as a function of the water content of the SAz-1 samples studied. We attribute these changes to strong electrostatic forces and mobility changes that occur when all of the water in the interlamellar space is associated with the metal ion. In order to perform such measurements, it was necessary to construct an IR cell where the hydration could be measured concurrently with the vibrational spectrum. This combination of macroscopic (weight or hydration level) and microscopic (vibrational spectroscopy) measurements is needed to obtain molecular level information on clay systems. The primary difficulties in obtaining molecular level information lie in the relatively low sensitivity of spectroscopic techniques. Once we had established the interesting hydration levels through the combined IR/gravimetric technique, we were able to prepare samples for NMR.

We believe that as water is removed from the metal-clay system we eventually get to the state where the only observable water vibrational modes are due to waters directly coordinated to the metal ion. At that point the effective pH of the water molecules changes because of the strong electrostatic interaction with the metal ion. This change in effective pH produces the observed shift in the ν_2 vibrational mode of the water.

The overall picture of Na^+ ion in montmorillonite obtained from the FTIR and NMR data presented is a rather simple one. We view the clay surface as having at least two distinct sites to which a hydrated Na^+ can bind. The present status of this work does not allow us to associate the sites with specific structural positions in the clay. We are working towards an understanding that will eventually allow us to be more specific about the structural associations. At very low levels of hydration, the Na^+ ion binds in these sites and cannot exchange between them. As water is inserted in the interlamellar space, the energy of solution is sufficient to increase the mobility of the Na^+ ion such that it begins to hop from site to site and to spend significant time as a solvated ion in the interlamellar space. At full hydration, the Na^+ ion is free to move in the interlamellar region and actually looks very much like Na^+ in bulk aqueous solution. This picture is consistent with the NMR data presented but it is clear that more work is needed to pin down many of the details of interest. For example, it would be interesting to assign the two peaks in the spectrum of the dry montmorillonite to well known sites in the clay. It would also be interesting to study both SAz-1 as a function of hydration and of temperature to slow down all chemical exchange processes and to obtain rate constants for the various exchanges. And finally, we are embarking on ^{17}O studies of water molecules associated with cations in the interlamellar spaces of clays.

In any case it is clear that NMR spectroscopy has much to offer in understanding the state of metals in soil components. It is also clear that it is useful to have different spectroscopic measurements to provide a more complete picture of the system under study. In our case the NMR measurements are indicative of chemical exchange while FTIR indicates that the water molecules are static. This is a result of the vastly different time scales probed by the two techniques. For NMR the time scales of 10^{-4} s are relatively fast while for vibrational spectroscopies, 10^{-12} s can be measured. It is also necessary to tie microscopic (spectroscopic) measurements to macroscopic data. We have combined our spectroscopies with desorption isotherms of water which allows us to tie the molecular effects to the measured hydration levels. This certainly influences the interpretation of the data.

REFERENCES

1. Mooney, R.W., A.G. Keenan and L.A. Wood, "Adsorption of Water Vapor by Montmorillonite. I. Heat of Desorption and Application of BET Theory," *J. Am. Chem. Soc.* 74:1367-1374 (1952).

2. Clementz, D.M., T.J. Pinnavaia and M.M. Mortland, "Stereochemistry of Hydrated Copper(II) Ions on the Interlamellar Surfaces of Layer Silicates. An Electron Spin Resonance Study," *J. Phys. Chem.* 77:196-200 (1973).

3. Clementz, D.M., M.M. Mortland and T.J. Pinnavaia, "Properties of Reduced Charge Montmorillonites: Hydrated Cu(II) Ions As a Spectroscopic Probe," *Clays and Clay Min.* 22:49-57 (1974).

4. McBride, M.B. and M.M. Mortland, "Copper (II) Interactions with Montmorillonite: Evidence From Physical Methods," *Soil Sci. Soc. Am. Proc.* 38:408-415 (1974).

5. Poinsignon, C., J.M. Cases and J.J. Fripiat, "Electrical-Polarization of Water Molecules Adsorbed by Smectites. An Infrared Study," *J. Phys. Chem.* 82:1855-1860 (1978).

6. McBride, M.B., T.J. Pinnavia and M.M. Mortland, "Electron Spin Resonance Studies of Cation Orientation in Restricted Water Layers on Phyllosilicate (Smectite) Surfaces," *J. Phys. Chem.* 79:2430-2435 (1975).

7. Prost, R., "Interactions Between Adsorbed Water Molecules and the Structure of Clay Minerals: Hydration Mechanism of Smectites," *Proc. Int. Clay Conf.* 351-359 (1975).

8. McBride, M.B., "Hydrolysis and Dehydration Reactions of Exchangeable Cu^{2+} on Hectorite," *Clays and Clay Min.* 30:200-206 (1982).

9. Sposito, G. and R. Prost, "Structure of Water Adsorbed on Smectites," *Chem. Rev.* 82:553-573 (1982).

10. Sposito, G., R. Prost and J.P. Gaultier, "Infrared Spectroscopic Study of Adsorbed Water on Reduced-Charge Na/Li Montmorillonites," *Clays and Clay Min.* 31:9-16 (1983).

11. Bidadi, H., P.A. Schroeder and T.J. Pinnavaia, "Dielectric Properties of Montmorillonite Clay Films: Effects of Water and Layer Charge Reduction," *J. Phys. Chem. Solids* 49:1435-1440 (1988).

12. Grandjean, J. and P. Laszlo, "Multinuclear and Pulsed Gradient Magnetic Resonance Studies of Na Cations and of Water Reorientation at the Interface of a Clay." *J. Mag. Res.* 83:128-137 (1989).

13. Kogelbauer, A., J.A. Lercher, K.H. Steinberg, F. Roessner, A. Soellner and R.V. Dmitriev, "Type, Stability, and Acidity of Hydroxyl Groups of HNaK-erionites," *Zeolites* 9:224-230 (1989).

14. Delville, A., J. Grandjean and P. Laszlo, "Order Acquisition By Clay Platelets In a Magnetic Field. NMR Study Of the Structure and Microdynamics of the Adsorbed Water Layer," *J. Phys. Chem.* 95:1383-1392 (1991).

15. Fripiat, J.J., A. Jelli, G. Poncelet and J. Andre, "Thermodynamic Properties of Adsorbed Water Molecules and Electrical Conduction in Montmorillonites and Silicas," *J. Phys. Chem.* 69:2185-2197 (1965).

16. Russell, J.D. and V.C. Farmer, "Infra-red Spectroscopic Study of the Dehydration of Montmorillonite and Saponite," *Clay Min. Bull.* 5:443-464 (1964).

17. Woessner, D.E., B.S. Snowden Jr. and G.H. Meyer, "Tetrahedral Model for Pulsed Nuclear Magnetic Resonance Transverse Relaxation: Application to the Clay-Water System," *J.Colloid Interface Sci.* 34:43-52 (1970).

18. Fripiat, J.J., J. Cases, M. Francois and M. Letellier, "Thermodynamic and Microdynamic Behavior of Water in Clay Suspensions and Gels," *J.Colloid Interface Sci.* 89:378-400 (1982).

19. Lipsicas, M., R. Raythatha, R.F. Giese,Jr. P.M. and Costanzo, "Molecular Motions, Surface Interactions, and Stacking Disorder in Kaolinite Intercalates," *Clays Clay Miner.* 34:635-644 (1986).

20. Laperche, V., J.F. Lambert, R Prost and J.J. Fripiat, "High-Resolution Solid-State NMR of Exchangeable Cations in the Interlayer Surface of a Swelling Mica: ^{23}Na, ^{111}Cd, and ^{133}Cs Vermiculites," *J.Phys.Chem.* 94:8821-8831 (1990).

21. Bank, S., J.F. Bank and P.D. Ellis, "Solid-State ^{113}Cd Nuclear Magnetic Resonance Study of Exchanged Montmorillonites," *J.Phys.Chem.* 93:4847-4855 (1989).

22. Weiss, C.A., R.J. Kirkpatrick and S.P. Altaner, "The Structural Environments of Cations Adsorbed onto Clays: ^{133}Cs Variable-Temperature MAS NMR Spectroscopic Study of Hectorite," *Geochim. Cosmochim. Acta* 54:1655-1669 (1990).

23. Weiss, C.A., R.J. Kirkpatrick and S.P. Altaner, Variations in Interlayer Cation Sites of Clay Minerals as Studied by ^{133}Cs MAS Nuclear Magnetic Resonance Spectroscopy," *Amer.Mineral.* 75:970-982 (1990).

24. Johnston, C.T., T. Tipton, D.A. Stone, C. Erickson and S.L. Trabue, "Chemisorption of p-dimethoxybenzene on Cu-Montmorillonite," *Langmuir* 7:289-296 (1991).

25. Johnston , C.T., T. Tipton, D.A. Stone, C. Erickson and S.L. Trabue, "C Vapor Phase Sorption of p-xylene on Co- and Cu-exchanged SAz-1 Montmorillonite," *Environ. Sci. Technol.* 26:382-390 (1992)

26. Van Olphen, H. and J.J. Fripiat, *Data Handbook For Clay Materials And Other Non-Metallic Minerals.* Pergamon Press, (Oxford, 1979).

27. Engelhardt, G. and D. Michel, *High-Resolution Solid-State NMR of Silicates and Zeolites.* John Wiley & Sons (New York, 1989).

28. Mooney, R.W., A.G. Keenan and L.A. Wood, "Adsorption Of Water Vapor By Montmorillonite. II. Effect of Exchangeable Ions and Lattice Swelling as Measured by X-ray Diffraction," *J. Am. Chem. Soc.* 74:1371-1374 (1952).

29. Fripiat, J.J., A. Jelli, G. Poncelet and J. Andre, "Thermodynamic Properties of Adsorbed Water Molecules and Electrical Conduction in Montmorillonites and Silicas," *J. Phys. Chem.* 69:2185-2197 (1965).

30. Lipsicas, M., R.H. Raythatha, T.J. Pinnavaia, I.D. Johnson, R. F. Giese, Jr., P.M. Costanzo and J.L. Robert, "Silicon and Aluminium Site Distributions in 2:1 Layered Silicate Clays," *Nature* 309:604-607 (1984).

31. Mortland, M.M. and K.V. Raman, "Surface Acidities of Smectites in Relation To Hydration, Exchangeable-cation and Structure," *Clays and Clay Min.* 16:393-398 (1968).

32. Lindman, B. and S. Forsen, *Chlorine, Bromine, and Iodine NMR: Physico-Chemical and Biological Applications.* Springer-Verlag, (Berlin, 1976).

10

TREATMENT OF SOILS CONTAMINATED WITH HEAVY METALS

Robert W. Peters and Linda Shem
Energy Systems Division
Argonne National Laboratory
Argonne, IL 60439

1. INTRODUCTION

Heavy metal contamination of subsurface soils and groundwaters results from a number of activities including the following: application of industrial waste, application of fertilizers and pesticides, mining operations, smelting operations, automobile battery production, metal plating/metal finishing operations, vehicle emissions and fly ash from combustion/incineration processes.

Metal concentrations in groundwater are largely governed by interactions with surrounding soils and geological materials. Many different mechanisms influence the partitioning of metals between the solid and solution phases (thereby affecting the leachability of metals from contaminated soils) including dissolution and precipitation, sorption and exchange, complexation, and biological fixation [1]. The major effect of complexation is the dramatic increase in solubility of heavy metal ions, especially for strong complexing agents such as ethylenediaminetetraacetic acid (EDTA) and nitrilotriacetic acid (NTA) [2-5]. Due to this increased solubility of metal ions in solution, complexing agents offer the potential to be effective extractants of heavy metals from contaminated soil.

Ranges of heavy metal concentrations reported in the technical literature for soils contaminated with various heavy metals are summarized in Table 1.

Table 1. Reported Heavy Metal Concentrations

Heavy Metal	Site Type	Reported Concentration Range, (ppm)	Reference
Zn	Smelting	6,000 to 80,000	[6]
Cd	Smelting	900 to 1,500	[6]
Cr	Chromium Production	500 to 70,000	[7,8]
Pb	Battery Reclamation	2 to 135,000 (Soil)	[9,10]
		2.16 to 43,700 (On Site Sediment)	[9,10]
		0 to 140 mg/L (Surface Waters)	[11]
Pb	Battery Recycling	210 to 75,850 mg/kg soil	[12]
		up to 211,300 mg/kg soil	[13,14]

2. BACKGROUND

Excavation and transport of heavy metal contaminated soil has been the standard remedial technique. This technique can hardly be viewed as a permanent solution. In addition, off-site shipment and disposal of the contaminated soil involves high expense, liability and appropriate regulatory approval. Further, recent U.S. EPA regulations require pretreatment prior to landfilling [15]. This has resulted in increased interest in technologies to treat contaminated soils either on-site or *in situ* [1].

For soils contaminated with organic pollutants, there are a number of techniques which can be considered for remediation of a particular site. For soils contaminated with organic pollutants, a variety of techniques can be considered, including thermal treatment, steam and air stripping, microbial degradation and chemical oxidation.

Fewer treatment techniques exist for remediation of metal-laden soils. Metals can be removed by either flotation or extraction. Process parameters affecting extraction technologies for cleaning up soils have been summarized by Raghavan *et al.* [16]. Migration of metals can also be minimized by solidifying or vitrifying the soil and fixing the metals in a nonleachable form. In solidification processes, lime, fly ash, cement-kiln dust, or other additives are added to bind the soil into a cement-like mass and immobilize the metallic compounds. In vitrification processes, the soil is formed into a glassy matrix by applying current across embedded electrodes. Solidification is very expensive because the waste must be thoroughly characterized to determine compatibility with the specific treatment processes. Many existing technologies result in a solid with less-than-optimal long-term stability. Solidification/stabilization also results in an increase in the waste volume. Following remediation, site reuse is limited, and long-term monitoring is generally required. For these reasons, solidification/stabilization is generally limited to radioactive or highly toxic wastes [17].

Techniques for removing metals from soil generally involve contacting the soil with an aqueous solution. Methods such as flotation and water classification are solid/liquid separation processes. Metal contamination is generally found on the finer soil particles. Because metals are often preferentially bound to clays and humic materials, separating out the finely-divided material may substantially reduce the heavy metal content of the bulk soil. Separating out the fines from the coarse fraction removed 90% of the metals but only 30% of the total dredged material [18]. In flotation processes, metallic minerals become attached to air bubbles which rise to the surface forming a froth and are thus separated from particles that are wetted by water [19]. While flotation has been used successfully in the mineral processing industry, it has not been widely used for treatment of contaminated soils [1]. Solid/liquid separation processes represent a preliminary step in the remediation of contaminated soils. The metals are still bound to a solid phase in a much more concentrated form. The advantage of this preliminary processing is that the mass of contaminated soil to be processed is considerably reduced and therefore reduces the treatment cost.

Pickering [20] identified the following four approaches whereby metals could be mobilized in soils: (1) acidity changes, (2) changes in system ionic strength, (3) changes in oxidation/reduction (REDOX) potential and (4) complex formation. In the latter technique, addition of complexing ligands can convert solid-bound heavy metal ions into soluble metal complexes. The effectiveness of complexing ligands in promoting metal release depends on the strength of bonding to the soil surface, the stability and adsorbability of the complexes formed, and the suspension pH [1]. From an application viewpoint, the type and concentration of the complexing ligand and the system pH are the operational parameters that can be controlled.

The ability of chelating agents to form stable metal complexes makes materials such as ethylenediaminetetraacetic acid (EDTA) and nitrilotriacetic acid (NTA) promising extractants for treatment of heavy metal-polluted soils [21]. Elliott and Peters [1] have noted that although complexation is the major mechanism responsible for the metal solubilization, the overall release process depends on the proton concentration and the system's ionic strength. Because hydrous oxides of iron and manganese can coprecipitate and adsorb heavy metals, they are believed to play an important role in the fixation of heavy metals in polluted soils [22]. Their dissolution under reducing conditions may weaken the solid-heavy metal bond and thereby promote solubilization of the metal ions. Elliott and Peters [1] noted that there are five major considerations in the selection of complexing agents for soil remediation. They are as follows:

1. Reagents should be able to form highly stable complexes over a wide pH range at a 1:1 ligand-to-metal molar ratio.
2. Biodegradability of the complexing agents and metal complexes should be low (especially if the complexing agent is to be recycled for reuse back into the process).
3. The metal complexes that are formed should be nonadsorbable on soil surfaces.
4. The chelating agent should have a low toxicity and potential for environmental harm.
5. The reagents should be cost-effective.

Elliott and Peters [1] note that while no compounds ideally satisfy all these criteria, there are several aminocarboxylic acids which form remarkably stable complexes with numerous metal ions. The properties of three chelating agents which have been employed as soil extractants (either for analytical or remediation purposes) are summarized in Table 2.

Extraction of heavy metals from contaminated soils can be performed either using *in situ* techniques or on-site extraction (following excavation). For the case of *in situ* soil flooding, the aqueous extractant is allowed to percolate through the soil to promote metal mobilization. For the case of on-site extraction following excavation, the operation can be performed either batchwise, semibatchwise, or continuously. The soil is first pretreated for size reduction and classification. The contaminated soil is then contacted with the extractant, and the soil is separated from the spent extractant. The eluant is further recycled to decomplex and precipitate the metals from solution, or alternatively it is treated using electrodeposition techniques to recover the metals. Elliott and Peters [1] schematically present the major components for a general soil washing process.

2.1 Previous studies involving extraction of heavy metals from contaminated soils

Ellis *et al.* [23] demonstrated the sequential treatment of soil contaminated with cadmium, chromium, copper, lead and nickel using ethylenediaminetetraacetic acid (EDTA), hydroxylamine hydrochloride, and citrate buffer. The EDTA chelated and solubilized all of the metals to some degree; the hydroxylamine hydrochloride reduced the soil iron oxide-manganese oxide matrix, releasing bound metals and also reduced insoluble chromates to chromium(II) and chromium(III) forms; and the citrate removed the reduced insoluble chromium and additional acid-labile metals. Using single shaker extractions, using a 0.1 M solution of EDTA was much more effective in metal removal than using a 0.01 M solution. A pH of 6.0 was chosen as optimum because it afforded slightly better chromium removal than that obtained at pH 7 or 8. EDTA was the best single extracting agent for all metals; however, hydroxylamine hydrochloride was more effective for removal of chromium. Results of the two-agent sequential extractions indicated that EDTA was much more effective in removing metals than the weaker agents. The results of the three-agent sequential extraction showed that, compared to bulk untreated soil, this extraction removed nearly 100% of the lead and cadmium, 73% of the copper, 52% of the chromium, and 23% of the nickel. Overall, this technique was shown to be better than three separate EDTA washes, better than switching the order of EDTA and hydroxylamine hydrochloride treatment, and much better than simple water washes. The EDTA washing alone can be effectively used, however, resulting in only a slight decrease in overall removal efficiency. Lead was easily removed by the EDTA and was also effectively removed by citrate. Cadmium was easily removed by EDTA and was also effectively removed by the hydroxylamine hydrochloride. However, copper was only removed by the EDTA. Although nickel removal was poor with EDTA alone, the treatment with all three agents showed no better removal.

Table 2. Properties of Chelating Agents

Chelating Agent	Molecular Weight (g/mol)	Acidity Constants					Metal Chelate log (Stability Constant)					
		pK_1	pK_2	pK_3	pK_4	pK_5	Cd	Zn	Cu	Pb	Mn^{2+}	Fe^{3+}
Nitrilotri-acetic Acid (NTA)	191	1.89	2.49	9.73	--	--	10.5	11.2	13.7	11.8	8.1	17.0
Ethylene-diaminetetra-acetic Acid (EDTA)	292	2.08	3.01	6.40	10.44	--	17.5	17.2	19.7	17.7	14.5	26.5
Diethylene-triamine-pentaacetic Acid (DTPA)	93	2.08	2.81	4.49	8.73	10.6	20.1	19.7	22.6	21.0	16.7	29.2

Hsieh *et al.* [7,8] studied soil washing for removal of chromium from soil. Chromium was selected for their study due to its prevalence in contaminated sites in northern New Jersey. In the first portion of their study, they investigated the effect of chromium concentration, the type of soil, and pH on chromium adsorption [7]. Sand did not adsorb Cr(III); pH and the quantity of sand had no effect on Cr(III) adsorption. Both Cr(III) and Cr(VI) adsorb onto kaolinite and bentonite clay, with Cr(III) being more prone to adsorption. The amount of chromium adsorbed was proportional to the concentration of chromium added to the soil. After reaching the maximum adsorption, the soil did not adsorb any more chromium. Kaolinite had less adsorption capacity for chromium compared with bentonite. Cr(VI) had a higher adsorption at low pH. Cr(III) precipitates above pH 5.5. Results from preliminary soil washing experiments indicated that the amount of chromium washed out from the soil was proportional to the number of washings performed and the amount of extracting agents used (sodium hypochlorite and EDTA were used as the extracting agents).

Hsieh *et al.* [8] observed that chromium washout was related to pH; the efficiency increased with increasing pH and then decreased. The optimum pH was approximately pH 10.4. They also noted that after some period of time, depending on pH and particle size, chromium would be released from the soil again. Approximately 20 to 50% of the chromium in the soil samples was in the free form and could be removed by washing with water alone. The researchers observed that the washing process for different size fractions of the soil followed second order kinetics. The rate constants for the various size fractions did not vary significantly, which they concluded indicated that the washing time was not dependent on particle size for the extractant used. Removal efficiency was observed to be related to particle size, with the -40 to +70 mesh size fraction giving the maximum, followed by the -70 to +200 mesh size fraction.

Hessling *et al.* [12] investigated soil washing techniques for remediation of lead-contaminated soils at battery recycling facilities. Three wash solutions were studied for their efficacy in removing lead from these soils: (1) tap water alone at pH 7, (2) tap water plus anionic surfactant (0.5% solution), and (3) tap water plus 3:1 molar ratio of EDTA to toxic metals at pH 7-8. Tap water alone did not appreciably dissolve the lead in the soil. Surfactants and chelating agents such as EDTA offer good potential as soil washing additives for enhancing the removal of lead from soils. There was no apparent trend in soil or contaminant behavior related to Pb contamination (predominant Pb species), type of predominant clay in the soil, or particle size distribution. The authors concluded that the applicability of soil washing to soils at these types of sites must be determined on a case-by-case basis.

Elliott *et al.* [13] performed a series of batch experiments to evaluate extractive decontamination of Pb-polluted soil using EDTA. Their study studied the effect of EDTA concentration, solution pH, and electrolyte addition on Pb solubilization from a battery reclamation site soil containing 21% Pb. The heavy metals concentrations in the soil were determined to be: 211,300 mg Pb/kg (dry weight); 66,900 mg Fe/kg; 1383 mg Cu/kg; 332 mg Cd/kg; and 655 mg Zn/kg. A nine-step chemical fractionation scheme was used to speciate the soil Pb and Fe. Results from their study indicated that increasing EDTA concentration resulted in greater Pb release. Recovery of Pb was generally greatest under acidic conditions and decreased modestly as the pH became more alkaline. Even in the absence of EDTA, a substantial increase in Pb recovery was observed below pH 5. As the pH became more alkaline, the ability of EDTA to enhance Pb solubility decreased, because hydrolysis was favored over complexation by EDTA. The researchers observed that EDTA can extract virtually all of the non-detrital Pb if at least a stoichiometric amount of EDTA is employed. When increased above the stoichiometric requirement, the EDTA was capable of effecting even greater Pb recoveries. However, the Pb released with each incremental increase in EDTA concentration diminished as complete recovery was approached. The researchers also investigated the release of Fe from the soil by EDTA. The Fe release increased markedly with decreasing pH. Despite the fact that the total iron was nearly 1.2 times the amount of lead in the soil, only 12% of the Fe was dissolved at pH 6 using 0.04 M EDTA, compared with nearly 86% dissolution of the Pb [14]. Little of the Fe was brought into solution during the relatively short contact time of the experiments (5-hrs). The iron oxides retained less than 1% of the total soil Pb [14].

Elliott *et al.* [13] observed that Pb recovery increased by nearly 10% in the presence of $LiClO_4$, $NaClO_4$, and NH_4ClO_4. They attributed this increase to an enhanced displacement of Pb^{2+} ions by the univalent cations and the greater solubility of Pb-containing phases with increased ionic strength. Below pH 6, calcium and magnesium salts also enhanced Pb recovery. Above pH 6, however, Pb recovery decreased due to a competition between Ca or Mg and Pb for the EDTA coordination sites. Their research [13,14] did not provide any evidence that the suspension pH must be raised to at least 12 to prevent Fe interference in soil washing with EDTA to effectively remove Pb.

3. GOALS AND OBJECTIVES

The primary goals of this project were to determine and compare the performance of several chelating agents for their ability to extract lead from contaminated soil. The objective of this study was to determine the removal efficiencies of the various chelating agents (and water alone) as functions of solution pH and chelating agent concentration.

4. EXPERIMENTAL PROCEDURE

The experiments were performed using a batch shaker technique in an effort to investigate the extraction removal efficiency of lead from contaminated soil. The procedures followed in this study are summarized below.

4.1 Batch shaker test

Soil from a previous study was prepped by grinding with a ceramic mortar and pestle to pass through an 850-μm sieve (ASTM mesh # 20). Characteristics of this soil are summarized in Table 3.

Four batches (~200 gm each) of uncontaminated soil were then spiked or artificially contaminated by soaking the soil with a solution of lead nitrate and left to air dry. Each batch of soil was "contaminated" with different lead nitrate concentrations (nominally 500, 1000, 5000, and 10,000 mg/kg soil). The contaminated soils were stored in glass jars with screw-on covers for approximately three months prior to conducting the screening studies to age the soil. The nominal experimental conditions are listed in Table 4.

Soil was weighed out using a top loading balance in 5-gm portions and placed in plastic shaker containers which had lids. To these containers, 45-mL of one of the following solutions was added:

- 0.01 M, 0.05 M or 0.1 M ethylenediaminetetraacetic acid (disodium salt) - EDTA-Na_2
- 0.01 M, 0.05 M or 0.1 M nitrilotriacetic acid (trisodium salt) - NTA-Na_3
- 0.01 M, 0.05 M or 0.1 M hydrochloric acid - HCl
- Deionized water

These conditions were employed to create a matrix of samples; each combination of lead-spiked soil and type and concentration of chelating agent were tested. Blanks of uncontaminated soil were also combined with each concentration of the various solutions of chelating agents and deionized water to detect any interferences in the analyses. Replicate experiments were performed in 25% of the experiments conducted.

Table 3. Soil Analysis of the Uncontaminated Soil Employed in Study

Parameter	Soil Analysis
pH	7.75
Electrical Conductivity, (mmhos/cm)	5.46
Cations (meq/L):	
Na^+	20.26
K^+	0.62
Ca^{2+}	37.88
Mg^{2+}	24.32
Anions (mg/l):	
NO_3-N (1:5 H_2O) extraction	2.7
PO_4 - P	10.3
Total Kjeldahl Nitrogen (TKN)	471
Fe (DTPA)	18.69
Sodium Absorption Ratio (SAR)	3.63
Calculated ESP	3.9
% H_2O at saturation	75.63
Calculated Exchange Capacity (CEC), (meq/100 g)	13.27
% Organic Matter (OM)	2.02
Texture	Clay Loam
Size Fractions (wt %):	
Sand	29.2
Silt	35.0
Clay	35.8

Table 4. Nominal Experimental Conditions

Parameter	Experimental Range
Heavy Metal (Lead) Concentration, (mg/kg soil)	500 to 10,000
Extractant Type and Concentration, (M)	
EDTA	0.01 to 0.10
NTA	0.01 to 0.10
HCl	0.01 to 0.10
Solution pH	4 to 12
Operation Temperature, °C	23.5
Preliminary Contact Time, (hr.)	0.5 to 6.0

The soil samples to which the extractant had been added were shaken for a period of nominally three hours at a low setting on an Eberbach shaker table. This time requirement was determined from preliminary experiments by monitoring the residual concentration as a function of time on the batch shaker table to be sufficient for chemical equilibrium to be achieved. Following this agitation, the samples were centrifuged in plastic Nalgene centrifuge tubes equipped with snap-on caps, filtered using No. 42 Whatman filter paper, and stored in glass vials maintained at $pH < 2$ (prepped using HNO_3) prior to AAS analysis. The AAS was calibrated using AAS lead standards. The analyses were performed in accordance with the procedures described in Standard Methods [24].

Data collected during the experiments included the following: operating temperature (~23.5°C), extractant type and concentration, lead concentration on the soil (before and after treatment), lead concentration in the extractant after treatment, pH of solution before and after treatment, and batch shaking time.

5. RESULTS AND DISCUSSION

5.1 Preliminary experiments

Prior to performing the more comprehensive experiments involving the extraction of lead from the contaminated soil using EDTA and NTA at various extractant concentrations (ranging from 0.01 M to

0.1 M) and at various pH conditions, preliminary experiments were conducted using soil which had been contaminated with lead at 100 mg/kg soil. The purpose of these preliminary experiments was to determine the contact time required to reach pseudo equilibrium conditions for the particular contaminated soil. The extractants studied included EDTA, NTA, and HCl (all at a concentration of 0.2 M). The pH quickly reached a constant value for each system, with pH corresponding to 4.2, 10.9 and 5.75 for the extractants of HCl, NTA, and EDTA, respectively. The approximate times for the HCl, NTA, and EDTA systems to reach pseudo equilibrium conditions were 0.5, 3 and 2 hrs., respectively. To facilitate the treatment with chelating agents, a contact time of 3.0 hours was selected to provide uniform treatment conditions. Based upon the very low solubilization of lead using hydrochloric acid as the extractant, this system was not studied further.

5.2 Batch extraction experiments

A series of batch extractions were conducted using water, NTA, and EDTA as the extractants. The lead concentrations on the soil were 500, 1,000, 5,000 and 10,000 mg/kg soil. The concentrations of EDTA and NTA ranged from 0.01 to 0.1 moles/L. The extractions were performed over the nominal pH range of 4 to 12.

Figure 1 presents the results for the extraction of lead using water alone for the 10,000 mg Pb/kg soil. The results for the other lead concentrations (ranging from 500 to 1,000 mg/kg soil) behaved in a similar fashion. The concentration of lead solubilized by the water extraction was pH sensitive; a significant increase in lead concentration in solution is observed for pH < 4. Generally, the amount of lead extracted from the soil by water alone is rather minimal; the amount extracted was generally less than 40 mg/kg soil. The maximum concentration extracted using water alone represented only 7.3% of the lead being solubilized.

Figure 2 presents the results for the extraction of lead from the 10,000 mg/kg soil using 0.1 M NTA. Again the results for the other lead concentrations behaved in a similar manner. Whereas, the amount of lead extracted from the soil using water as the extractant was as high as 730 mg/kg soil, the lead extracted using NTA as the extractant was significantly higher, with the amounts of lead extracted being as high as 1910 mg/kg soil at the lower pH range. The lead concentration in solution is observed to be much more pH sensitive than the water system. Fairly small amounts of lead were extracted (indicating little lead solubilization) over the pH range of 6 to 9. Significant increases in lead concentration are observed on both sides of this pH range. The maximum amounts of lead solubilized over the pH range of 4 to 12 was 19.1%.

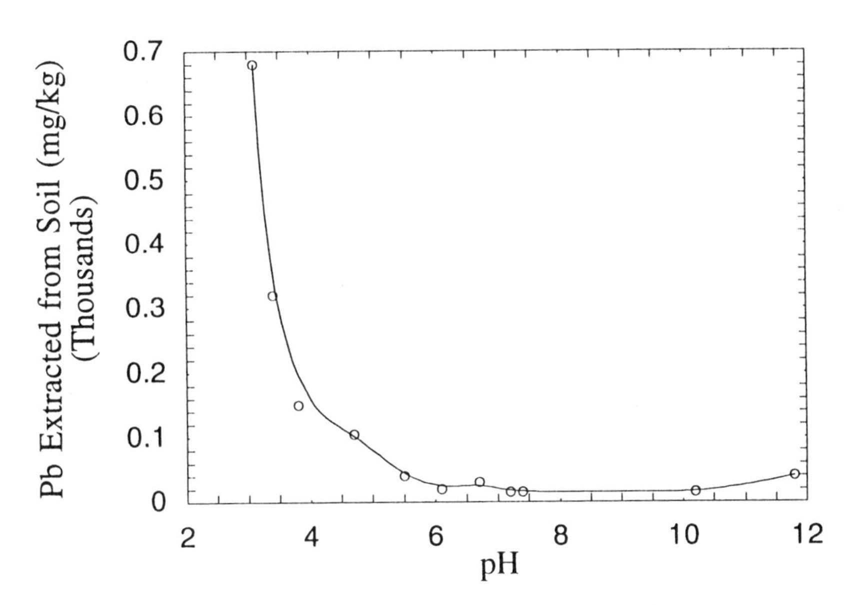

FIGURE 1. ***Lead Extracted from Contaminated Soil using Water as the Extractant. Initial Pb concentration = 10,000 mg/kg soil.***

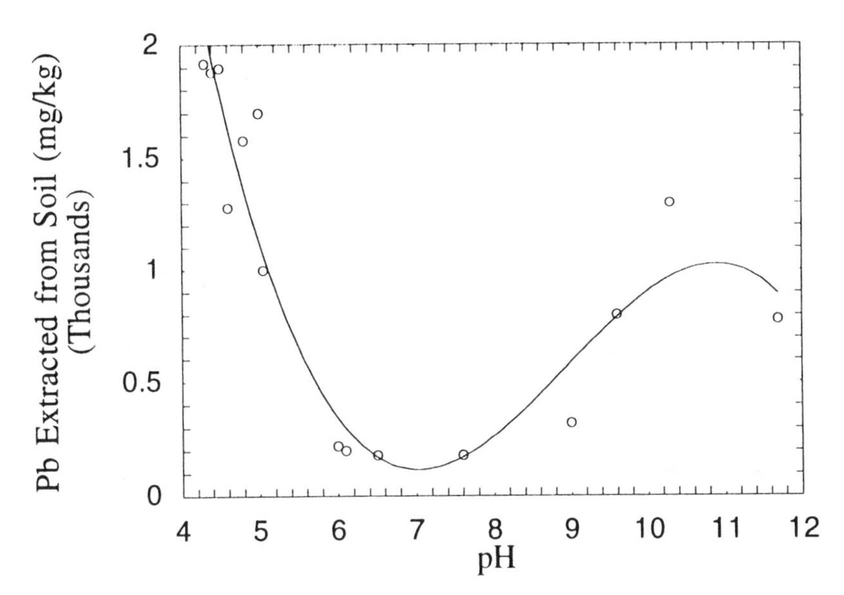

FIGURE 2. ***Lead Extracted from Contaminated Soil using NTA as the Extractant. Initial Pb concentration = 10,000 mg/kg soil; NTA concentration = 0.1 mole/L.***

Figure 3 presents the analogous set of data for the extraction of lead from the 10,000 mg/kg soil using 0.1 M EDTA. Once again, the results for the other lead concentrations behaved in a similar manner. In striking contrast to the other two extraction systems (employing water and NTA) which exhibited a significant pH dependency, the EDTA extraction system was strongly pH insensitive. The amount of lead extracted from the soil ranged from 5730 to 6420 mg/kg soil over the pH range of 4.9 to 11.3, corresponding to lead removals ranging from 57.3 to 64.2%.

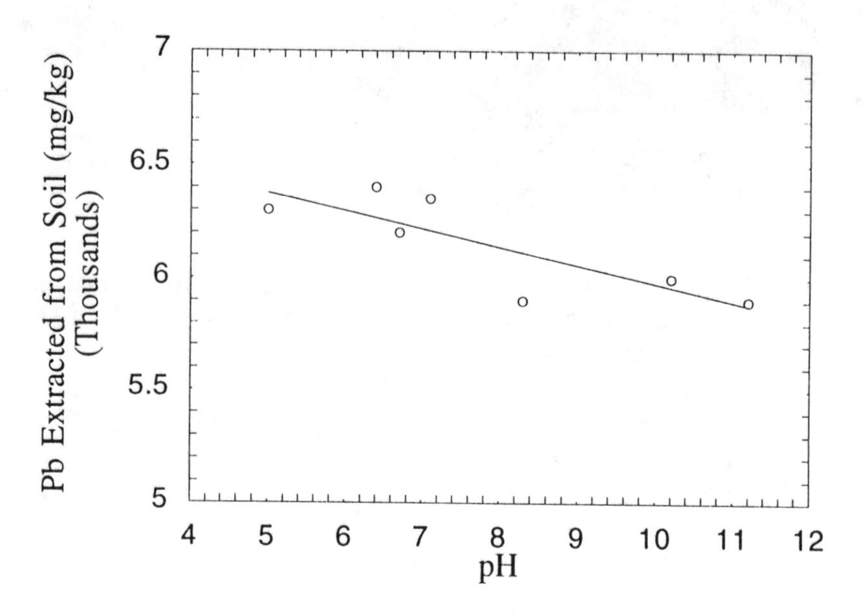

FIGURE 3. Lead Extracted from Contaminated Soil using EDTA as the Extractant. Initial Pb concentration = 10,000 mg/kg soil; EDTA concentration = 0.1 mole/L.

Figure 4 compares these three extractants in terms of the amount of lead extracted from the soil (expressed as a percentage of the nominal initial lead concentration of 10,000 mg/kg applied to the soil). The NTA extraction shows the largest pH dependency, while the EDTA extractant is strongly pH insensitive. For the EDTA extraction system, the removal of lead remained relatively constant at approximately 61% over the entire pH range.

Figure 5 shows the results of the lead extraction removal efficiency plotted as a function of pH for various initial lead concentrations on the soil. For these experiments, the EDTA

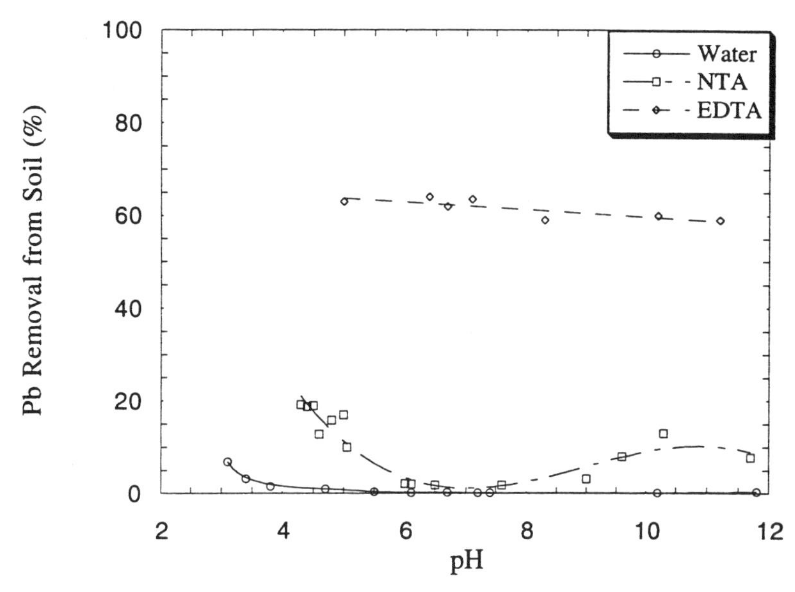

FIGURE 4. ***Lead Removal Efficiency from Contaminated Soil using Different Extractants Versus pH. Initial Pb concentration = 10,000 mg/kg soil; concentration of EDTA and NTA = 0.1 mole/L.***

concentration was 0.10 M. The results indicate that the removal efficiency of lead using EDTA as the extractant was fairly insensitive to the initial lead concentration applied to the soil over the range of 500 to 10,000 mg/kg soil. The lead removal efficiency ranged from a low of 56.19% to a high of 68.67%.

The effect of varying the EDTA concentration for removal of lead from contaminated soil is shown in Figure 6. The pH for this EDTA extractant ranged from 5.9 to 6.9. The removal efficiency is shown for four separate lead concentrations on the soil, ranging from 500 to 10,000 mg/kg soil. The removal efficiency was greatest for the case of a lead contamination level of 5000 mg/kg, although all the removal efficiencies were fairly constant, being in the range of 54 to 68%.

Figure 7 shows the effect of varying the EDTA concentration for treating the four lead contamination levels (ranging from 500 to 10,000 mg/kg soil). The EDTA concentration ranged from 0.01 to 0.1 moles/L and the solution pH ranged from 5.9 to 6.9. As shown in Figure 8, nearly all the EDTA concentration studied resulted in removal efficiencies in the 60% range.

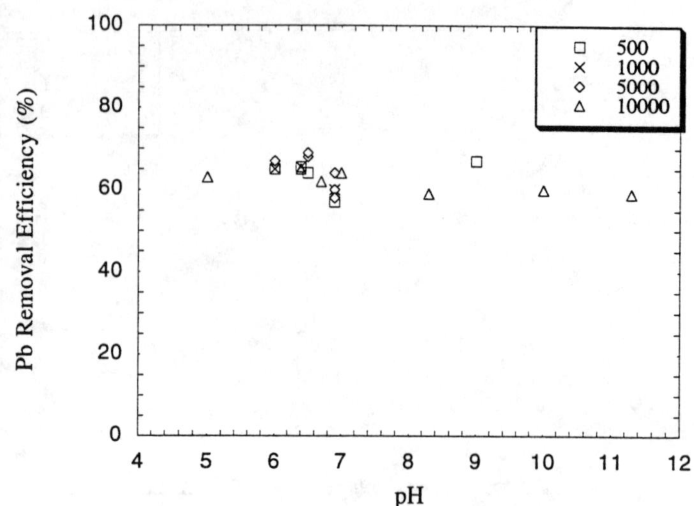

FIGURE 5. Lead Removal Efficiency from Contaminated Soil Versus pH for Different Initial Pb Concentrations using EDTA as the Extractant. Initial Pb concentration = 500, 1000, 5000, or 10,000 mg/kg soil; EDTA concentration = 0.10 mole/L.

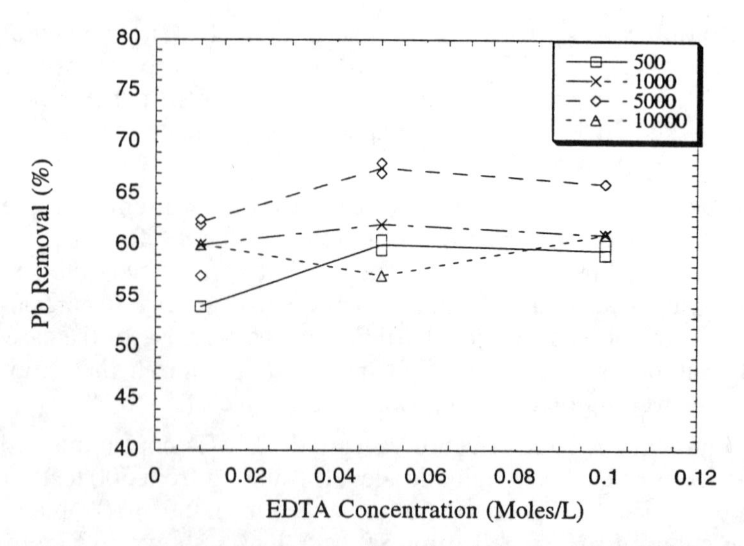

FIGURE 6. Comparison of the Lead Removal Efficiency for EDTA Extraction of Lead as a Function of EDTA Concentration. Initial Pb concentrations = 500, 1000, 5000 and 10,000 mg/kg soil.

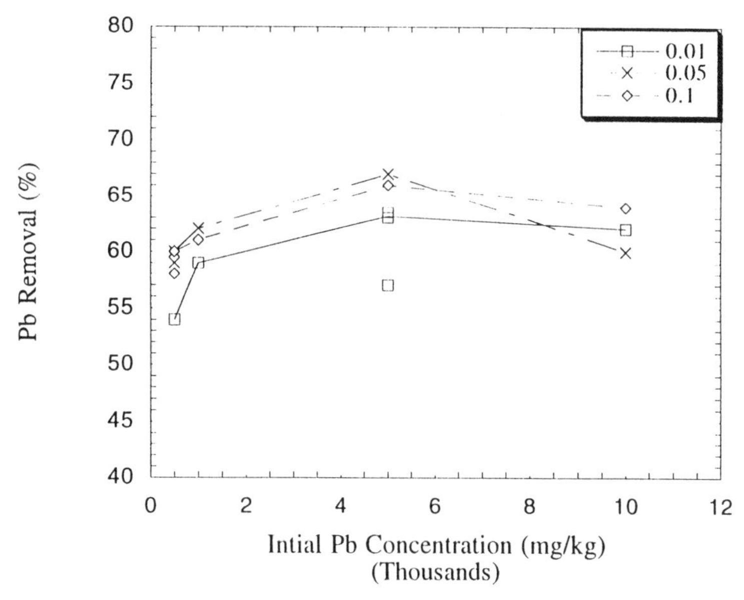

FIGURE 7. Comparison of the Lead Removal Efficiency for EDTA Extraction of Lead as a Function of the Initial Lead Concentration. EDTA concentrations = 0.01, 0.05 and 0.1 mole/L.

6. SUMMARY AND CONCLUSIONS

The results from other research studies [13,14] indicate that removal of lead from contaminated soil can exceed 80%. This study, which involved extraction of lead from a soil with a very high clay and silt content (~71%), indicated that lead removal exceeds 60%, using EDTA concentrations in the range of 0.01 to 0.1 M for lead concentrations on the soil ranging from 500 to 10,000 mg/kg. Extraction of lead using water alone removed a maximum of 7.3% for pH ~ 4. The removal of lead using water and NTA as extractants were both pH dependent, whereas the removal of lead using EDTA was strongly pH insensitive. The removal of lead ranged from 57 to 69% over the entire range of initial lead concentrations in the soil for the pH range of 4.9 to 11.3. The initial lead content had very little effect on the removal efficiency of lead for the EDTA system (for initial lead concentrations in the range of 500 to 10,000 mg/kg soil). The applied EDTA concentration over the range of 0.01 to 0.1 moles/L also had little effect on the removal efficiency of lead from the soil.

Extraction of lead with EDTA was rapid, reaching equilibrium within a contact time of 1.0 hr. Extraction of lead with NTA was slower; a contact time of approximately 3.0 hr was required to reach equilibrium.

The order of lead removal efficiency for the various extractants was in the order: EDTA >> NTA >> water. The maximum lead removals observed for this high clay and silt soil were 68.7, 19.1, and 7.3%, respectively, for the cases of EDTA, NTA, and water used as the extractant.

Acknowledgments: The authors acknowledge the support of Argonne National Laboratory's Energy Systems Division for funding of this research project.

REFERENCES

1. Elliott, H.A. and R.W. Peters, "Decontamination of Metal-Polluted Soils Using Chelating Agents," *Removal of Heavy Metals from Groundwaters*, CRC Press, Inc., (Boca Raton, FL in press).

2. Ku, Y. and R.W. Peters, "The Effect of Weak Chelating Agents on the Removal of Heavy Metals by Precipitation Processes," *Environ. Prog.,* 5:147-153 (1986).

3. Ku, Y. and R.W. Peters, "The Effect of Complexing Agents on the Precipitation and Removal of Copper and Nickel from Solution," *Particulate Sci. Technol.,* 6:441-466 (1988).

4. Peters, R.W. and Y. Ku, "The Effect of Citrate, a Weak Complexing Agent, on the Removal of Heavy Metals by Sulfide Precipitation," in *Metals Speciation, Separation, and Recovery*, J.W. Patterson and R. Passino, Eds., pp. 147-169. Lewis Publishers, (Chelsea, MI 1987).

5. Peters, R.W. and Y. Ku, "The Effect of Tartrate, a Weak Complexing Agent, on the Removal of Heavy Metals by Sulfide and Hydroxide Precipitation," *Particulate Sci. Technol.,* 6:421-439 (1988).

6. Oyler, J.A., "Remediation of Metals-Contaminated Site Near a Smelter Using Sludge/Fly Ash Amendments," *Proc. 44th Purdue Indus. Waste Conf.,* 44:75-82 (1989).

7. Hsieh, H.N., D. Raghu, J.W. Liskowitz and J. Grow, "Soil Washing Techniques for Removal of Chromium Contaminants from Soil," *Proc. 21st Mid-Atlantic Indus. Waste Conf.*, 21:651-660 (1989).

8. Hsieh, H.-N., D. Raghu and J. Liskowitz, "An Evaluation of the Extraction of Chromium from Contaminated Soils by Soil Washing," *Proc. 22nd Mid-Atlantic Indus. Waste Conf.*, 22:459-469 (1990).

9. Trnovsky, M., J.P. Oxer, R.J. Rudy, G.L. Weinstein and B. Hartsfield, "Site Remediation of Heavy Metals Contaminated Soils and Groundwater at a Former Battery Reclamation Site in Florida," in *6th Internat. Conf. on Heavy Metals in the Environ., Vol. I,* S.E. Lindberg and T.C. Hutchinson, Eds., pp. 88-90. (New Orleans, LA 1987).

10. Trnovsky, M., J.P. Oxer, R.J. Rudy, M.J. Hanchak and B. Hartsfield, "Site Remediation of Heavy Metals Contaminated Soils and Groundwater at a Former Battery Reclamation Site in Florida," in *Hazardous Waste, Detection, Control, Treatment*, R. Abbou, Ed., pp. 1581-1590. Elsevier Science Publishers, B.V., (Amsterdam, The Netherlands 1988).

11. Trnovsky, M., P.M. Yaniga and R.S. McIntosh, "Lead Removal - The Main Remedial Design Parameter for Two Florida Superfund Sites," in *7th Internat. Conf. on Heavy Metals in the Environ., Vol. II*, J.-P.Vernet, Ed., pp. 233-236. (Geneva, Switzerland 1989).

12. Hessling, J.L., M.P. Esposito, R.P. Traver and R.H. Snow, "Results of Bench-Scale Research Efforts to Wash Contaminated Soils at Battery-Recycling Facilities," in *Metals Speciation, Separation, and Recovery, Vol. II*, J.W. Patterson and R. Passino, Eds., pp. 497-514. Lewis Publishers, Inc., (Chelsea, MI, 1989).

13. Elliott, H.A., G.A. Brown, G.A. Shields and J.H. Lynn, "Restoration of Pb-Polluted Soils by EDTA Extraction," in *7th Internat. Conf. on Heavy Metals in the Environ., Vol. II*, J.-P. Vernet, Ed., pp. 64-67. (Geneva, Switzerland 1989).

14. Elliott, H.A., J.H. Linn and G.A. Shields, "Role of Fe in Extractive Decontamination of Pb-Polluted Soils," *Haz. Waste Haz. Mater.*, 6:223-229 (1989).

15. Winslow, G., "First Third Land Disposal Restrictions: Consequences to Generators," *Haz. Waste Manage. Mag.*, Nov.-Dec., (1988).

16. Raghavan, R., E. Coles and D. Dietz, *Cleaning Excavated Soil Using Extraction Agents: A State-of-the-Art Review*, EPA/600/2-89/034, U.S. Environ. Protection Agency, Risk Reduction Engineering Laboratory, (Cincinnati, OH 1989).

17. EPA, *Handbook for Remedial Action at Waste Disposal Sites*, EPA-625/6-82-006, U.S. Environ. Protection Agency, Municipal Environmental Research Laboratory, (Cincinnati, OH 1982).

18. Werther, J., R. Hilligardt and H. Kroning, "Sand from Dredge Sludge-Development of Processes for the Mechanical Treatment of Dredged Material," in Contaminated Soil, J.W. Assink and W.J. Van Der Brink, Eds., pp. 887. Martinus Nijhoff, (Dordrecht, The Netherlands 1986).

19. Aplan, F., "Flotation," in *Encyclopedia of Chemical Technology*, *Vol. 10, 3rd Edition,* M. Grayson and D. Eckroth, Eds., pp. 523-547. John Wiley & Sons, (New York, NY 1980).

20. Pickering, W.F., "Metal Ion Speciation - Soils and Sediments - (A Review)," *Ore Geol. Rev.,* 1:83-146 (1986).

21. Wagner, J.-F., "Heavy Metal Transfer and Retention Processes in Clay Rocks," in *7th Internat. Conf. on Heavy Metals in the Environment, Vol. I*, J.P. Vernet, Ed., pp. 292-295. (Geneva, Switzerland 1989).

22. Slavek, J. and W.F. Pickering, "Extraction of Metal Ions Sorbed on Hydrous Oxides of Iron (III)," *Water, Air, Soil Pollut.,* 28:151-162 (1986).

23. Ellis, W.D., T.R. Fogg and A.N. Tafuri, "Treatment of Soils Contaminated with Heavy Metals," in *Land Disposal, Remedial Action, Incineration and Treatment of Hazardous Waste, 12th Ann. Res. Sympos.*, EPA 600/9-86/022, pp. 201-207. (Cincinnati, OH 1986).

24. American Public Health Association, *Standard Methods for the Examination of Water and Wastewater,* 15th Ed., American Public Health Association, (Washington, D.C. 1981).

11

EFFECT OF MINE WASTE ON ELEMENT SPECIATION IN HEADWATER STREAMS

Michael C. Amacher[1], Ray W. Brown[1], Roy C. Sidle[1] and Janice Kotuby-Amacher[2]
[1]Intermountain Research Station
Forest Service
U.S. Department of Agriculture
Logan, UT 84321
[2]Department of Plants
Soils and Biometeorology
Utah State University
Logan, UT 84322

1. INTRODUCTION

National forest lands provide numerous valuable resources such as timber and minerals, but one of the most valuable and often overlooked resources is water. National forest lands as watersheds are especially vital during drought periods such as that now occurring throughout much of the western United States.

The Forest Service, U.S. Department of Agriculture, is committed to a policy of multiple resource use on national forest lands. This presents problems to resource managers because they must often accommodate conflicting and disparate uses of the national forest. Some uses of forest resources, such as mining and exploration, are potentially destructive and may limit the use of other resources. In addition, water quality on national forest lands may be severely degraded. To complicate matters further, the cumulative impacts of multiple uses (e.g., mining and associated activities, grazing and recreation) can be more severe than the impact of any single disturbance [1].

Our overall research goal is to assess the cumulative effects of multiple land uses on national forests and to develop and evaluate remedial treatments where previous uses have degraded the impacted area. As part of this overall objective, we are evaluating the effect of mining activities and mine waste on water quality in headwater streams originating in affected areas. In this paper we report the effect of mine waste on water quality in two headwater streams.

The first study area is at Maybe Canyon near Soda Springs, Idaho, on the Caribou National Forest in the southeastern Idaho phosphate mining district. A cross-valley fill of overburden material was constructed across the channel of Maybe Canyon Creek, a small headwater stream flowing into the Blackfoot River. The second study area is at Daisy and Fisher Creeks, headwater tributaries of the Stillwater and Clark's Fork of the Yellowstone Rivers, respectively, near Cooke City, Montana, on the Custer and Gallatin National Forests. Acid mine drainage from the abandoned McLaren and Glengary gold and copper mines flow into Daisy and Fisher Creeks, respectively.

Past research at Maybe Canyon has focused on the hydrology of the cross-valley fill [2], whereas past research at the McLaren and Glengary Mines has focused on developing and evaluating revegetation methods for the mine spoils in the area [3,4,5]. Little attention has been focused on potential impacts on water quality despite the obvious visual stream degradation that has occurred on Daisy and Fisher Creeks. For this phase of research at the two study sites the specific objectives are to determine:

1. The effect of mine waste on the chemical composition of the headwater streams.
2. The equilibrium chemical speciation of the headwater streams.
3. Possible solid phase controls of water composition at the study sites.
4. The relative importance of hydrologic and geochemical controls on the chemical composition of the headwater streams.
5. The solid phase associations of trace elements in sediments at the study sites.

Bencala and McKnight [6] and Bencala *et al.* [7] studied concentration changes in small streams as a result of both hydrologic and instream geochemical processes and showed that the spatial and temporal variations of concentrations resulting from hydrologic processes may be as great as variations resulting from geochemical processes. Their findings have important implications for designing stream surveys to assess the impact of land use practices on stream water quality. Sampling must be done on a sufficient scale to assess variability due to hydrologic as well as geochemical processes. The approach that we are using in the initial assessment of Maybe Canyon, Daisy, and Fisher Creeks is to use the concentration normalization techniques of Bencala and McKnight [6] and Bencala *et al.* [7] to estimate the effects of dilution by incoming waters on instream element concentrations and to use MINTEQA2, an equilibrium chemical speciation program, to assess potential geochemical controls on stream water composition.

2. MATERIALS AND METHODS

2.1 Study site descriptions

Maybe Canyon lies at an elevation of about 2134 m northeast of Soda Springs, Idaho in the Caribou National Forest. Economic deposits of phosphatic shale occurring along ridges are mined by surface strip mine methods, and the overburden material consisting primarily of chert (silica with carbonates) and mudstone [8] has been deposited in Maybe Canyon as a cross-valley fill. Included with the overburden is phosphatic shale, which had a phosphorus content too low for economic recovery.

A small headwater tributary of the Blackfoot River flows through the canyon, and the fill material was placed across the stream channel. The fill material acts somewhat like a porous dam in that the stream flows through the fill. The channel through which the stream flows was covered primarily with chert, and the remainder of the fill material was placed on top until the fill covered an area of about 1216 m long by 274 m wide by 152 m high. The areal extent of the fill is about 33 ha. The north-facing downstream side of the fill "dam" is capped with chert on the west side and mudstone on the east side.

The stream above the fill area is seasonal and has water in the channel only during snowmelt and shortly thereafter in May and June. The stream below the fill area contains water throughout the year and is fed by groundwater springs and seeps. A spring under the fill was piped to a storage tank downstream during construction of the fill. The storage tank is the only means of sampling the piped spring. A small pond is located about 1900 m downstream. A flume was installed about 70 m downstream from where the stream exits the fill to provide discharge measurements.

Daisy and Fisher Creeks, two headwater tributaries of the Stillwater and Clark's Fork of the Yellowstone Rivers, respectively, lie at an elevation of about 2900 to 3000 m, a few miles north of Cooke City, Montana, on the Gallatin and Custer National Forests. Daisy Creek begins in the saddle below Daisy Pass between Henderson Mountain and Crown Butte, whereas Fisher Creek begins in the saddle just below Lulu Pass between Fisher and Scotch Bonnet Mountains. Acid mine drainage from two abandoned gold and copper mines (McLaren and Glengary) on opposite sides of Fisher Mountain flows into the two streams.

The McLaren and Glengary Mines consist of old open pit areas with exposed minespoil in the pits and on surrounding hill slopes. Acid mine drainage from the McLaren Mine runs downslope in several channels to intersect the headwaters of Daisy Creek. Acid mine drainage from an old adit on the Glengary Mine comprises a significant fraction of the base flow of Fisher Creek, which begins further upslope at Lulu Pass where the open pit area is located.

The geology of the Fisher Mountain area is described in the Environmental Baseline Study Plan for the New World Project (Crown Butte Mines, Inc. and Noranda Minerals Corp.[1]). Sedimentary rock units in the Fisher Mountain area include the Cambrian units (from oldest to youngest): Wolsey shale, Meagher limestone, Park shale, and the Gallatin unit (Pilgrim limestone).

Numerous Tertiary igneous intrusions in the area have altered the sedimentary rocks in the contact zone. Intrusive phases include andesite, dacite, rhyolite, quartz-latite , and monzonite. Gold-copper-skarns occur primarily in the Cambrian Meagher limestone and in the upper part of the Cambrian Wolsey shale. Sulfide minerals (pyrite-chalcopyrite) are associated with the skarns. Weathering of the sulfide ore deposits produces the acid mine drainage that contaminates Daisy and Fisher Creeks. Annual precipitation in the area is in excess of 150 cm with most of that falling as snow in the winter months. Snowmelt begins in May and lasts through July.

2.2 Water and sediment sampling

At Maybe Canyon water samples were collected from snowmelt runoff above the cross-valley fill on 5/12/89 and 5/25/90. A composite

---

[1]The use of trade or firm names in this paper is for reader information and does not imply endorsement by the U.S. Department of Agriculture of any product or service.

snowpack sample was also collected on 5/25/90. The piped spring was sampled on 7/6/90 at the storage tank about 2000 m downstream from the fill. The stream below the fill was sampled on 5/12/89, 8/28/89, 5/25/90, 7/6/90, and 9/4/90 at various distances downstream. Downstream distances were determined using a Walktax meter. A 2-L grab sample of unfiltered water was collected at each location in acid-washed polypropylene bottles rinsed three times with sample water prior to sample collection. Fresh sediment fines were collected to a depth of up to 15 cm on either 5/25/90 or 7/6/90. Composite samples of phosphatic shale, mudstone, and chert were collected to a depth of 15 cm on 5/1/89. Composite samples of hillslope soil and alluvial soil along the stream were collected to a depth of 15 cm on 5/1/89 and 5/25/90, respectively.

At Daisy and Fisher Creeks, water and sediment samples were collected on 8/13/89, 7/18-19/90, 8/18-19/90, and 9/9-10/90. Water and sediment samples were collected at four drainage locations on the McLaren Mine, nine locations along Daisy Creek, and eleven locations along Fisher Creek including tributaries and Glengary Mine drainage. Not all stations were sampled on each date. A 2-L grab sample of unfiltered water was collected at each station in acid-washed polypropylene bottles rinsed three times with sample water before collection. Sediment samples consisted of fresh sediment fines collected to a depth of up to 15 cm. Distances between sample stations were measured from aerial photographs and represent straight line distances.

2.3 Field measurements

Discharge at Maybe Canyon was calculated from the water depth in the flume, which was measured on the sampling dates. Discharge at several locations along Daisy and Fisher Creeks was calculated from stream cross-section length and depth measurements and current velocity measured with a current meter. These measurements were made on 8/13/89 and 9/9-10/90. Temperature, conductivity, and pH at both study sites were measured *in situ* at each sample station using portable meters with automatic temperature compensation.

2.4 Sample preparation and analysis

Water samples were filtered through 0.4 μm polycarbonate membrane filters using 1-L polycarbonate filter units. Filters and units were preconditioned and rinsed by filtering and discarding 50 to 100 mL of each sample prior to filtering the remaining sample. A 500 mL subsample of each water sample was acidified with select grade concentrated HNO_3 to an acid content of 0.5%. The remaining samples were stored at < 4°C in the dark until analysis. Sediment samples were air dried, crushed, and passed through a 0.5-mm stainless steel sieve using a vibration sieve shaker.

The acidified water samples were analyzed for Na, K, Mg, Ca, Sr, Cr, Mo, Mn, Fe, Co, Ni, Cu, Zn, Cd, B, Al, Pb, P, As, S, and Se by inductively coupled plasma emission spectrometry (ICPES) [9]. Trace levels of Cr(VI), Mo, Mn, Fe, Co, Ni, Cu, Zn, Cd, Pb, As(III), and Se(IV) in the acidified water samples were determined by extraction into pyrrolidine dithiocarbamic acid in chloroform, evaporation of the chloroform, acid digestion of the residue, and analysis by ICPES [10]. The unacidified water samples were analyzed for the elements listed in Table 1. EPA recommended QA/QC procedures were followed where applicable.

Table 1. Water Analysis Methods

Analyte	Method	Reference
V	Gallic acid	[11]
Fe(II)	Bipyridine	[12]
Alkalinity	Titration	[9]
DOC	Mn(III) pyrophosphate	[13]
Si	Silicomolybdic acid	[14]
Nitrate	Cd reduction	[15]
P	Asorbic acid	[16]
F	SPADNS	[17]
Cl	Hg(II) thiocyanate	[17]

Sediment samples were sequentially extracted to determine elements associated with specific solid phases. For Maybe Canyon fill material, soil, and sediment samples, elements associated with the exchangeable, carbonate, chelate (EDTA) extractable, Mn oxide, amorphous Fe oxide, crystalline Fe oxide, organic matter (OM) +

sulfide, and residual phases were determined by sequential extraction of 0.5-g samples and analysis of the extracts by ICPES. Extractants and extraction conditions are listed in Table 2. The extraction sequence is similar to that used by Salomons and Förstner [18] and includes extractants and procedures for carbonates, Mn oxides, amorphous Fe oxides, and crystalline Fe oxides developed by Tessier *et al.* [19], Chao [20], Chao and Zhou [21], and Shuman [22], respectively. For Daisy and Fisher Creek sediments, elements associated with exchangeable, Mn oxide, amorphous Fe oxide, crystalline Fe oxide, and organic matter + sulfide + residual phases were determined by sequential extraction of 0.5-g samples and analysis of the extracts by ICPES. Extractants and extraction conditions are listed in Table 2.

Table 2. Sequential Extraction Conditions

Solid Phase	Extractant	Conditions
Exchangeable	0.1 M $Mg(NO_3)_2$	1:20 ratio, shake 2h.
Carbonates	pH 5, 1 M NaOAc	1:20 ratio, shake 5h.
Chelate extractable	0.05 M EDTA	1:20 ratio, shake 2h.
Mn oxides	pH 2, 0.1 M $NH_2OH:HCl$	1:50 ratio, shake 30min.
Amorphous Fe oxides	0.25 M NH_2OH HCl + 0.25 M HCl	1:50 ratio, shake 2h. at 50°C
Crystalline Fe oxides	pH 3, 0.2 M ammonium oxalate + 0.2 M oxalic acid + 0.1 M ascorbic acid	1:50 ratio, heat 30min. in boiling H_2O bath
Organic matter +sulfide	pH 2, 30% H_2O_2	1:50 ratio, heat 30min. in boiling H_2O bath
Residual	conc. HF + sat. H_3BO_3	1:50 ratio, add 10 mL conc. HF, shake overnight, add 2 g H_3BO_3, dil. to 25 mL

2.5 Data analysis

To determine the effect of inflow from tributaries, springs, and seeps on element concentrations in stream water, the concept of a conservative solute was used. Following the treatment of Bencala and McKnight [6] and Bencala *et al.* [7], conservation of flow in the absence of evaporation or other losses requires that when an inflow enters a stream that:

$$Q_A + Q_I = Q_B \tag{1}$$

where Q_A is stream discharge above the inflow, Q_I is inflow discharge, and Q_B is stream discharge below the inflow. Conservation of mass of a conservative solute requires that:

$$C_A^C Q_A + C_I^C Q_I = C_B^C Q_B \tag{2}$$

where C_A^C is the concentration of conservative solute above the inflow, C_I^C is concentration of conservative solute in the inflow, and C_B^C is concentration of conservative solute below the inflow. Substituting Equation (1) into (2) and rearranging leads to the definition of a normalized or relative concentration of a conservative solute:

$$\frac{Q_A}{Q_B} = \frac{C_B^C - C_I^C}{C_A^C - C_I^C} = C_N^C \tag{3}$$

where C_N^C is the normalized concentration of a conservative solute. The normalized or relative concentration of any solute is

$$C_N = \frac{C_B - C_I}{C_A - C_I} \tag{4}$$

The predicted downstream concentration of any solute, assuming a concentration change results only from mixing of inflow and instream waters, can be calculated from the normalized concentration of the conservative solute:

$$C_B^P = C_A C_N^C + C_I\left(1 - C_N^C\right) \tag{5}$$

Equations (3) and (4) are for a single inflow. In the case of multiple inflows it is best to normalize concentrations with respect to maximum and minimum inflow concentrations:

$$C_N = \frac{C - C^I_{min}}{C^I_{max} - C^I_{min}} \qquad (6)$$

where C^I_{max} is the maximum concentration observed in the inflows and C^I_{min} is the minimum concentration observed in the inflows.

To examine possible geochemical controls on element concentrations in Maybe Canyon and Daisy and Fisher Creeks, the equilibrium chemical speciation program MINTEQA2 [23] was used to analyze the stream data. Details on the use of this program can be found in the MINTEQA2 reference manual [23]. To determine equilibrium chemical speciation, solid phase precipitation was not imposed in each case. However, pH sweeps were run allowing certain likely solids to precipitate to compare predicted concentrations with data. In addition, the diffuse layer model (DLM) was used to predict potential Cu adsorption on hydrous Fe oxides in Daisy and Fisher Creeks. Details on the theory and use of this model can be found in Dzombak and Morel [24]. An extensive database on element adsorption reactions with hydrous Fe oxide is included in MINTEQA2 and makes adsorption modeling extremely convenient with the DLM.

3. RESULTS AND DISCUSSION

3.1 Maybe Canyon

An initial stream survey showed relatively high levels of several elements (e.g., Na, Mg, Ca, Sr, V, Mo, Ni, Zn, Cd, NO_3-N, P, SO_4-S, F, and Cl) downstream from the fill (Table 3). Because these results were unexpected, a more detailed survey of other water sources in the canyon was conducted, and the results are also presented in Table 3. Clearly snowpack, snowmelt runoff, and the piped spring are not the source of these elevated element concentrations, although the snowpack and piped spring contained similar concentrations of P and NO_3-N, respectively, compared to stream waters below the fill. By summer the stream above the fill dries up, but water exits the fill throughout the year even during the driest months. Springs and groundwater seeps occur underneath the fill. One of these springs was piped during construction of the fill dam, and the pipe leads to a storage tank about 2000 m downstream from the fill.

Table 3. Selected Water Analysis Data From Maybe Canyon

Sample Station	Date	Discharge cms	Temp deg C	Field pH	Na mg/L	K mg/L	Mg mg/L	Ca mg/L	Sr mg/L	V µg/L	Mo µg/L	Mn µg/L
Snowpack above cross-valley fill	5/25/90				0.18	2.45	0.09	0.47		1.5	<3	6
Runoff above cross-valley fill	5/12/89		2.3	8.10	2.13	3.76	13.23	48.83	0.10	1.0	1	2
Runoff above cross-valley fill	5/25/90				3.24	2.38	15.05	52.44	0.15	1.3	<3	1
Spring runoff at tank	7/6/90				4.09	1.81	15.10	75.69	0.22	0.4	<3	3
Below flume	5/12/89	0.0266	5.8	7.15	7.22	2.91	36.95	157.90	0.33	10.0	20	2
Below flume	8/28/89	0.00652	8.2	6.70	9.64	1.88	50.79	213.80	0.49	15.4	18	1
Below flume	5/25/90	0.0119	7.7	7.68	8.40	3.64	43.98	185.20	0.43	12.4	21	1
Below flume	7/6/90	0.00453	8.8	7.30	9.50	3.64	51.86	214.50	0.50	14.7	20	1
Below flume	9/4/90	0.00255	7.2	7.32	10.04	3.61	53.17	222.70	0.51	16.2	21	19

Sample Station	Date	Fe µg/L	Ni µg/L	Zn µg/L	Cd µg/L	HCO^- mg/L	Si mg/L	NO_3-N mg/L	P µg/L	SO_4-S mg/L	F mg/L	Cl mg/L
Snowpack above cross-valley fill	5/25/90	5	<6	5	2	8	0.02	0.17	119	0.22	0.04	0.56
Runoff above cross-valley fill	5/12/89	10	<3	5	<1	207	3.6	<0.5	70	2.68	<0.1	2.4
Runoff above cross-valley fill	5/25/90	3	<6	3	<1	226	4.50	0.13	40	3.04	0.12	0.82
Spring runoff at tank	7/6/90	5	<6	3	<1	269	7.09	2.14	5	10.55	0.06	2.80
Below flume	5/12/89	11	57	216	4	212	4.8	1.7	83	115.90	0.4	4.4
Below flume	8/28/89	38	73	276	3	263	5.1	2.2	97	159.20	0.3	11.0
Below flume	5/25/90	3	54	183	3	229	5.56	2.96	110	135.10	0.40	4.47
Below flume	7/6/90	3	83	316	4	266	7.53	3.54	101	161.40	0.43	5.02
Below flume	9/4/90	6	85	333	3	272	6.48	4.06	94	187.10	0.42	5.74

Analysis of this spring showed that it is not a source of the high element concentrations, so it seems reasonable to conclude that other springs and seeps under the fill also do not contribute to the contaminants observed downstream from the fill.

A charge distribution graph of water exiting the fill (Figure 1) shows that Mg, Ca, HCO_3, and SO_4 are the major charge components.

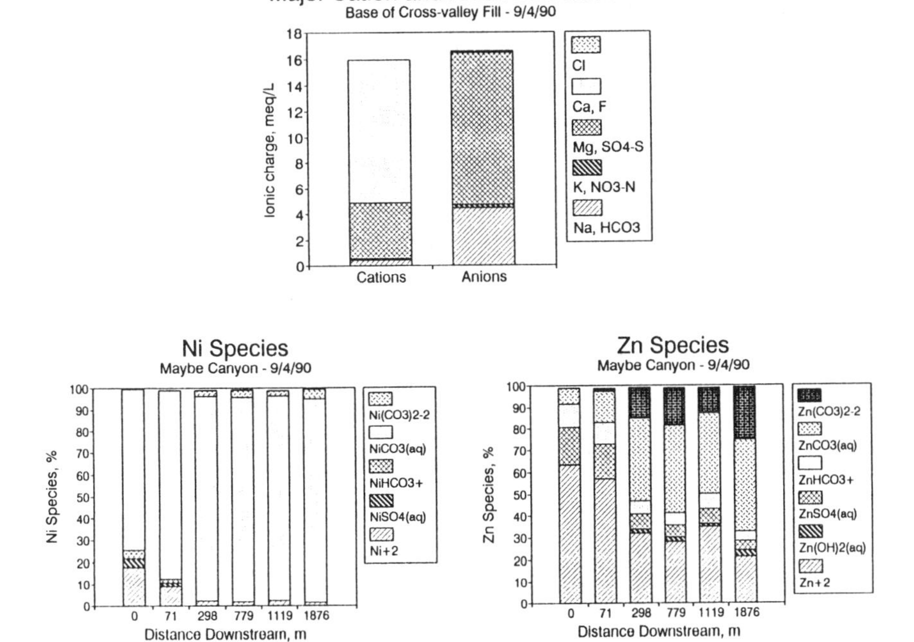

FIGURE 1. Charge Distribution (top) and Ni and Zn Speciation Versus Distance Downstream (bottom) at Maybe Canyon Creek below the Cross-valley Fill.

The source of Ca and SO_4 is unknown but is believed to be gypsum within the fill material. The source of HCO_3 is the carbonate rock formations underlying the drainage system.

MINTEQA2 calculations revealed significant carbonate and sulfate ion pair formation for Mg, Ca, Ni, and Zn. Nickel and Zn speciation is also shown in Figure 1. Speciation changes downstream were largely the result of increasing pH with distance downstream.

Saturation index calculations (Figure 2) showed that the runoff above the fill is supersaturated with respect to calcite. Supersaturation with respect to calcite in groundwater is a common occurrence [25], and because runoff in the spring probably includes shallow groundwater seeping into the stream, the observed supersaturation is not surprising. At the point where the stream exits the fill (0 m in Figure 2), stream water is near equilibrium with respect to both calcite and gypsum. The likely explanation is that stream water flowing through the fill dissolves gypsum, which increases the concentration of Ca in solution resulting in an even greater degree of supersaturation. Calcite precipitation then occurs to move the solution closer to equilibrium. As the stream exits the fill and proceeds downslope, outgassing of CO_2 occurs, pH and pCO_2 increase, and thus the calcite saturation index increases with distance downstream (Figure 2). The gypsum saturation index becomes more negative (Figure 2) as Ca and SO_4 concentrations decrease downstream because of dilution (discussed below).

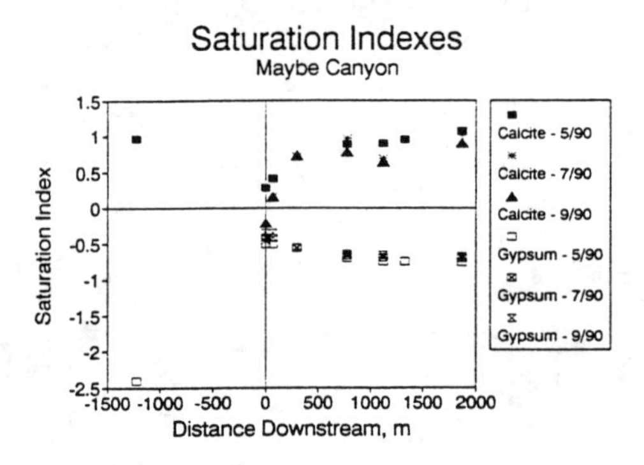

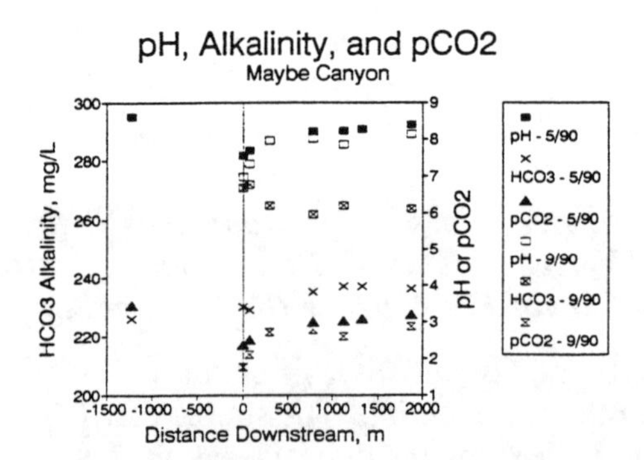

FIGURE 2. ***Saturation Indexes of Calcite and Gypsum (top) and pH, pCO_2 and HCO_3 (bottom) Versus Distance Downstream at Maybe Canyon Creek. 0 m refers to the downstream base of the cross-valley fill where stream water emerges.***

A kinetic constraint on calcite precipitation downstream seems likely, because Ca concentration decreases can be accounted for by dilution (see below) and are not as large as would be expected for calcite equilibrium to be restored. For example, MINTEQA2 predicted Ca concentrations at the last sample station just above the pond about 1876 m downstream from the cross-valley fill were 0.902, 1.03, and 1.12 mM for the 5/90, 7/90, and 9/90 samplings assuming calcite precipitates to maintain equilibrium. Actual concentrations were 1.93, 2.14, and 2.16 mM, more than double the predicted values obtained by assuming calcite precipitates to maintain equilibrium. Furthermore, there was no significant decrease in HCO_3 concentrations downstream. The cause of the observed supersaturation with respect to calcite is unknown, but is not because of organic matter, because dissolved organic matter concentrations were below the detection limit of about 1 mg/L for the Mn(III) pyrophosphate method.

Concentration profiles for the major and trace elements as the stream water moves downslope are shown in Figure 3 for the 5/90 samples. Concentration profiles for the 7/90 and 9/90 samplings are similar except that no sample could be collected above the fill because the stream had already dried up at that location. Large increases in the concentrations of the ions indicated in Figure 3 occur as stream water moves through the fill. Moving downstream from the fill, ion concentrations decrease, except for Mn, which increases as the concentrations of the other ions decrease. The increase in Mn concentration may be the result of adsorption by the other trace elements (e.g., Ni and Zn) onto Mn oxides in the sediments and subsequent release of displaced Mn originally adsorbed to the Mn oxide surfaces. Sequential extraction analysis revealed that the sediment samples contained from 0.14 to 0.38% total extractable Mn with more than 80% in the EDTA and $NH_2OH{\cdot}HCl$ extractable fractions. Most of the extractable Ni and Zn was in these fractions as well.

The source of the surprisingly high concentrations of NO_3 in stream water exiting the fill is unknown. Fertilizer sources are ruled out because this site is in the mountains well away and upslope from agricultural activities. Furthermore, most of the agriculture in the area consists of raising cattle. No high N demand crops are grown. One possibility is N mineralization and subsequent oxidation in the buried soils underneath the fill, or from the fill material itself. This hypothesis requires verification, however, and must await subsequent experimental testing.

The similarity of the concentration profiles of the major elements indicates that dilution is probably responsible for the observed decreases downstream from the fill. This conclusion is supported by the relative

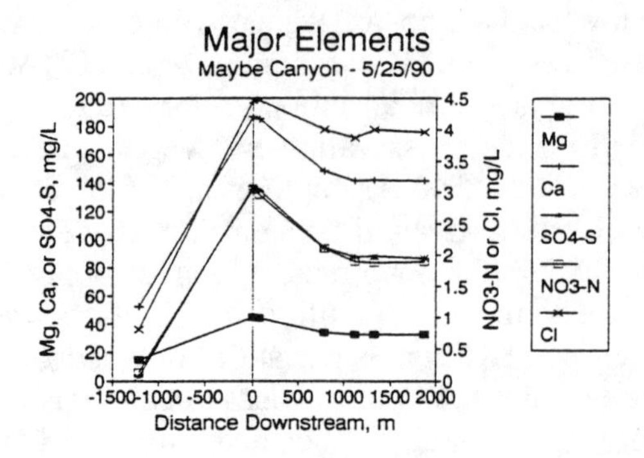

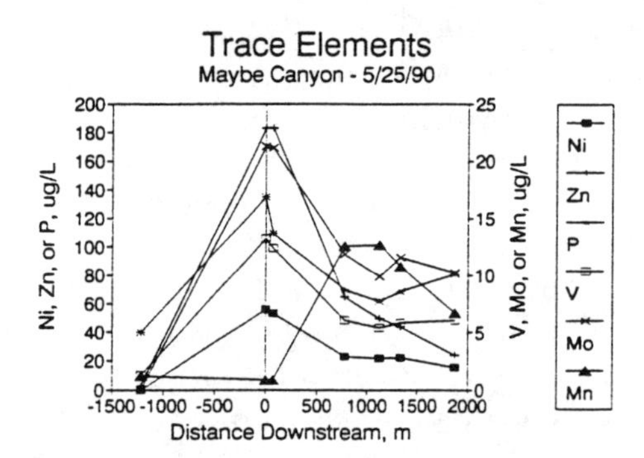

FIGURE 3. Major (top) and Trace (bottom) Element Concentrations Versus Distance Downstream at Maybe Canyon Creek.

or normalized concentration profiles for samples taken in 5/90 shown in Figure 4. Relative (normalized) concentrations were calculated using Equation 4 assuming that element concentrations for runoff above the fill could be used as inflow element concentrations representative of springs and seeps below the fill that were not sampled. The relative concentration profiles for Mg, Ca, NO_3, and SO_4 are so similar that it is difficult to conceive of any process other than dilution that could produce the same result. The relative concentrations of Cl were significantly larger than those for Mg, Ca, NO_3, and SO_4. It is expected that Cl would show the same conservative behavior observed for the other major elements, but the Cl concentration in runoff above the fill may not be representative of that in downstream springs and seeps. It is necessary to inventory and analyze those springs and seeps to ensure maximum possible accuracy in evaluating relative concentrations.

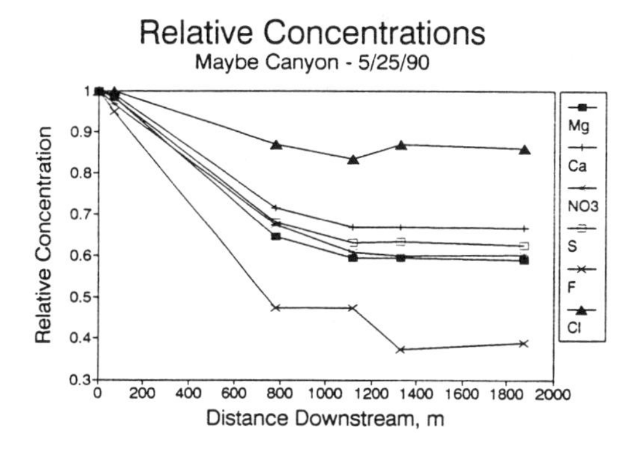

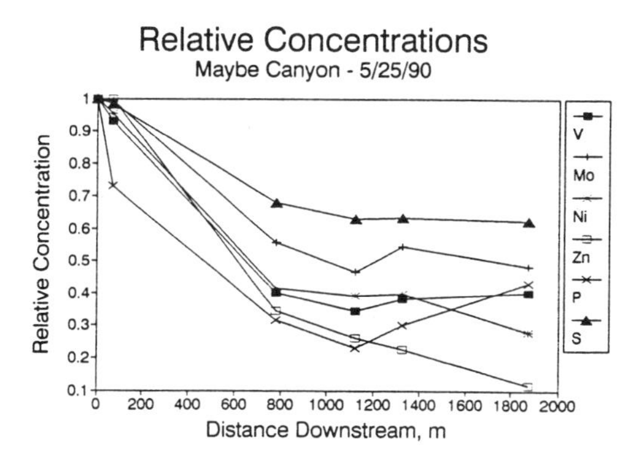

FIGURE 4. Relative (normalized) Concentrations of Major (top) and Trace (bottom) Element Concentrations Versus Distance Downstream at Maybe Canyon Creek.

Further evidence for dilution as a major controlling factor of instream concentrations of the major elements is given in Figure 5, where element concentrations are shown as a function of discharge. Linear relationships are seen for Mg, Ca, HCO_3, and SO_4. Thus, the observed concentration changes are largely the result of physical mixing of inflows and instream waters.

The greater downstream decreases in the relative concentrations of trace elements (V, Mo, Ni, Zn, P, and F) compared to major elements (e.g., S) indicate that other processes besides dilution control their concentrations in solution (Figure 4).

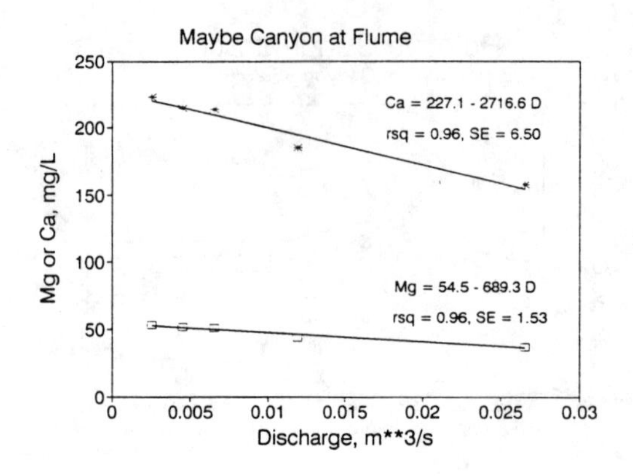

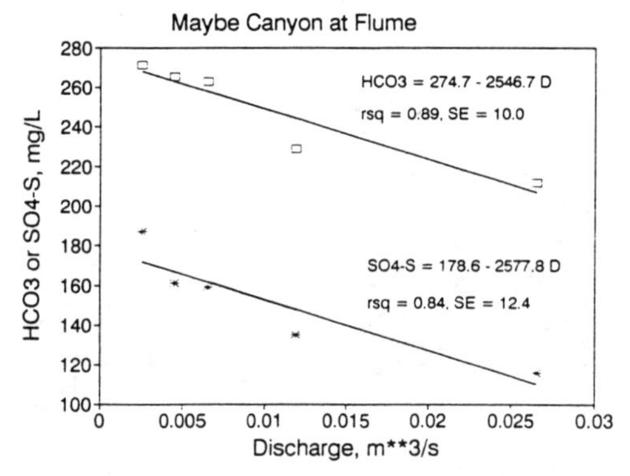

FIGURE 5. Major Cation (top) and Anion (bottom) Concentrations Versus Discharge at Maybe Canyon Creek.

MINTEQA2 calculations did not indicate that any particular pure trace element solid phase might be controlling their concentrations. Adsorption onto sediment solid phases is the most likely explanation of the observed downstream decreases that could not be accounted for by dilution. Sequential extraction of the sediments showed that the sediments contain substantial quantities of trace elements (e.g., Ni and Zn) in readily extractable fractions (e.g., exchangeable, EDTA extractable, carbonate-bound) (Figure 6).

Sequential extractions confirmed that the fill materials (phosphatic shale, mudstone, and chert) and not the native hillslope and alluvial soils are the source of trace element contaminants in the stream (Figure 6). Also, Connors [8] presented data to show that phosphatic shale contains high concentrations of trace elements. Most of the Ni and Zn in the sediments could be removed with weaker extractants such as NaOAc, EDTA, and $NH_2 \cdot HCl$, and thus most of these elements are associated with carbonates, organic matter, and Mn and amorphous Fe oxides. The EDTA extraction step is not as specific as the others and probably includes elements associated with carbonates not previously removed by NaOAc, organic matter, and metal oxides [26]. Only the fill materials contained appreciable quantities of trace elements in the crystalline Fe oxide, organic matter plus sulfide, and residual fractions.

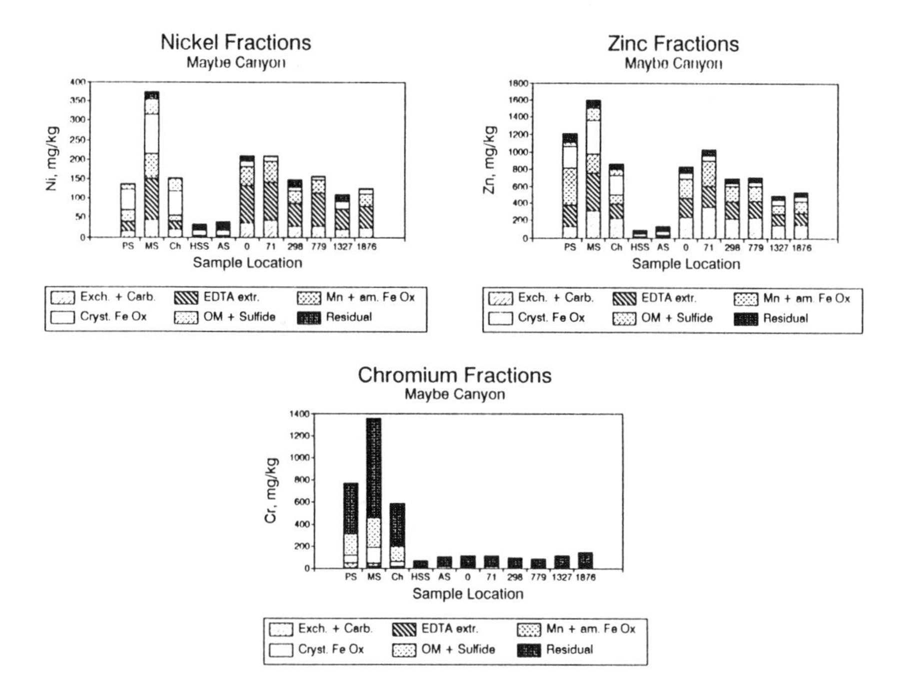

FIGURE 6. Nickel, Zinc, and Chromium Fractions in Maybe Canyon Fill Material, Soils and Sediments. The sample location designations are PS = phosphatic shale, MS = mudstone, Ch = chert, HSS = hillslope soil, AS = alluvial soil, and 0, 71, 298, 779, 1327 and 1876 refer to sediment locations in meters downstream from the base of the cross-valley fill.

The sequential extraction analysis indicates that weathering and not erosion of the fill material is responsible for the contaminants in the sediments below the fill. The relatively high concentrations of Ni and Zn in the easily extractable fractions, and their relative absence in the crystalline Fe oxide, organic matter + sulfide, and residual phases when compared to the fill materials indicates that these elements were likely leached from the fill materials and readsorbed on sediment components.

If erosion of the fill had distributed fill material particles downstream in the sampled fresh sediment fines, then the distribution of elements in the sediments should be similar to the original fill materials.

Confirmation of the weathering hypothesis can be obtained by examining the sequential extraction data for Cr (Figure 6). The fill materials contain relatively high concentrations of Cr, but the sediments do not. The Cr concentrations in the sediment samples are more nearly like those in the native hillslope and alluvial soils in the canyon bottom. Erosion of hillslope and alluvial soils during spring snowmelt runoff is probably the source of fresh sediment fines in the stream channel. Mineralogical analysis of fill materials, native soils, and sediments is necessary to confirm this hypothesis and will be the subject of future research at this site.

Our initial assessment of hydrologic and geochemical controls of water composition at Maybe Canyon is:

1. Water exiting the cross-valley fill contains higher concentrations of Na, Mg, Ca, Sr, V, Mo, Ni, Zn, Cd, NO_3, P, SO_4, F, and Cl compared to snowpack, surface runoff upstream from the fill, and a piped spring buried under the fill. Sequential extraction data confirmed the fill materials to be the primary source of the higher concentrations of many of the elements in stream water exiting the fill.

2. Stream water within the fill is near equilibrium with respect to calcite and gypsum.

3. As stream water exits the fill and continues downstream outgassing of CO_2 occurs with a corresponding increase in pH. Supersaturation of calcite occurs, but there appears to be a kinetic constraint on calcite precipitation.

4. MINTEQA2 calculations indicated that significant ion pairing of cations with carbonate and sulfate ions occurs in stream waters below the cross-valley fill.

5. MINTEQA2 calculations revealed that trace element concentrations are probably not being controlled by any particular pure solid phase.

6. Downstream decreases in concentrations of major elements (e.g., Mg, Ca, NO_3, SO_4) are the result of dilution by uncontaminated springs and seeps contributing to stream flow. This is supported by

the relative concentration data and the linear relationship between discharge and concentration.

7. Some trace elements (e.g., Ni and Zn) are adsorbed to downstream sediments. This conclusion is supported by the relatively greater decrease in downstream concentrations of these trace elements as compared to the major elements, and by the relatively high concentrations of carbonate-bound and EDTA extractable metals in downstream sediments.

8. Nickel and Zn are relatively mobile in the geochemical environment of Maybe Canyon despite the high pH and presence of carbonates, whereas Cr is relatively immobile. This is supported by the presence of relatively high concentrations of Ni and Zn in stream water, whereas soluble Cr was not detected. Also, sediments contained relatively high levels of readily extractable Ni and Zn compared to source materials. In contrast, sediments contained low concentrations of Cr compared to source materials in the fill.

A number of findings at Maybe Canyon require further investigation, however, to fully elucidate the geochemical behavior of elements at this site and to complete our assessment of the impact of mining to the watershed. Future studies include:

1. Inventory and chemical analysis of springs and seeps that influence the chemical composition of downstream water (dilution effects).

2. Tracer study to determine hydrologic parameters for modeling element transport in stream flow.

3. Possibility of calcite precipitation in the pond about 2000 m downstream from the fill.

4. Trace element adsorption on stream sediments.

5. Rates of element release from fill materials.

6. Source of high levels of NO_3 in stream waters.

7. Mineralogy of fill materials, native soils and stream sediments.

8. Development and evaluation of a model to account for the reactions and transport of elements in Maybe Canyon stream waters.

Despite the incompleteness of the study it is possible to make a few tentative recommendations regarding the handling of waste materials from phosphate rock mining in southeastern Idaho. Waste rock should be stockpiled in dry valleys without springs or streams. Because the phosphatic shale and mudstone in the area contain high levels of trace elements that could create serious contamination problems if sufficiently high concentrations migrated into surface and ground waters, it is ill advised to remove such material from a geochemical environment where they are slowly weathered, and place them in an environment where more rapid weathering can lead to contamination of local streams and aquifers. Placement in dry valleys would at least reduce weathering of fill material. Similarly, runoff should be diverted away from waste rock stockpiles. Backfilling of mined areas along the ridges with overburden would be a preferred alternative. There seems to be little remediation that can be done now at Maybe Canyon since the fill dam is in place, but continued monitoring of the stream is recommended for the near future.

3.2 Daisy and Fisher Creeks

Weathering of sulfide minerals at the McLaren and Glengary Mines has resulted in contamination of Daisy and Fisher Creeks with high levels of Mn, Fe, Cu, and Al and low stream water pH levels. Concentrations of Mn, Fe, Cu, and Al in stream water were strongly associated with S concentrations (Figure 7). The close association between Mn and S would indicate that these elements are likely controlled by the same instream process, probably dilution. The associations of Fe, Cu, and Al to S, although strong, are not as close as that between Mn and S and indicate that additional instream processes may control the concentrations of Fe, Cu, and Al. McKnight and Bencala [27] showed that Mn and SO_4 were closely correlated with each other in the Snake River, CO, an acidic, metal-rich stream, but that Fe and SO_4 were not. Bencala *et al.* [7] also found that Mn and SO_4 behaved conservatively in this stream and were largely controlled by variations in inflow.

The presence of orange-red precipitates on the streambeds of Daisy and Fisher Creeks indicates that hydrous Fe oxides are controlling the concentrations of Fe in these streams. MINTEQA2-generated Fe(III) activities for those stream and acid mine drainage samples in contact with hydrous Fe oxides were plotted against pH and are shown in Figure 8. There is some scatter in the data, but the close correspondence to a straight line is remarkable considering that the data are from a field site and not from a controlled laboratory experiment.

The scatter in the data may have several causes including lack of equilibrium between stream waters and hydrous Fe oxides, differences in particle sizes, ages, and chemical composition of the Fe oxides in different stream locations, differences in temperature, colloidal Fe not removed by membrane filtration, and inaccurate Fe(II) concentrations because the analyses were done much later in the lab instead of onsite.

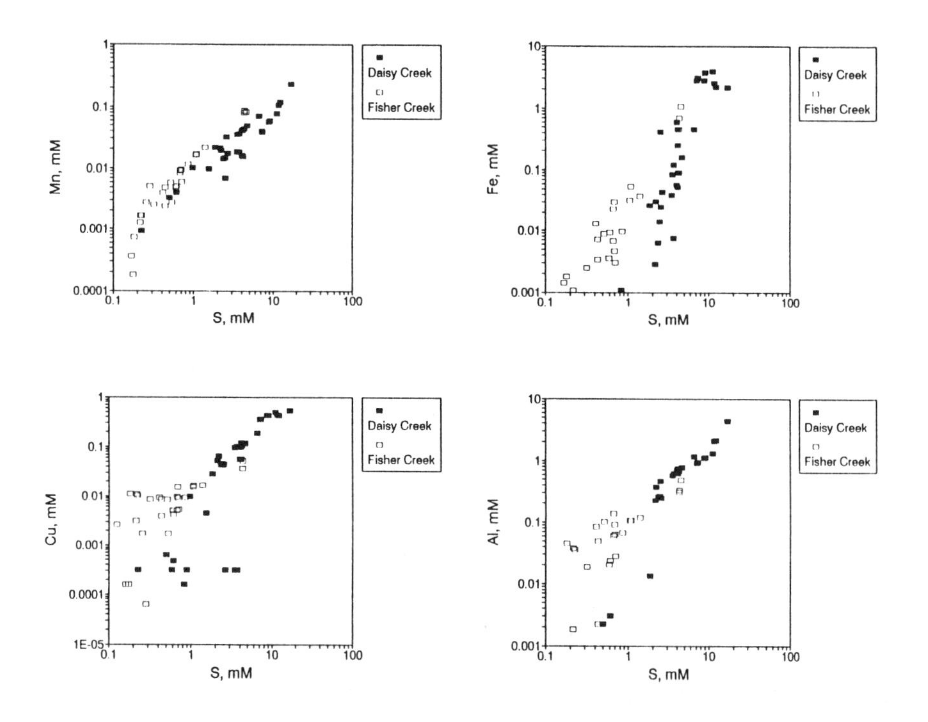

FIGURE 7. Mn, Fe, Cu and Al Versus S in McLaren and Glengary Mines Drainages and Daisy and Fisher Creeks.

McKnight and Bencala [28] showed that hydrous Fe oxide equilibria in the Snake River, Colorado, responded quickly to pH perturbations, indicating that equilibrium between freshly precipitated $Fe(OH)_3$ and dissolved Fe(III) is probably attained fairly rapidly. Thus, Daisy and Fisher Creek waters are probably near or at equilibrium with respect to amorphous $Fe(OH)_3$ and possibly microcrystalline goethite. Temperature variations in Daisy and Fisher Creek waters ranged from 3 to 19°C over the periods sampled and may account for some of the observed scatter in Figure 8.

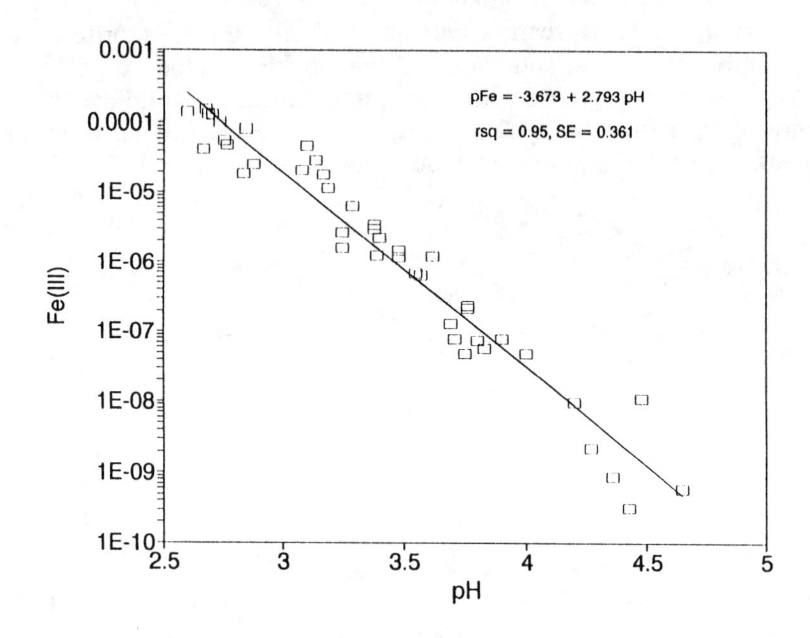

FIGURE 8. Fe(III) Activities as a Function of pH for McLaren and Glengary Mine Drainages and Daisy and Fisher Creeks.

Total Fe(II) and Fe(III) input into MINTEQA2 may also have contained some error because of colloidal Fe(III) passing through the membrane filters and changes in Fe(II) during storage, even though samples were kept at < 4°C and in the dark. Probable sources of variation are differences in particle sizes, ages, and chemical composition of hydrous Fe oxides at different stream locations. Much of the variation in ion activity products for amorphous $Fe(OH)_3$ can be accounted for by age and particle size variations [29]. Stream waters nearer the source of acid mine drainage probably contained more freshly precipitated $Fe(OH)_3$ with smaller particle sizes.

The stoichiometry of the reaction

$$Fe^{3+} + 3OH^{-} = Fe(OH)_3 \qquad (7)$$

indicates that the slope of Figure 8 should be 3. The experimental value of 2.79 ± 0.12 is close to the theoretical value but indicates that some anion substitution has occurred during precipitation. Slopes of less than

3 have been obtained by others where anion substitution was found to occur during $Fe(OH)_3$ precipitation [29].

The mean ion activity product (pFe(III) - 3pOH) of $Fe(OH)_3$ for Daisy and Fisher Creek waters is 39.0 ± 0.1 and compares well with similar values obtained by others for hydrous Fe oxides [29], despite any sources of variability discussed above. To obtain more accurate values for Fe(III) and Fe(II), onsite analysis of reactive filterable Fe(III) and Fe(II) are needed, and the presence of colloidal material could be confirmed by ultrafiltration studies.

The mean pK value for the reaction

$$Fe(OH)_3 + 3H^+ = Fe^{3+} + 3H_2O \quad (8)$$

obtained from MINTEQA2-generated ion activity products is -4.38 for Daisy and Fisher Creeks. MINTEQA2 uses a value of -4.891 in its database for this reaction. Using the original MINTEQA2 value and the new value, which best fits the Daisy and Fisher Creek data, pH sweeps were run using MINTEQA2 to calculate predicted Fe(III) concentrations at different pH values. The objective was to simulate expected downstream Fe(III) concentrations as pH increases with distance away from the acid mine drainage source. Initial values for the pH sweeps were the composition of acid mine drainage from the McLaren Mine entering Daisy Creek and acid mine drainage from the adit at the Glengary Mine flowing into Fisher Creek on the 9/90 sampling dates. These MINTEQA2-generated predictions and actual field data are shown in Figure 9 for Daisy and Fisher Creeks. The pK value of -4.38, which was generated from ion activity products for the field sites, provides better predictions of actual field values than does the original MINTEQA2 pK for $Fe(OH)_3$ formation.

The data at pH levels above 5 show considerable scatter. All the field data from all sample locations and dates in which soluble Fe could be measured are plotted in Figure 9 and include sites above pH 5 from tributaries and nonacid mine drainages where little or no hydrous Fe oxide precipitates were observed. Thus, Fe in these waters was not controlled by equilibrium with hydrous Fe oxides. The six data points above pH 6 from Daisy Creek that show the most departure from the MINTEQA2 predicted curve were taken from groundwater seeps with no obvious hydrous Fe oxide precipitates in their drainages. The form of Fe initially present in these samples was likely Fe(II), but since analyses were done much later in the lab, oxidation of Fe(II) (likely at the observed pH's) probably occurred and erroneous Fe(II) and Fe(III) values were input into MINTEQA2 for these samples. Field measurement of Fe(II) would settle this question.

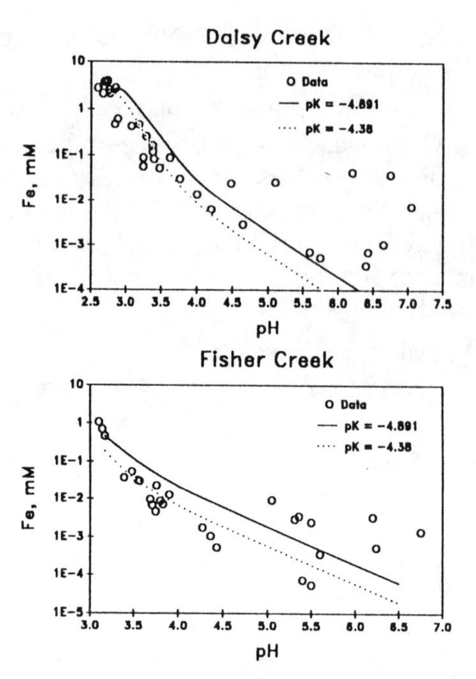

FIGURE 9. MINTEQA2-predicted Fe(III) Concentrations for Daisy (top) and Fisher (bottom) Creeks as a Function of pH at Two pK Values for the Reaction $Fe(OH)_3 + 3H^+ = Fe^{3+} + 3H_2O$. Compositions of Acid Mine Drainages from McLaren and Glengary Mines for 9/90 Sampling were used as Initial Values for Daisy and Fisher Creeks, Respectively.

A plot of Cu concentrations versus pH for study site waters at all sampling dates showed a sharp decrease as pH increased and suggested an adsorption edge. The possibility of modeling Cu adsorption by particulate hydrous Fe oxides using the diffuse layer model of Dzombak and Morel [24], which is included in MINTEQA2 along with an extensive database of adsorption reactions onto hydrous Fe oxide, was therefore investigated. Simulated Cu adsorption edges at different particulate Fe concentrations using initial element concentrations for McLaren and Glengary Mine drainages flowing into Daisy and Fisher Creeks on the 9/90 sampling dates, respectively, are shown in Figure 10 along with the actual Cu concentrations in mine drainages and stream waters for all sampling dates. It is clear that Cu adsorption would not be significant until stream pH levels increased above about pH 4.5 or particulate Fe levels were high. Unfortunately, suspended particulate Fe data are not yet available for Daisy and Fisher Creeks, but they probably show considerable spatial and temporal variability depending on stream

conditions. Thus, a range of values was used to obtain the simulations shown in Figure 10.

The 0.12 and 0.025 g/L particulate Fe values for Daisy and Fisher Creeks, respectively, were obtained by allowing MINTEQA2 to calculate the total amount of Fe that would precipitate as pH increased, and it is assumed that this particulate Fe remained in solution. The simulation for 0.025 g/L particulate Fe in Fisher Creek does come close to the two data points above pH 6, where significant Cu adsorption is expected to occur, but the Fisher Creek Cu concentrations do not show the sharp decrease with increasing pH expected of an adsorption edge. Particulate Fe levels of 0.5 to 2.0 g/L were needed to approximate the Cu concentrations for Daisy Creek, which show a sharp decrease above pH 4.5 as predicted by the adsorption model.

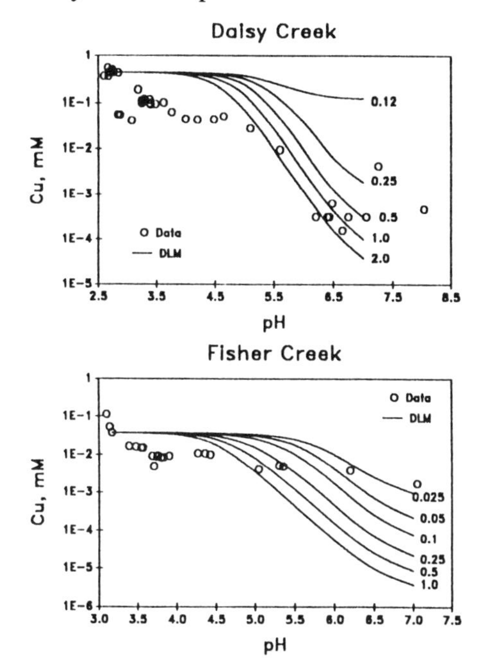

FIGURE 10. MINTEQA2-DLM Predicted Cu Concentrations for Daisy (top) and Fisher (bottom) Creeks as a Function of pH. Compositions of Acid Mine Drainages from McLaren and Glengary Mines for 9/90 Sampling were used as Initial Values for Daisy and Fisher Creeks, Respectively. Numbers next to DLM Predicted Lines Refer to Concentrations of Particulate Fe in g/L.

The 0.5 to 2.0 g/L particulate Fe values are greater than the 0.12 g/L level expected if all solution Fe precipitated. Of course, the expected concentration of particulate Fe assuming all solution Fe precipitates does not account for resuspended particulate Fe that might already be in solution, nor does it account for Fe oxide coatings on sediments that come into contact with stream water. Partitioning of Fe in the stream between suspended particulates, freshly precipitated gels, and older hard coatings on streambed rocks might improve our estimate of the actual amount of solid phase Fe available for adsorbing trace metals such as Cu.

The pH sweep simulations permit approximate predictions of the effect of pH on Fe and Cu behavior in study area waters, but they cannot calculate the effect of inflows on instream element concentrations along a given reach of stream. Because initial surveys have shown that Daisy and Fisher Creeks have multiple inflows of varying element concentrations, Equation 6 was used to calculate relative or normalized concentrations for the two streams. Thus, instream relative concentrations are constrained between the maximum and minimum values observed for inflows along a given stream reach. The relative concentrations for Mn, Fe, Cu, Al, and S in Daisy and Fisher Creeks for the 9/90 sampling are shown in Figure 11. Similar results were obtained on the other sampling dates. It is apparent that all the elements except Fe have similar relative concentrations and are thus mostly influenced by inflows rather than instream reactions. Iron is mostly controlled by precipitation of hydrous Fe oxides as already discussed rather than by inflows.

To assess the relative importance of inflows versus adsorption of Cu or precipitation of Fe as controls of instream metal concentrations, additional simulations were run. To assess the potential effect of dilution on downstream Fe and Cu concentrations, Equation 5 was used to calculate expected downstream Fe and Cu concentrations resulting from mixing of instream and inflow waters using relative concentrations of S as the conservative element. In addition, MINTEQA2 pH sweep simulations were run using downstream observed pH levels and the composition of inflow acid mine drainages as initial values. Precipitation of hydrous Fe oxide was allowed, using a pK of -4.38 for the solubility product of hydrous Fe oxide, and the diffuse layer model was used to predict Cu adsorption. The results of the simulations along with the stream data for the 9/90 dataset are shown in Figures 12 and 13. Although the dataset is meager, it is clear that dilution alone cannot account for the observed downstream Fe concentrations when acid mine drainage flows into the streams. The precipitation of hydrous Fe oxide as predicted by MINTEQA2 does approximate the data, but there are insufficient data over a wide enough pH range to determine how well MINTEQA2 can predict downstream Fe concentrations. Better

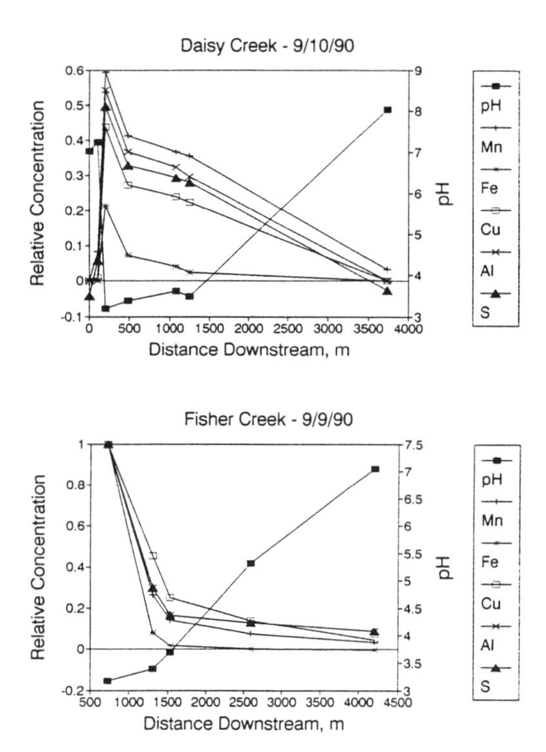

FIGURE 11. Relative (normalized) Concentrations of Mn, Fe, Cu, Al and S in Daisy (top) and Fisher (bottom) Creeks Versus Distance Downstream for the 9/90 Sampling. The main AMD channel from the McLaren Mine enters Daisy Creek about 207 m from Daisy Creek's source. The main AMD source for Fisher Creek, the Glengary Mine adit, enters the Creek about 729 m from the creek's source.

predictions were obtained in Figure 9, but that figure included data from all study site waters including acid mine drainages. More instream data are needed for both Daisy and Fisher Creeks between the pH extremes to fully assess the capability of MINTEQA2 to predict downstream Fe concentrations. Nevertheless, Figures 9 and 12 make it clear that Fe concentrations in study site waters are controlled by precipitation of hydrous Fe oxides and not by dilution.

By contrast, downstream Cu concentrations are well predicted by the simple mixing model of Equation 5. At higher pH levels, however, the diffuse layer model is as good as the mixing model at predicting instream Cu concentrations. The dataset is too meager to determine which approach is more applicable. More data between the observed pH extremes are needed. Additional potential sampling points have been identified and will be sampled during subsequent field investigations.

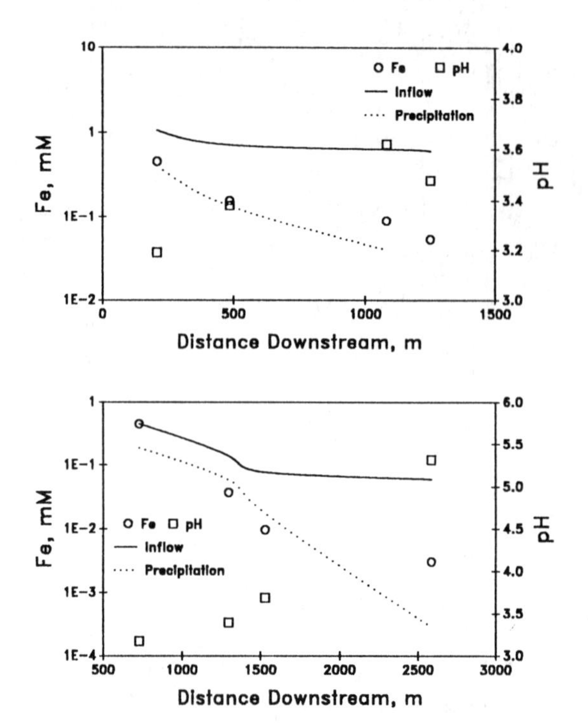

FIGURE 12. Predicted Downstream Fe(III) Concentrations in Daisy (top) and Fisher (bottom) Creeks for the 9/90 Sampling Assuming Dilution or Precipitation as Processes Responsible for Instream Concentration Changes.

Sequential fractionation data for Daisy and Fisher Creek sediments collected in 8/89 are shown in Figures 14 and 15. Although a detailed analysis of this and subsequently collected data will be the subject of another paper, one finding will be stressed. Extractable Fe and Cu per kilogram of sediment are shown in Figure 14. Extractable Cu in the amorphous Fe oxide fraction per kilogram of extractable Fe in that fraction is shown in Figure 15. This approach normalizes extractable Cu with respect to the amorphous Fe oxide content of the sediment. The numbers shown above the bars are the pH values of the stream water at the point where the sediment samples were collected. It is apparent that the amorphous Fe oxide fraction in the most acid sediments is relatively depleted in Cu, although the most acid sediments contain the highest amounts of amorphous Fe oxide. Sediments

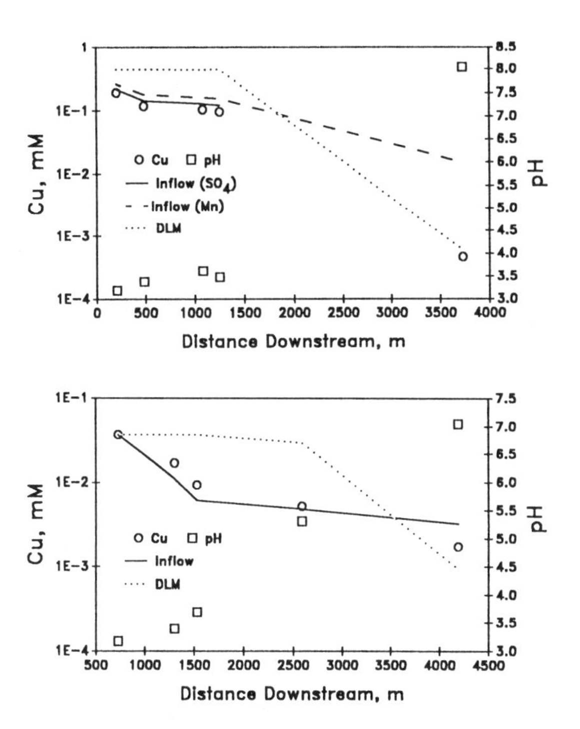

FIGURE 13. Predicted Downstream Cu Concentrations in Daisy (top) and Fisher (bottom) Creeks for the 9/90 Sampling Assuming Dilution or Adsorption on Hydrous Fe Oxides as the Processes Responsible for Instream Concentration Changes.

underneath waters with higher pH levels contain relatively higher amounts of Cu in the amorphous oxide fraction. Because the diffuse layer model predicts that metal adsorption increases as pH increases, the sequential extraction data support the validity of the diffuse layer model predictions of Figures 10 and 13.

The initial assessment of hydrologic and geochemical controls of element concentrations in Daisy and Fisher Creeks leads to the following conclusions:

1. Mn and S concentration changes are primarily the result of inflows mixing with stream waters. The similarity of relative concentrations and close correlation between the concentrations of these two elements support this conclusion. This finding is also in accord with the findings of Bencala and McKnight [6] and Bencala *et al.* [7] for the Snake River, Colorado.

2. Fe concentrations are controlled by precipitation of hydrous Fe oxide. This conclusion is supported by MINTEQA2 calculations and sediment analysis.

3. Cu concentration changes are the result of mixing inflows with stream waters at pH < 4.5, but adsorption on hydrous Fe oxides is possible at pH > 4.5. The relative concentration data, DLM calculations, and relatively high levels of Cu associated with amorphous Fe oxides in sediments at pH > 4.5 support this conclusion.

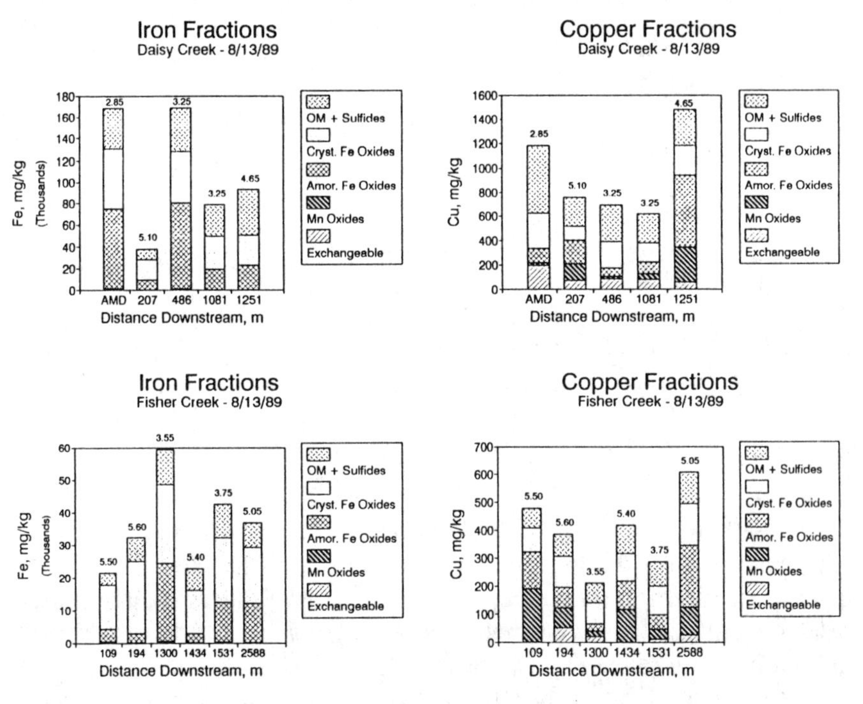

FIGURE 14. Sequential Extraction Data for Fe and Cu in Daisy and Fisher Creek Sediments from 8/89. AMD refers to sediment in the McLaren mine drainage which enters Daisy Creek between the 207 and 486 m samples. Acid mine drainage from the Glengary adit enters Fisher Creek between the 194 and 1300 m samples at about 729 m. The 1434 m sample is from an uncontaminated tributary into Fisher Creek. The numbers above the bars are the pH levels of stream waters at the sample locations.

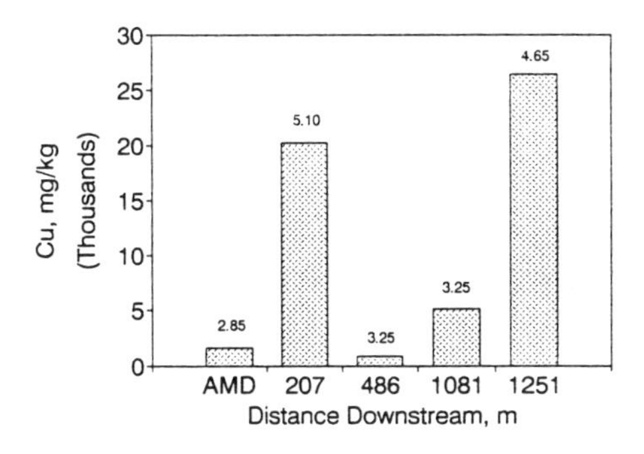

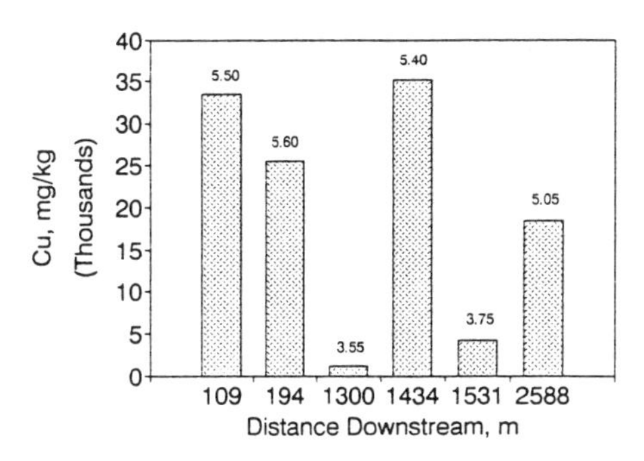

FIGURE 15. Copper Sorbed to the Amorphous Fe Oxide Fraction for Daisy (top) and Fisher (bottom) Creek Sediments from 8/89. The numbers above the bars are pH levels of stream waters at the sample locations.

Future investigations at Daisy and Fisher Creeks will be designed to confirm or refute these conclusions. Such studies include:

1. Inventory and chemical analysis of inflows that cause significant spatial variability in chemical composition of Daisy and Fisher Creeks.

2. Tracer study to determine hydrologic parameters needed to model element transport by stream flow.

3. Temporal variation in pH, pCu, Fe(II), and reactive Fe(III) caused by photoreduction of Fe(III) oxides and microbially mediated oxidation of Fe(II).

4. Rate of hydrous metal oxide formation in solution and rate of deposition on the streambed.

5. Partitioning metals between suspended particulates, freshly precipitated hydrous metal oxide gels, and older hard metal oxide coatings on stream sediments.

6. Development and evaluation of a model to account for reactions and transport of metals in Daisy and Fisher Creeks.

The temporal variation in solution Fe caused by photoreduction and microbially mediated oxidation was studied in the Snake River, Colorado, by McKnight and Bencala [27,28] and could also be important in Daisy and Fisher Creeks.

The area around Fisher Mountain is currently undergoing intense exploration drilling by Noranda Minerals Exploration, Inc. for possible mining in the future. A mining and reclamation plan has been submitted to the Montana State Lands Board and to the Gallatin National Forest for review and does include remediation proposals that could lead to partial restoration of Daisy and Fisher Creeks. However, because uncertainty surrounds the actual extent of future disturbances, only tentative recommendations regarding remediation can be made at this time. Suggested remediation efforts should include:

1. Sealing of all adits.

2. Burying high acid generating potential materials underneath low acid generating potential materials and revegetating surface materials.

3. Backfilling pits with low acid generating potential material.

4. Lining drainage channels with limestone rock available in the area.

The large amounts of fresh water produced each year during snowmelt would help flush existing hydrous metal oxide coatings downstream. However, the sequential extraction data show substantial amounts of heavy metals in the organic matter + sulfide + residual fraction in the sediments. Apparently, past erosion, because of a lack of vegetation in the old mined areas, has resulted in sulfide minerals being moved downstream and deposited in sediments. These sulfide minerals serve as a source of Cu that can be leached into stream waters far into the future until natural weathering has reduced their presence to small background levels.

REFERENCES

1. Sidle, R.C. and A.N. Sharpley, "Cumulative Effects of Land Management on Soil and Water Resources: An overview," *J. Environ. Qual.* 20:1-3 (1991).

2. Farmer, E.E., "Phosphate Mine Dump Hydrology," in *Symposium on Watershed Management 1980.* pp. 846-854. Amer. Soc. Civ. Eng. (New York, 1980).

3. Brown, R.W., R.S. Johnston and J.C. Chambers, "Responses of Seeded Native Grasses to Repeated Fertilizer Applications on Acidic Alpine Mine Spoils," in *Proceedings: High Altitude Revegetation Workshop No. 6.* Information Series No. 53. T.A. Colbert and R.L. Cuany, Eds. pp. 200-214. Colorado State Univ., Water Resources Research Institute, (Colorado, 1984) .

4. Brown, R.W., J.C. Chambers and R.C. Sidle, "Temporal Changes in Pyritic Spoil Properties Following Revegetation of An Abandoned High Elevation Mine Site," *Agron. Abstr.* pp. 32 (1989).

5. Brown, R.W. and J.C. Chambers, "Reclamation Practices in High Mountain Ecosystems," in *Proceedings: Symposium on Whitebark Pine Ecosystems.* W.C. Schmidt and K.J. McDonald, Eds. pp. 329-334. USDA Forest Service Gen. Tech. Rep. INT-270 (Ogden, UT 1990).

6. Bencala, K.E. and D.M. McKnight, "Identifying Instream Variability: Sampling Iron in An Acidic Stream," in *Chemical Quality of Water and the Hydrologic Cycle.* R.C. Averett and D.M. McKnight, Eds. pp. 255-269. Lewis Publishers, Inc. (Chelsea, MI 1987).

7. Bencala, K.E., D.M. McKnight and G.W. Zellweger, "Evaluation of Natural Tracers in An Acidic and Metal-Rich Stream," *Water Resour. Res.* 23:827-836 (1987).

8. Connors, E.B., "Phosphate Mining in Southeast Idaho," in *Symposium on Watershed Management 1980.* pp. 867-876, Amer. Soc. Civ. Eng., (New York, 1980)

9. *Standard Methods. 16th Ed.* American Public Health Association, American Water Works Association and Water Pollution Control Federation, (Washington, D.C. 1985).

10. *Methods for Chemical Analysis of Water and Wastes.* EPA-625-/6-74-003a. U.S. Environmental Protection Agency, (Cincinnati, OH 1976).

11. Weiguo, Q., "Determination of Trace Vanadium in Water by a Modified Catalytic-Photometric Method," *Anal. Chem.* 55:2043-2047 (1983).

12. Skougstad, M.W., M.J. Fishman, L.C. Friedman, D.E. Erdman and S.D. Duncan, Eds., "Methods for Analysis of Inorganic Substances in Water and Fluvial Sediments," U.S. Geological Survey Open File Report 78-679, (Reston, VA 1978)

13. Bartlett, R.J. and D.S. Ross, "Colorimetric Determination of Oxidizable Carbon in Acid Soil Solutions," *Soil Sci. Soc. Am. J.* 52:1191-1192 (1988).

14. Hallmark, C.T., L.P. Wilding and N.E. Smeck, "Silicon," in *Methods of Soil Analysis, Part 2, 2nd Ed.* A.L. Page, Ed. pp. 263-273, American Society of Agronomy (Madison, WI 1982)

15. Keeney, D.R. and D.W. Nelson, "Nitrogen-Inorganic Forms," in *Methods of Soil Analysis, Part 2. 2nd Ed.* A.L. Page, Ed. pp. 643-698. American Society of Agronomy, (Madison, WI 1982).

16. Olsen, S.R. and L.E. Sommers, "Phosphorus," in *Methods of Soil Analysis*, *Part 2, 2nd Ed.* A.L. Page, Ed. pp. 403-430. American Society of Agronomy, (Madison, WI 1982).

17. Adriano, D.C. and H.E. Doner, "Bromine, Chlorine and Fluorine," in *Methods of Soil Analysis*, *Part 2, 2nd Ed.* A.L. Page, Ed. pp. 449-483. American Society of Agronomy, (Madison, WI 1982) .

18. Salomons, W. and U. Förstner, *Metals in the Hydrocycle.* Springer-Verlag (Berlin, 1984).

19. Tessier, A., P.G.C. Campbell and M. Bisson, "Sequential Extraction Procedure for the Speciation of Particulate Trace Metals," *Anal. Chem.* 51:844-851 (1979).

20. Chao, T.T., "Selective Dissolution of Manganese Oxides From Soils and Sediments with Acidified Hydroxylamine Hydrochloride," *Soil Sci. Soc. Am. Proc.* 36:764-768 (1972).

21. Chao, T.T. and L. Zhou, "Extraction Techniques For Selective Dissolution of Amorphous Iron Oxides From Soils and Sediments," *Soil Sci. Soc. Am. J.* 47:225-232 (1983).

22. Shuman, L.M., "Separating Soil Iron- and Manganese-Oxide Fractions for Microelement Analysis," *Soil Sci. Soc. Am. J.* 46:1099-1102 (1982).

23. Allison, J.D., D.S. Brown and K.J. Novo-Gradac, "MINTEQA2 /PRODEFA2, A Geochemical Assessment Model for Environmental Systems: Version 3.0 User's Manual," U.S. Environmental Protection Agency (Athens, GA 1990).

24. Dzombak, D.A. and F.M.M. Morel, *Surface Complexation Modeling: Hydrous Ferric Oxide*, John Wiley & Sons, (New York 1990).

25. Suarez, D.L., "Ion Activity Products of Calcium Carbonate in Waters Below the Root Zone," *Soil Sci. Soc. Am. J.* 41:310-315 (1977).

26. Pickering, W.F., "Selective Chemical Extraction of Soil Components and Bound Metal Species," *CRC Critical Rev. Anal. Chem.* Nov: 233-266 (1981).

27. McKnight, D. and K.E. Bencala, "Diel Variations in Iron Chemistry in an Acidic Stream in the Colorado Rocky Mountains," *Arctic Alpine Res.* 20:492-500 (1988).

28. McKnight, D.M. and K.E. Bencala, "Reactive Iron Transport in An Acidic Mountain Stream in Summit County, Colorado: A Hydrologic Perspective," *Geochim. Cosmochim. Acta* 53:2225-2234 (1989).

29. Macalady, D.L., D. Langmuir, T. Grundl and A. Elzerman, "Use of Model-Generated Fe^{3+} Ion Activities to Compute Eh and Ferric Oxyhydroxide Solubilities in Anaerobic Systems," in *Chemical Modeling of Aqueous Systems II.* ACS Symposium Series No. 416. D.C. Melchior and R.L. Bassett, Eds. pp. 350-367. American Chemical Society, (Washington, D.C 1990).

12

ASSESSMENT OF METAL CONTAMINANTS DISPERSED IN THE AQUATIC ENVIRONMENT

Jingyi Liu, Hongxiao Tang, Yuhuan Lin and Meizhou Mao
Research Center for Eco-Environmental Sciences
Academia Sinica
Beijing, China

1. INTRODUCTION

Metal contamination in the aquatic environment is an environmental problem in China. Rivers, coastal waters and soils were contaminated by industrial and mining activities. Metals discharged from industries have been controlled to a great extent in recent years, yet heavy metals dispersed in river sediments still need to be dealt with. Characterization, transformation, transport and fate of metal contaminants in aquatic ecosystems and the measures to alleviate and control their ecological effects need to be studied.

China is rich in nonferrous metal resources, particularly multi-elemental minerals. Exploitation and utilization of resources discharges heavy metals into the environment and contaminates neighboring aquatic systems. Weathering of sulfide minerals generates highly acidic mine drainage which further releases large amounts of metals and deserves special concern.

The distribution of metal species and their transformation in the environment is largely dependent on their eco-environmental conditions [1,2]. This paper reports briefly some of our work in the past decade on the assessment of metal contaminants dispersed in different eco-environmental conditions:

- Mercury dispersed in Ji Yun River sediments;
- Chemical stability of heavy metals in Xiang River; and

0-87371-697-3/95/$0.00+$.50

- Assessment of metal contamination at the Dexing mining area.

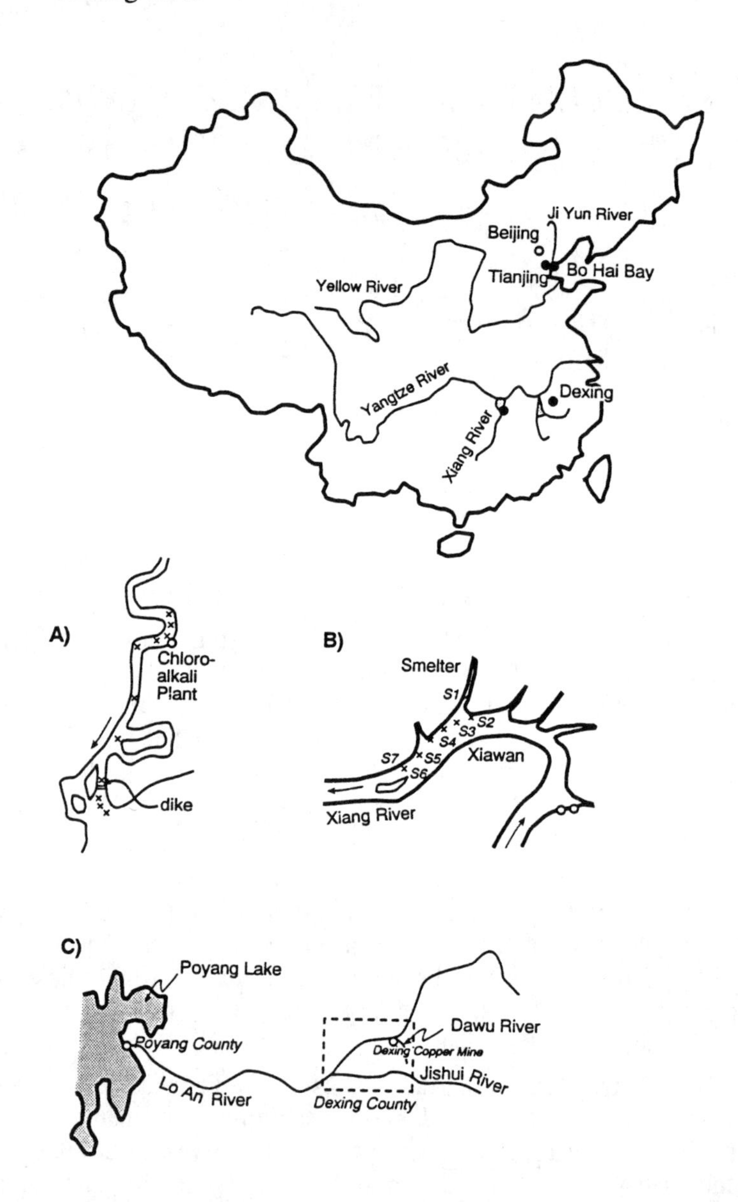

FIGURE 1. (a) Location and Sketches of the Ji Yun River: Bo Hai Bay Area, China. (b) Location and Sketch of the Xiang River: Xiawan Section, China, and (c) Location and Sketch of the Dexing Copper Mine - Poyang Lake Area, China.

Identification, speciation of metal contaminants, some physico-chemical processes of metal transformation and transport in aquatic systems together with some simulation experiments and modeling work are reported.

2. MERCURY DISPERSED IN RIVER-SEDIMENTS

The Ji Yun River is a Hg heavily contaminated river in North China; it runs through a chemical, industrial area near Tianjin to Bo Hai Bay [Figure 1(a)]. Hg was discharged from a chloro-alkali plant and accumulated mostly in river sediments. Studies have been made on the characterization of sediments, Hg speciation, stability of HgS, complexation and adsorption of Hg, etc.

2.1 Speciation and distribution of Hg

The composition and grain size of the sediments, humic acid and some other organics in the sediments were characterized by various physico-chemical techniques [3,4]. Sequential chemical extraction for Hg speciation has been developed following Eganhouse's procedure [5] with slight modifications [6]. A relatively high percentage (>80%) of Hg was found to be associated with humic acids and other organics (Figure 2a and 2b)[7].

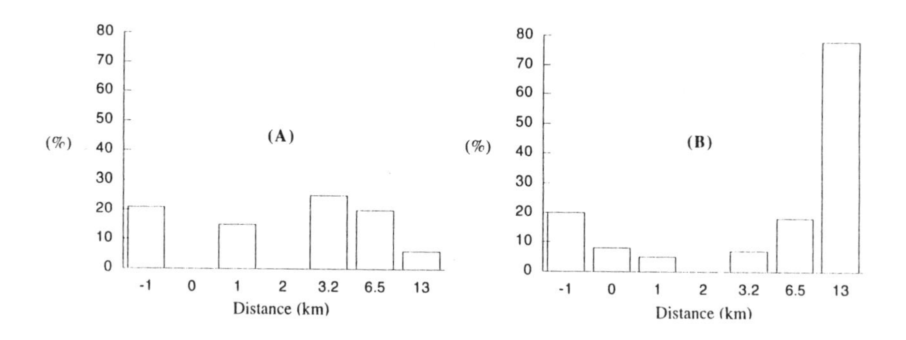

FIGURE 2. Distribution of Hg in Ji Yun River Sediment: Bonded with (a) Humic Acids and (b) Easily Degradable Organics.

A comparative study of the horizontal and vertical distribution of Hg in river sediments was made from the data obtained from 1976-77 [8] and 1980-81 [9]. It was found that the river was contaminated heavily by Hg. Concentrations of Hg in the sediments were lowered considerably in 1981 (pollution source being controlled), and Hg was transported downwards into the Bay (Figure 3).

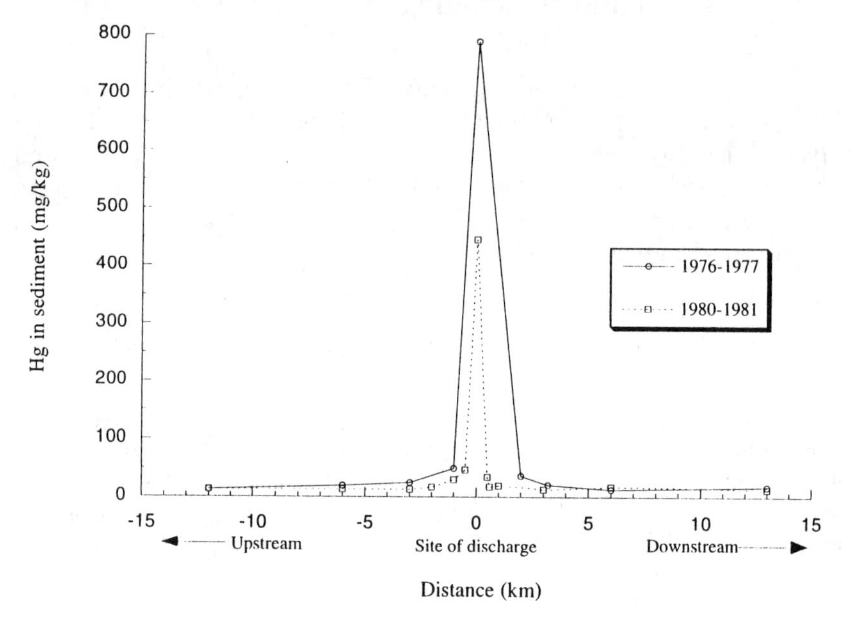

FIGURE 3. ***Distribution of Mercury Content in Ji Yun River Sediment.***

2.2 Organic associated Hg and stability of HgS

The relatively high value of total acidic functional groups of humic acids and the presence of some organic compounds shown in GC-MS spectrogram indicated the complexing capacity of Hg in the river sediments. From the horizontal and vertical distribution of Hg along the river, it was found that the relative percentage of humic acids associated with Hg and the easily degradable organic species would decrease downstream.

Stability of synthesized HgS (amorphous form) showed that the presence of humic acids was able to dissolve HgS in the sediments by complexation with Hg and release Hg into the aqueous phase. The released Hg was adsorbed on the sediments and then gradually redissolved [10].

2.3 Complexation, adsorption, transport and fate of Hg

Hg in the sediments existed mostly as organic associated species and as sulfides which were capable of transforming into soluble species and being transported along the river. The complexing capacity of humic acids, determined by the gel-filtration technique, was in the order [7]: $Water_{FA}$ > $Sediment_{FA}$ >$Sediment_{HA}$ where FA and HA stand for fulvic acid and humic acid respectively. The apparent stability constants of the complex, Hg-humate, were calculated to be greater than that of the other Hg complexes (with Cl^-, OH^-, SO_4^{2-} and PO_4^{3-}) in the river.

The humic acid or chlorides in the interstitial water of the sediment also dissolve Hg slowly from the sediment, thus enhancing the transport of Hg in the river. It was found that the sediments from Ji Yun River were mainly composed of clay minerals, quartz and sand [7] and the predominant fraction (>80%,<60μm) contained 80% of the adsorbed Hg. However, the adsorption of Hg on fine particles (<5μm) was found to be the strongest. Adsorption of Hg is still the most important process enriching Hg in the river-sediments. The adsorptive capacity of Hg on different adsorbents were found to be: [11]

$$HA > MnO_2 > \text{Clay minerals} > Fe_2O_3 > SiO_2$$

The relatively high content of S and N in the humic acids extracted from river sediments would likely enhance Hg adsorption.

If chloride is less than 10^{-4} M, adsorption of Hg on different adsorbents is still relatively strong, and if it reaches 0.56 M (chloride concentration in seawater), then the adsorbed Hg is relatively easily desorbed due to the formation of stable Hg-chloro-complexes [12]. Thus, it is inferred that complexation, dissolution, adsorption, desorption and degradation of organics may be the main physico-chemical processes participating intermittently and interactively in the transport of mercury in the river.

As the flow rate of Ji Yun River in 1980-81 was low, the aforementioned physico-chemical processes in the presence of a considerable amount of organic matter might be the important factors for the transport of Hg in Ji Yun River at that period. With much higher flow rate in 1977-79, the physical process was considered to be the most important process for the transport of Hg. By analyzing the Hg content on the suspended sand samples near Bo Hai Bay (1979-81), the amount of Hg dispersed into Bo Hai Bay can roughly be estimated.

3. CHEMICAL STABILITY OF HEAVY METALS IN THE XIANG RIVER

The Xiang River, situated in mid-southern China, is one of the largest tributaries of the Yangtze River with abundant nonferrous metal resources in its vicinity. Large amounts of metal contaminants from mining and metallurgical activities were discharged into the Xiang River and accumulated in the river sediments. More than 10 institutions have worked together for seven years in this aquatic environment for the assessment of metal contamination, pollution control and water quality management. Studies in our laboratory on chemical speciation and chemical stability of the metals are briefly reported here.

3.1 Field work and direct measurements

Large amounts of samples of river water, suspended matter, surface sediments and core samples along the river were collected and analyzed (1978-81). Different analytical methods, e.g. AAS, ASV, XRF, TOC were used. Sampling at Xiawan section near a large smelter (S1, see Figure 1b]) was emphasized at S1-S5 (about 5 Km) in 1982-84.

Metal speciation of water, suspended matter, surface sediments and core samples was carried out using different procedures including separating by membrane, Chelex resin, and ultrafiltration-ASV determination for water and ultrasonic sieves, density-gradients, and sequential chemical extraction for sediments and suspended matter, [13,14]. It was found that in river sediments metals were enriched in 20-60 μ fractions. Carbonate and Fe and Mn hydrous oxide (reducible) bound fractions were rather high in sediments along the river (Figure 4a and 4b).

The six fractions species were classified into three categories: direct bioavailable (water soluble + cation exchangeable), intermediate fraction (carbonate bound + reducible + organic/sulfide bound) and inert fraction. The direct bioavailable fraction for Cd was estimated to reach 30-40%, while for Cu, Pb, and Zn it was 5% of the total metal content at site 4 as shown in Figure 5.

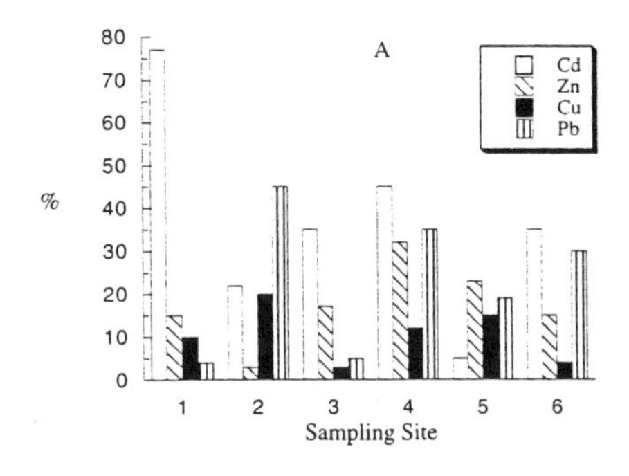

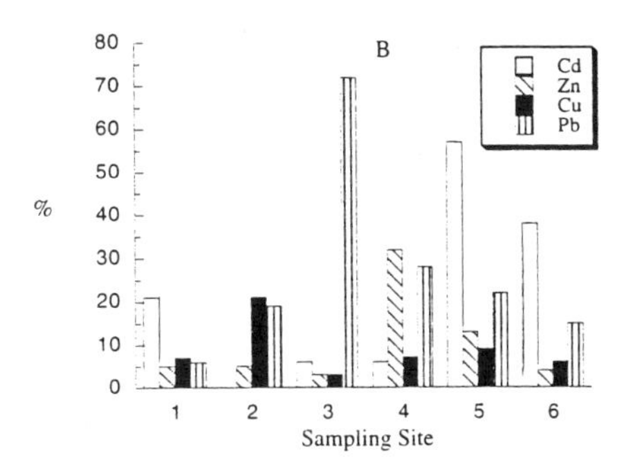

FIGURE 4. Metal Speciation of Surface Sediment along the Xiang River. Distribution of (a) Carbonate-bound Fraction and (b) Fe/Mn Hydrous Oxide-bound Fraction.

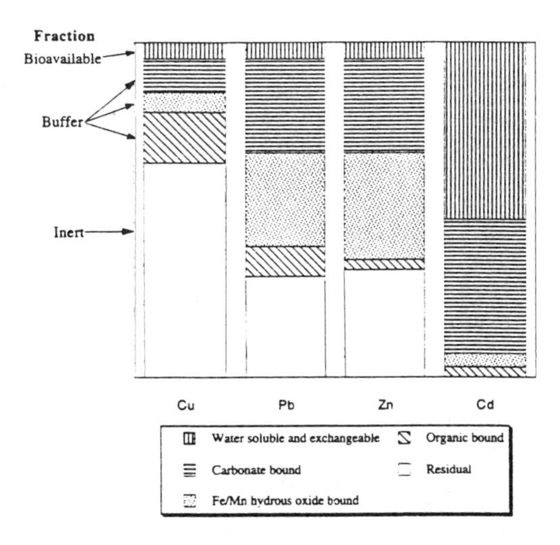

FIGURE 5. Distribution and Classification of Pb, Zn and Cd Species at Site 4 on the Xiang River.

At Xiawan, near site 3, it was found that the direct bioavailable fraction was about 1-1.5% for Cu and Pb and 10-20% for Cd; the intermediate fraction contained 10-30% of the metals and the inert fraction contained 50-90%. The high solid particles with crystalline structure may originate from ore and from the smelter slags [14].

3.2 Chemical equilibrium modeling

The suspended sediments from Xiang River were fractionated and various components (clay, hydrous metal oxides, carbonates, organic matter) were isolated. The adsorption isotherms of Cd on each of these components were systematically studied and these experimental data were used to calculate the contributions of the sediment fractions to the total adsorption employing a multi-component adsorbent model. The low content of organic ligands was characterized by the complexation capacity and stability constants which was determined by the ASV method [15].

With a view to predict the species distribution of the components at the equilibrium condition in that aquatic system, REDEQL-2 Program involving 12 metals, 49 possible precipitation and 5 redox reactions with hydrological data over the years were studied.

All the free ion concentrations calculated are below 10^{-8} M and would not be precipitated under current conditions. The effects of pH and the concentration of metals were calculated. Figure 6 shows pH-pM diagrams of Cu, Cd and Pb species.

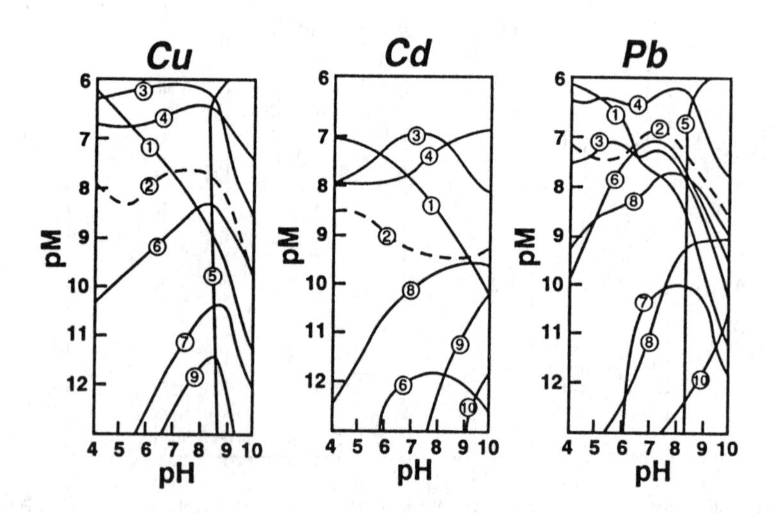

FIGURE 6. pH-pM Diagrams of Cu, Cd and Pb Species.

3.3 Release of metals from sediments

Static and dynamic extraction experiments were carried out to simulate the release of metals from the contaminated sediments. All experimental results indicated that the release rate was rather low and the amount of metal released was not large (Figure 7).

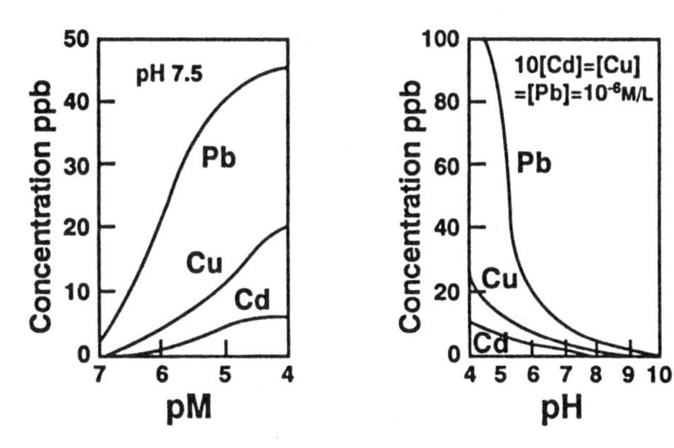

FIGURE 7. Concentrations of Directly Bioavailable Species under Various Conditions.

3.4 Chemical stability of metal contaminants in the Xiang River

Based on the results from chemical speciation by direct measurements, chemical equilibrium modeling and extraction experiments of contaminated sediments; it was shown that the direct bioavailable fraction of metal contaminants was rather limited. The metals bound to organic ligands, adsorbed particles and precipitated species presented an intermediate fraction for formation of soluble species. Most of the metal contaminants were dispersed in the residue. Thus it was inferred that the metal contaminants exhibited a rather high chemical stability in the Xiang River [16].

4. ASSESSMENT OF METAL CONTAMINATION AT MINING AREA

The Dexing Copper Mine is the largest open-pit copper mine in Jiangxi province, southeastern China. Dawu River (14 Km) runs through it and flows into Lo An River (279 Km) and finally into Poyang Lake, the largest lake in China [see Figure 1 (C)]. Weathering of this multi-elemental sulfide mineral generates acid mine drainage (Table 1), and a large amount of alkaline wastewater together with concentrated suspension of ore tailings are discharged into the Dawu River. Thus, the acid mine drainage and the alkaline wastewater mix and seriously deteriorate the Dawu River and contaminate Lo An River-Lake.

Table 1. Chemical Composition of Acid Mine Drainage in the Dexing Copper Mine (December 1987-December 1988).

Species	Range (mg/L)	Species	Range (mg/L)
pH	2.24-2.60	Ca	180-375
Total Fe	2130-2420	Mg	362-376
Fe(II)	120-270	Na+K	32-147
Mn	30-60	SO_4	4400-5022
Al	650-870	S	3025-4680
Cu	80-120	P	20-154
Zn	2.5-4.2	F	10-20
Cd	0.2-0.4		

4.1 Field survey and chemical analysis

Field surveys along the Dawu River (1987-88) and the Lo An River (1987-90) were carried out both in dry and rainy seasons. Riverwater and suspended sediments were sampled from upstream through Gukou, the converging point of the rivers, to downstream at Caijiawan near the Lake, and a few sediment cores at Caijiawan were taken (1988-90). River water samples were filtered through a 0.45 μm membrane filter, acidified *in situ* and analyzed for metals by AAS and ICP, and for anions by ion chromatography. Sediments and suspended matter were analyzed by X-ray fluorescence spectrometery and minerals were identified by X-ray diffractometery.

Geochemical speciation of river sediment/suspended matter for the Lo An River was carried out by sequential chemical extraction following Tessier's [17] and Förstner's [18] procedures.

4.2 Distribution of metals along the aquatic system

It was found that Fe, Cu and S contents increased gradually from upstream to downstream along the Dawu River at the mining area. The dissolved Fe and S were decreased by oxidation and precipitation, but Cu remained in solution. A large amount of Fe, S and trace metals accumulated in the sediments (Figure 8).

After receiving the alkaline wastewater, precipitation of Fe and Al was nearly complete, and metals together with ore tailings had settled down into the river basin. In the dry seasons, the concentrations of dissolved metal species were lower than those in rainy seasons. As the acidic and alkaline wastewater could not mix well at increasing flow rate in rainy seasons, neutralization and precipitation reactions extended in the Lo An River for a longer distance [19].

From the Lo An River downwards, it was found that the riverwater was slightly contaminated and the suspended matter together with the tailings were transported along the river. The trend of the metal contamination by Cu, Zn and Pb in the sediments was found to be slightly higher in dry seasons than in rainy seasons. Copper in sediments amounts to several thousands ppm at Gukou, and a few tens to one hundred ppm at Caijiawan, near the Lake; Zn and Pb concentrations are less [20].

4.3 Simulation experiments

Metals from acid mine drainage are transformed and transported in Dawu-Lo An River, depending on the concentration and the chemical behavior of the elements and the seasonal variation. The main physico-chemical processes occurring at the mining area along the rivers, e.g. neutralization, precipitation, flocculation of Fe and Al in acidic solution, oxidation of Fe(II) in concentrated Fe(III) solution at low pH, adsorption of metals on Fe and Al oxides, clays and ore tailings and flocculation in systems from acid mine drainage with alkaline tailings, were simulated [19,21].

It was found that Fe(III), Al and Cu precipitated completely at pH > 3.0, > 5.0 and 6-7, respectively. Cu, Pb and Zn coprecipitated at pH 5-6 and Cd coprecipitated at pH 7-8. Removal of Cu, Zn and Pb by coprecipitation is shown in Figure 9. Amorphous Fe hydroxides and ore tailings settle down rapidly at pH 5-9 and form coatings with strong sorption capacity for Cu, Pb and Cd (Figure 10).

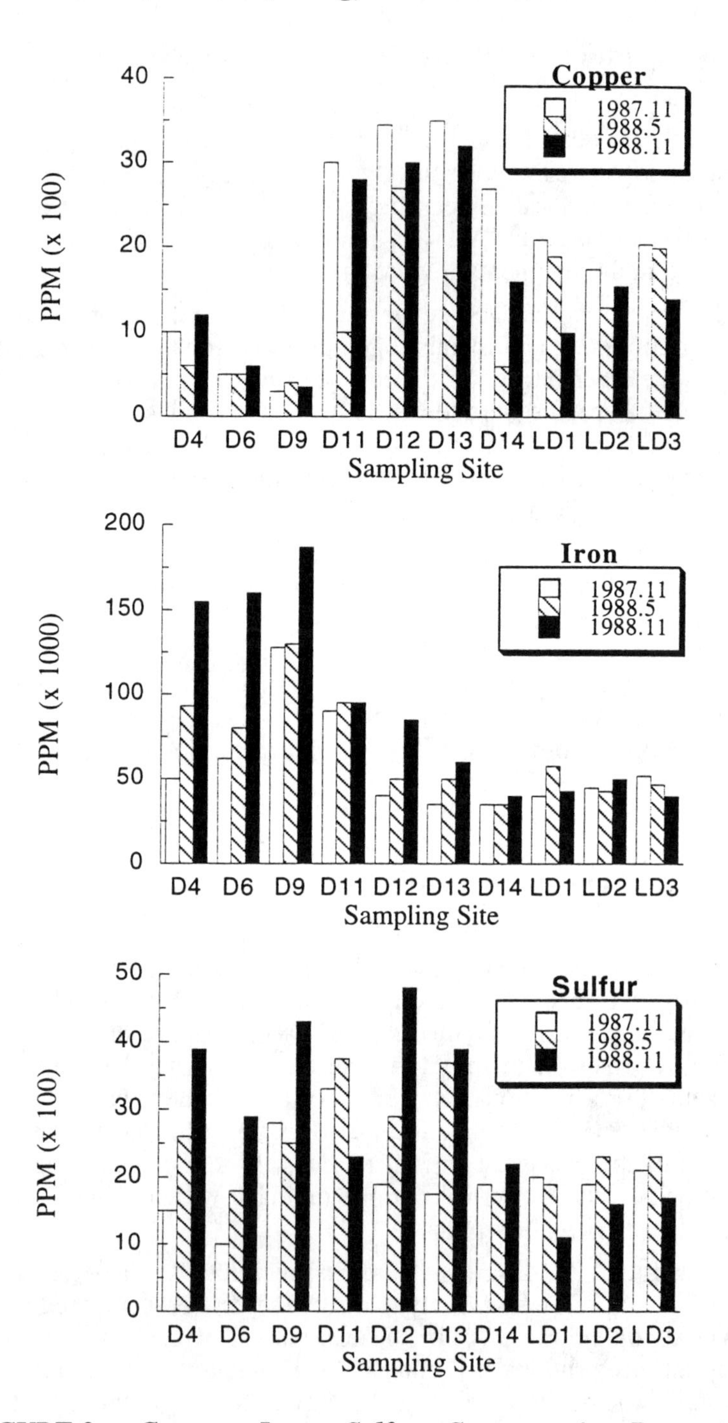

FIGURE 8. Copper, Iron, Sulfur Contents in Dawu River Sediments.

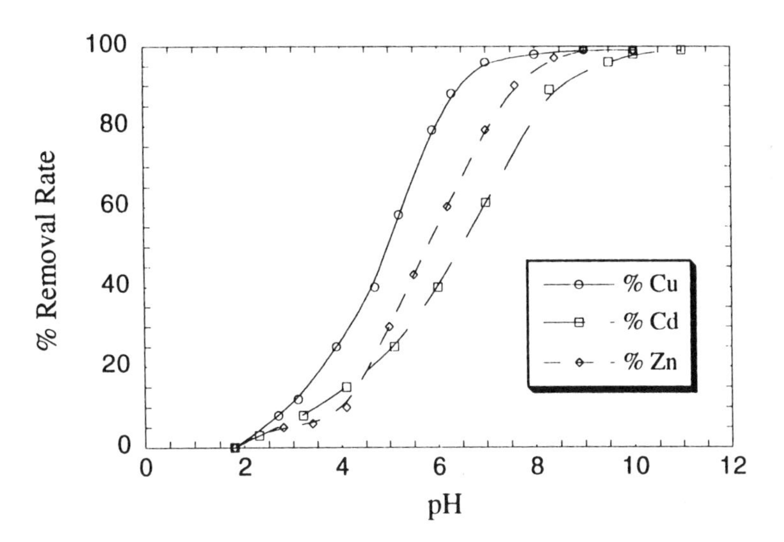

FIGURE 9. Removal of Cu, Zn and Cd.

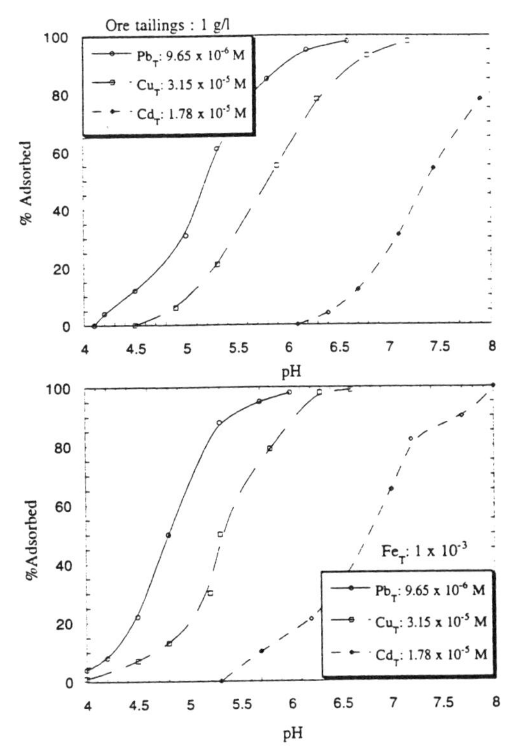

FIGURE 10. Fractional Adsorption of Cu, Pb and Cd on a) Tailings and b) Fe Hydrous Oxide.

Thus, most of the physico-chemical processes would be complete in the Dawu River in dry seasons, while in rainy seasons, these processes would stretch to the Lo An River due to the direct impact of unmixed acid mine water and metals could be released from the sediments containing active fractions of Fe/Al and adsorbed metals. Mixing of acid and alkaline water plays the major role in determining the concentration of the dissolved Fe, Al, Mn and Cu at the lower reach of the Dawu-Lo An River. The adsorption to surface coatings of Fe and Al hydroxides on ore tailings regulates the concentration of trace metals entering into the Lo An River.

4.4 Preliminary study on water quality modeling

In order to evaluate and predict the water quality under different eco-environmental conditions, a chemical equilibrium model for the Dawu River and a river model for the Lo An River were studied [22,23].

The chemical equilibrium computer program MINTEQA2 (U.S. EPA Environmental Research Laboratory Athens) was used for chemical speciation of contaminants involving Dawu River water, acid mine drainage, alkaline wastewater and the mixed waters. The effect of changing water quality and the effect of precipitation, adsorption and flocculation on metal species distribution were calculated. It was shown that Fe, Al, and Cu precipitate completely at pH greater than 4, 5 and 6 respectively. After receiving wastewater, total Cu, Cd and Fe would be precipitated completely at mixing ratios of 2, 4 and 6, respectively (Figure 11) [22].

The river model (assisted by the Institute of Soil Fertility, The Netherlands) involves simple metal speciation in the soluble and articulate fractions of river water and the chemically active and soluble tal fractions in the sediments. The geochemical speciation data was ned by Tessier's procedure (Figure 12) [20].

he water soluble phase, the exchangeable cations, the carbonate nd reducible phases were combined as the chemically active d the sum of the sulfide/organic bound and the residual phases fractions. In the model, the water fraction is divided into suspended matter fractions. Both suspended matter and in chemically active and inert fractions. Using a mass and hydrological data for three segments on the Lo An of Cu and Zn was calculated for different seasons. e to be 50, 80-100 and 500-1200 m^3/s respectively rainy seasons, it was found that Cu and Zn in t near the mining area in dry seasons, while in rainy seasons, Cu and Zn in sediments were largest at Caijiawan near the lake. Cu and Zn amounted to 100 ppm and 300 ppm respectively [23]. These results are in agreement with the surface metal concentration of the core sample obtained at Caijiawan in April, 1989.

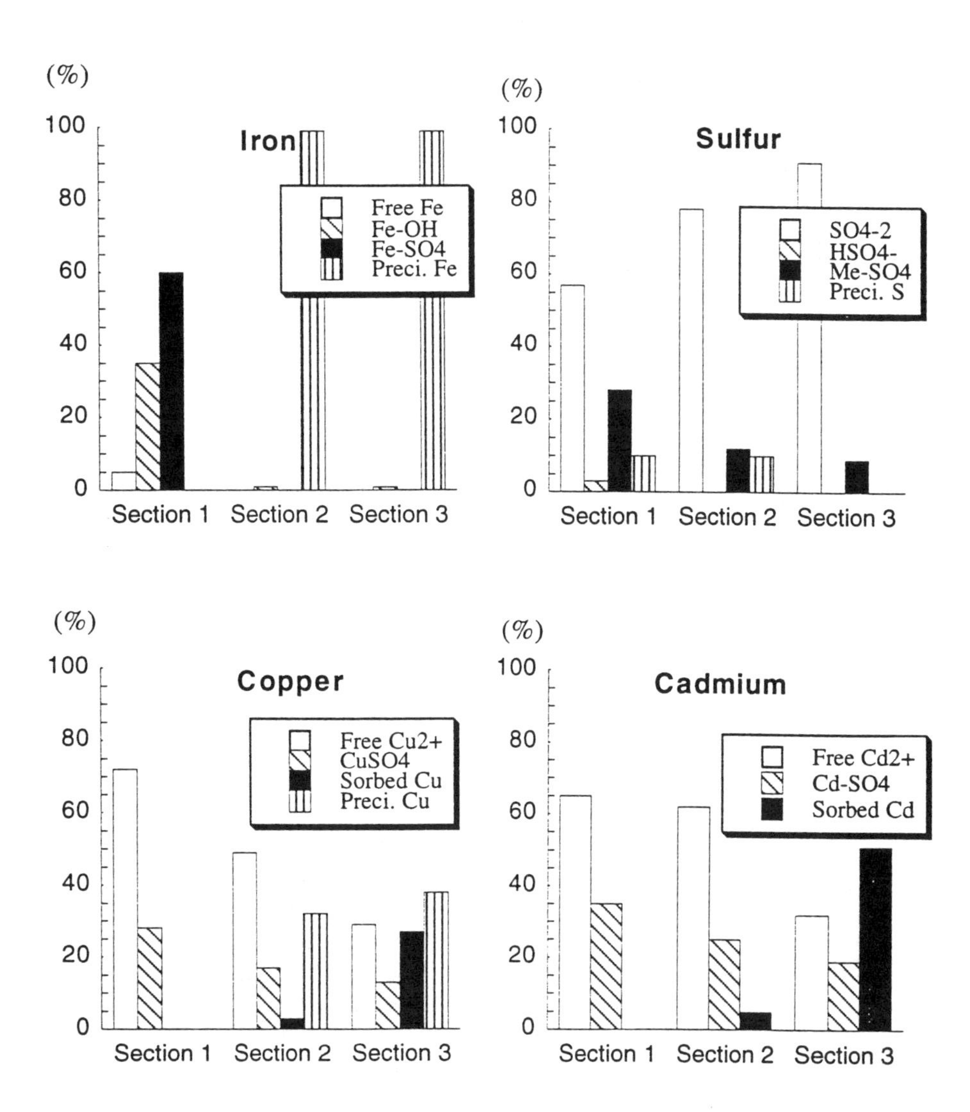

FIGURE 11. Speciation of Fe, S, Cu and Cd in the Dawu River. Sections 1 and 2 Refer to the Section of the Dawu River Before and After Receiving Alkaline Wastewater. Section 3 Refers to the Section of the Dawu River Before Entering into the Lo An River.

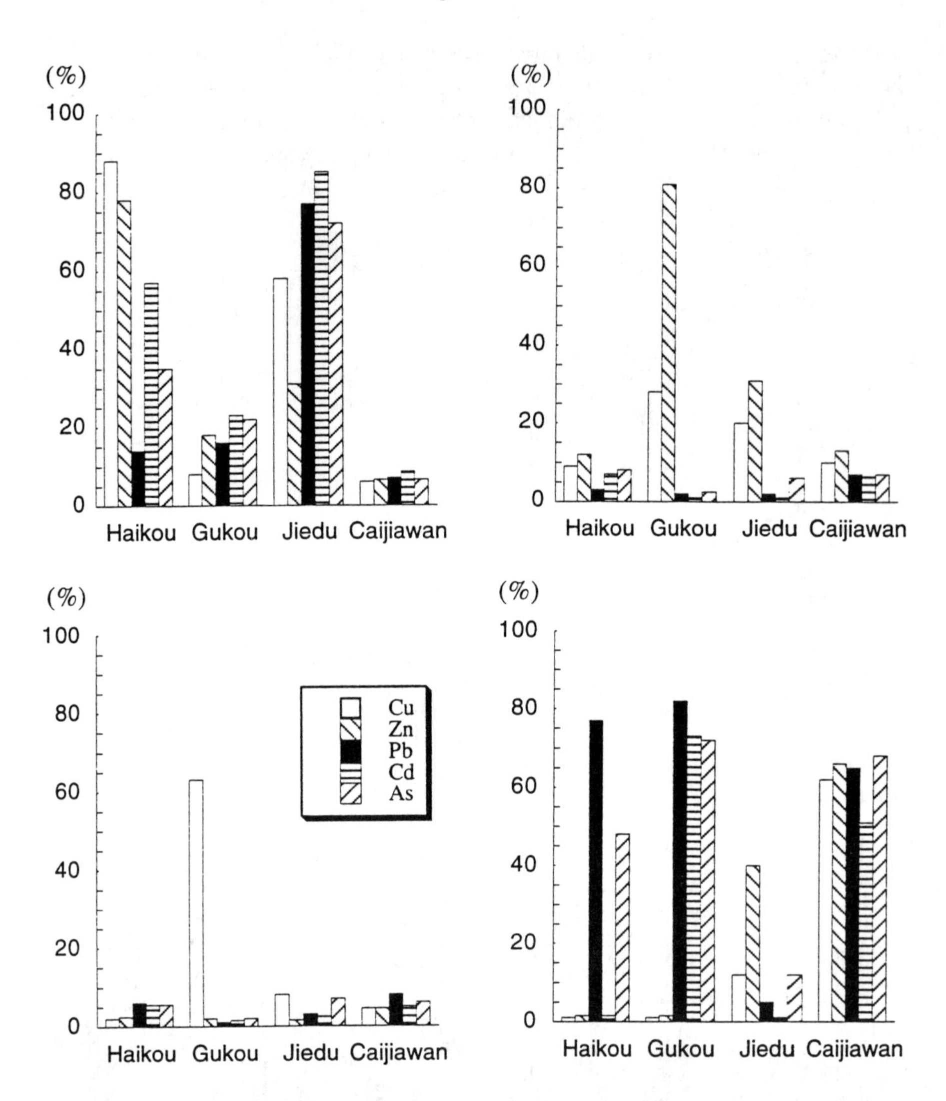

FIGURE 12. Metal Speciation of Sediments in the Lo An River [February 1987] (a) Carbonate Bound Phase, (b) Reducible Phase, (c) Organic-sulfide Bound Phase, and (d) Residual Phase.

4.5 Site-specific sediment quality assessment

An approach to evaluate transport, an inventory of historical input was developed and metal contamination was predicted by interpreting horizontal distribution and accumulation rates in river sediments.

Sediments were collected, using a Van Veen grab sampler at 40 stations along the Lo An River from the entry of wastes from the Dexing Copper Mine to the lake. A few vertical core profiles were taken, using a Phlegar valve corer. Surface sediment fraction <20 μm (top 10-15 cm) and core profiles (1 cm slices) were digested by HNO_3. Cu, Zn, Pb and Cd of total digests and leachate were determined by AAS. Pb-210 and Cs-137 techniques were used for dating the sediment core. The Index of Geoaccumulation (Igeo) [24] was used for assessing metal pollution in aquatic sediments:

$$Igeo = \log_2 C_n / 1.5 B_n \tag{1}$$

where C_n is the measured concentration of the element 'n' in the sediment fraction (<20 μm in this study). B_n is the geochemical background value from unpolluted sediments of the lake.

This index consisted of 7 grades, starting at background level (Igeo-class 0), doubling in each Igeo-class, whereby the highest grade reflects 100-fold enrichment above the background values.

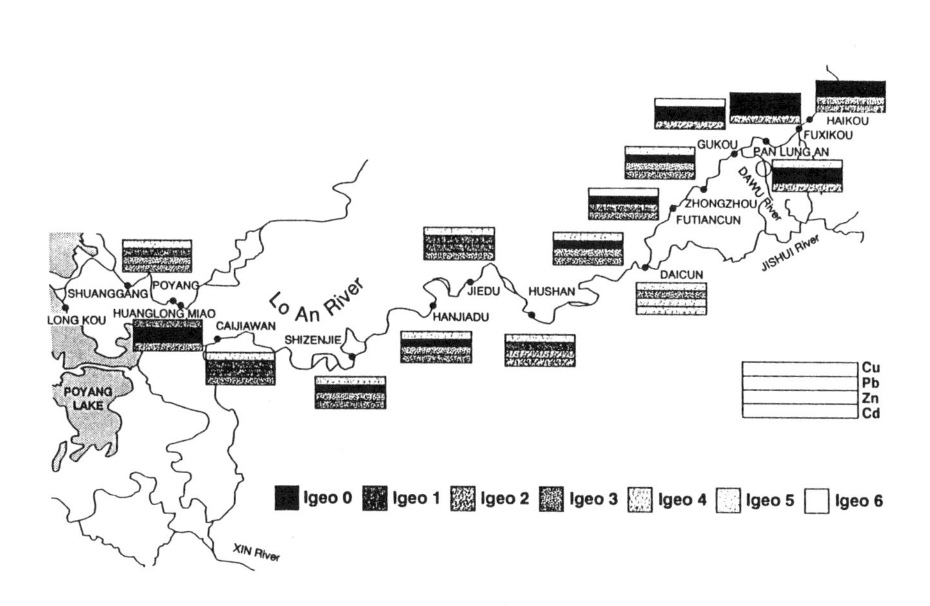

FIGURE 13. Sediment Quality: April 1990.

It was found that near the mine area (at Gukou), Cu increases the highest grade 6, indicating heavy contamination. The indices for Zn and

Cd increase by one grade, as compared with the upstream background values. A high percentage of Cu is partitioned in the sulfide/organic fraction, suggesting the effect of floatation agents. A high percentage of Zn has been found in the reducible fraction, indicating that Fe and Mn hydrous oxides are the main carriers for Zn. The pollution indices for Cu, Zn, Pb and Cd remain 3-4, 2-3, 1-2 and 1-2, respectively in Caijiawan sediment, indicating the sediment near the lake is still moderately contaminated by Cu and Zn, and slightly contaminated by Pb and Cd. Preliminary results from the sediment quality of Lo An River-Poyang Lake area are shown in Figure 12 [25].

Acknowledgments: Dexing Project was a part of the Cooperative Ecological Research Project (CERP) on "Metal Pollution and Its Ecological Effects" (1987-90). This was a joint project of UNESCO (MAB) Programme, Chinese Academy of Sciences and Federal Ministry for Research and Technology, Germany. It was carried out by Research Center for Eco-Environmental Sciences, Academia Sinica, China, in cooperation with Institute of Sedimentology, University of Heidelberg, Germany. Also collaborating were Nanchang Institute of Aeronautical Technology, Environmental Protection Bureau of Jiangxi Province, China and Technical University of Hamburg-Harburg, Germany.

REFERENCES

1. Liu, C.I. and H.X. Tang, "Chemical Studies of Aquatic Pollution by Heavy Metals in China," in *Environmental Inorganic Chemistry*, J.K. Irgolic and A.E. Martell, Eds. 359 pp. VCH Publishers, (1985).

2. Liu, J.Y., H.X. Tang and G. Müller, "UNESCO MAB Cooperative Ecological Research Project on Heavy Metal Pollution and Its Ecological Effects," *J. Environ. Sci.* (China) 4:4-13 (1992).

3. Batley, G.E. and T.M. Florence, "Determination of the Chemical Forms of Dissolved Cadmium, Lead and Copper in Seawater," *Marine Chem.* 4:347-363 (1976).

4. Florence, T.M., "Trace Metal Species in Fresh Water," *Water Res.* 11:681-687 (1977).

5. Eganhouse, R.P., D.R. Young and J.N. Johnson, "Geochemistry of Mercury in Palos Verdes Sediment," *Environ. Sci. Technol.* 12:1151-1157 (1978).

6. Pang, S.W., Q.K. Qu and J.F.Sun, "Sequential Chemical Extraction for Speciation of Mercury in River," *Acta Scient. Circum.* 1:234-241 (In Chinese 1981).

7. Peng, A. and W.H. Wang, "Extraction and Characterization of Humic Acid from Ji Yun River," *Acta Scient. Circum.* 1:126-139 (In Chinese 1981).

8. Zhang, S., Y.J. Tang and W.L. Yang, "Characteristics of Mercury Polluted Chemical Geography of Ji Yun River," *Acta Scient. Circum.* 1:349-362 (In Chinese 1981).

9. Lin, Y.H., T.M. Kang and C.I. Liu, "Speciation and Distribution of Mercury in Ji Yun River Sediment," *Environ. Chem.* 2:10-19 (In Chinese 1983).

10. Peng, A. and W.H. Wang, "Influence of Humic Acid on the Transport and Transformation of Mercury in Ji Yun River," *Environ. Chem.* 2:33-38 (In Chinese 1982).

11. Reimers, R.S. and P.A. Krenkel, "Kinetics of Mercury Adsorption and Desorption in Sediments," *J. Water Pollut. Control Fed.* 46:352-365 (1974).

12. Li, C.S. and C.I. Liu, "Adsorption of Mercury in Freshwater," *Environ. Chem.* 1:304-313 (In Chinese 1982).

13. Mao, M.Z., Z.H. Liu and J.X. Wei, "Chemical Speciation of Heavy Metals of the Surface Sediments in Xiang River," *J. Environ. Sci.* 2:355-361 (China, 1981).

14. Mao, M.Z., Z.H. Liu and H.R. Dong, "Distribution of Trace Metals in the Surface Sediments of Xiang River by Density Gradient Method," *Environ Chem.* 1:168-175 (In Chinese 1982).

15. Tang, H.X., H.B. Xue, B.Z. Tian, H.R. Dong and P.S. Lei, "Study on Multielement Adsorption Model of Aquatic Sediments with a Sequential Chemical Separation Procedure," *Acta Scient. Circum.* 2:279-292 (In Chinese 1982).

16. Tang, H.X., H.B. Xue, M.Z. Mao and Z.K. Luan, "The Chemical Stability of Heavy Metals in a Natural Water System," *J. Environ. Sci.* 1:55-64 (1989).

17. Tessier, A., P.G.C. Camphell and M. Bisson, "Sequential Extraction Procedure for the Speciation of Particulate Trace Metals," *Anal. Chem.* 51:844-851 (1979).

18. Förstner, U. and G. Wittmann, *Metal Pollution in the Aquatic Environment*, p. 498 Springer, (Berlin, 1983).

19. Tang, H.X., Z.K. Luan, F.C. Cao and B.Z. Tian, "General Survey of Metal Pollution in Dawu River at Dexing Copper Mine," *J. Environ. Sci.* 4:36-41 (China, 1992).

20. Mao, M.Z., Z.H. Liu and H.R. Dong, "Distribution and Speciation of Heavy Metals Along Lo An River," *J. Environ. Sci.* (China) 4:72-81 (1992).

21. Luan, Z.K., H.X. Tang and F.C. Cao, "Chemical Processes of Acid Mine Drainage in the Aquatic System of Copper Mine Area," *J. Environ. Sci.* (China) 4:42-48 (1992).

22. Chen, M., H.X. Tang and Z.K. Luan, "Chemical Speciation of Pollutants in Waters of Dexing Copper Mine Area by Using Chemical Equilibrium Model-MINTEQA2," *J. Environ. Sci.* 4:58-67 (China, 1992).

23. Lin, Y.H., and Q. Li, "Study on the Metal Speciation Modelling on Lo An River," *J. Environ. Sci.* (China) 4:100-108 (1992).

24. Müller, G., "Schwermetalle in den Sedimenten des Rheins - Veranderungen seit 1971," Umschau 79:778-783 (1979).

25. Schmitz, W., M.Z. Mao, N. Ramezani and Z.H. Wang, "Heavy Metal Pollution of Aquatic Sediment in the Lo An River-Poyang Lake Area, Jiangxi Province, China," *J. Environ. Sci.* (China) 4:88-90 (1992).

APPLICATION OF INORGANIC-CONTAMINATED GROUND WATER TO SURFACE SOILS AND COMPLIANCE WITH TOXICITY CHARACTERISTIC (TCLP) REGULATIONS

Chris L. Bergren[1], Mary A. Flora[1], Jeffrey L. Jackson[2] and Eric M. Hicks[2]
[1]Westinghouse Savannah River Company
Savannah River Site
Aiken, SC 29808
[2]Sirrine Environmental Consultants
P.O. Box 24000
Greenville, SC 29616

1. ABSTRACT

The Westinghouse Savannah River Company (WSRC) is currently implementing a Purged Water Management Program (PWMP) at the Savannah River Site (SRS) near Aiken, South Carolina. A variety of constituents and disposal strategies are being considered. Constituents investigated in the PWMP include radionuclides, organics, and inorganics (As, Ba, Cd, Cr, Pb, Hg, Se, and Ag). One practical disposal alternative is to discharge purged water (all constituents below regulatory levels) to the ground surface near the monitoring well that is being purged. The purpose of this investigation is to determine if long-term application of purged water that contains inorganic constituents (below regulatory levels) to surface soils will result in the accumulation of inorganics such that the soil becomes a hazardous waste according to the Toxicity Characteristic regulations (40 CFR Part 261.24). Two study soils were selected that encompass the range of soils found at the SRS: Lakeland series and Orangeburg series. Laboratory batch equilibrium studies indicate that the soils, although able to retain a large

0-87371-697-3/95/$0.00+$.50

amount of inorganics, will not exceed Toxicity Characteristic concentrations when subjected to the TCLP. Field studies are underway to confirm this.

2. BACKGROUND

The Westinghouse Savannah River Company (WSRC), operator of the Savannah River Site (SRS) near Aiken, South Carolina, is currently investigating the hydrogeology of the SRS for environmental compliance purposes. Over 1200 monitoring wells have been installed for this investigation. The water removed from a monitoring well prior to sampling is referred to as "purged water." Purged water is normally discharged on the ground surface away from the well from which it was removed. Because purged water may have an impact on the local environment, SRS has, for specific parameters, established specific concentration limits above which purged water may require special handling. These concentration limits are referred to as trigger levels.

SRS has established a preliminary plan for managing purged ground water [1]. This plan includes recommended trigger levels for constituents in purged water. Trigger levels were selected using health-based criteria and regulatory definitions of hazardous waste (e.g., Toxicity Characteristics Leaching Procedure or TCLP). The selection and justification of trigger levels is further discussed in the Purged Water Management Plan (PWMP) [1].

There are three general categories of contaminants in the PWMP: radionuclides, organics, and inorganics. Specific parameters in each category have trigger levels that govern their disposal. Purged ground water that contains any constituents above the trigger level will be contained and treated to an acceptable level. Purged ground water that does not exceed any trigger levels is discharged directly on the ground surface near the monitoring well.

This project is concerned with proper disposal of inorganic constituents in purged ground water. Recent revisions to 40 CFR Part 261.24 (RCRA) specify eight regulated TCLP inorganic constituents, primarily metals. These inorganics are arsenic (a metalloid), barium, cadmium, chromium, lead, mercury, selenium (a nonmetal), and silver. The TCLP is used to determine the amount of these inorganic constituents that can leach out of the parent material (e.g., soils, sludges, solids). The concentration in the extracted leachate is determined and then compared to the regulatory level (40 CFR Part 261.24; Table 1).

Table 1. Summary of DCB-1A Well Water Data (PWMP)*

Parameter	Initial Concentrations: Historical SRS Data Range	n	Mean	September 1990 Data	Conc. After Spiking	Trigger Level 40 CFR, 261.24
TCLP Inorganics (mg/L)						
As (‡)	0-0.270	4	0.070	<0.005	5.1	5
Ba (‡)	0.010-0.050	7	0.018	0.015	0.35(1)	100
Cd (‡)	0.018-0.091	11	0.033	0.019	1.0	1
Cr (‡)	0.020-0.240	11	0.140	0.084	4.7	5
Pb (‡)	0.010-0.310	11	0.080	<0.150	0.5(1)	5
Hg (‡)	0.0001-0.112	11	0.0003	0.0005	0.14(1)	0.2
Se (‡)	0.002-0.027	10	0.008	<0.025	0.89	1
Ag (‡)	0.0001-0.039	10	0.012	<0.025	4.0	5
Other Metals (mg/L)						
Aluminun	500	1	500	340	300	NA
Beryllium	0.03-0.11	9	0.07	<0.1	<0.1	NA
Calcium	ND	ND	ND	210	190	NA
Copper	0.2-1.8	10	0.8	1	0.59	NA
Iron	10-1750	11	420	100	100	NA
Lithium	0.55	1	0.55	<0.5	<0.5	NA
Magnesium	70-425	3	230	180	160	NA
Manganese	10-50	11	26	25	22	NA
Nickel	0-7.5	10	2.7	2.3	2.1	NA
Potassium	0.66-2.5	2	21	<2.0	3.4	NA
Sodium	12-34	7	6	20	18	NA
Zinc	0-12	10	-	5.3	4.8	NA
Anions (mg/L)						
Chloride	1.5-18	7	7.3	ND	ND	NA
Fluoride	<0.1-14.2	8	3.9	ND	ND	NA
Nitrate as (N)	0.27	1	0.27	ND	ND	10,000(2)
Phosphate ‡	0.04	1	0.04	ND	ND	NA
Sulfate	540-8300	15	4700	ND	ND	NA

Notes:

*PWMP - Purged Water Management Program
NA - not applicable
ND - not determined
(1) - average of data from two laboratories
(2) - proposed trigger level for nitrate [1]
(‡) - total

If a constituent equals or exceeds the regulatory level, the parent material is considered a characteristic hazardous waste under the RCRA program, and must be handled as a hazardous waste.

The purpose of this study is to determine or predict the long-term effect of applying purged water (in which all eight TCLP inorganics are below regulatory levels) to the soil surface. Because soil has the ability to retain inorganics (e.g., adsorption and precipitation), it is possible that long term application of ground water with low levels of TCLP inorganics could result in the buildup of inorganic constituents at the soil surface. Eventually, the soil could retain enough inorganics such that it failed to meet the TCLP regulatory levels and hence become a hazardous waste.

This study is being conducted in two phases, laboratory and field. The laboratory study was a batch study to determine the capacity of two study soils to retain TCLP inorganics. The field study is being conducted using purged water at two study sites at the SRS. We did not intend to:

(a) perform adsorption isotherms and determine equilibrium distribution coefficients (Kd).

(b) discern the retention mechanism (e.g., adsorption, precipitation, complexation).

(c) determine the transport rate of TCLP inorganics through the vadose zone.

(d) predict the impact on ground water of TCLP inorganics.

3. METHODS

Following is a discussion of the (a) selection and sampling of the soil study sites, (b) selection and sampling of the purged water monitoring well, (c) laboratory study, and (d) field study.

3.1 Selection and sampling of study sites

Two soil series were selected that generally encompassed the range of physical and chemical properties of the soil series found at the SRS. The results from an extensive soils study conducted at the SRS was examined to aid in selection of study soils [2].

One of the more significant parameters affecting fate of inorganics (i.e., metals) in soils is cation exchange capacity (CEC). CEC is the sum total of exchangeable cations that a soil can adsorb and is a function of minerology, soil texture, pH, organic matter content, and within a given soil texture, the amount and type of clay [3]. Other factors being equal, soils with a high clay content (>35% particles by mass <0.002 mm per USDA soil classification system) typically have a medium to high CEC. Soils that are predominantly sand (>90% particles by mass from 0.05 to 2.0 mm) have a low CEC. Both types of soil profiles are found at SRS. The range of responses of soils to inorganics-bearing water (e.g., the TCLP metals) were met by selecting a soil profile that contained a significant amount of clay and a profile that contained a significant amount of sand.

General soil horizon descriptions, based on USDA soil classification, were examined to aid in the selection of study soils. In addition, the general occurrence and location of the soils at SRS were also considered. After review of SRS soil data, the two soil types selected were the Lakeland series (sand) and Orangeburg series (sand/sandy clay) [2]. Soil descriptions are provided in Tables 2 and 3. Because of the extremely sandy, highly permeable, and pedogenically undeveloped character of the Lakeland soil profile, it is highly infertile and droughty. As a result, Lakeland soils generally support an ecosystem tolerant of these conditions, such as scrub oaks and lichens at our site. Soil was collected from 0-8 inches (0-20 cm) below the surface and at 30-36 inches (76-91 cm) below the surface at the Lakeland site. The Orangeburg site is in a planted pine forest. In contrast to the Lakeland profile which is sandy throughout, the Orangeburg series typically has a sandy surface but an argillic (subsoil) horizon of sandy clay to sandy clay loam texture. Because of this horizonation, the Orangeburg soils are better for agricultural and silvaculture use because of the better fertility and availability of water, both of which are attributable to the higher clay content and concomitant CEC in the subsoil. Soil was collected from 0-6 inches (0-15 cm) and at 30-36 inches (76-91 cm). Chemical and physical characteristics of the study soils are provided in Tables 4 and 5.

Approximately 20 kg of each soil type (10 kg of surface soil, 10 kg of subsurface soil) was collected at each site. Soil was collected and transported in one gallon plastic bags. A sample of each soil horizon was analyzed for the following: pH, moisture content, bulk density, particle density, total CEC, total As, Ba, Cd, Cr, Pb, Hg, Ag, Se, Al, Be, Ca, Cu, Fe, Li, Mg, Mn, Ni, K, Na, Zn, and TCLP inorganics after the TCLP extraction.

Table 2. Lakeland Series*

The Lakeland series consists of excessively drained, rapidly permeable soils that formed in sandy marine sediment on the Coastal Plain and in areas intermingled with the Sand Hills. These soils are on broad ridges and side slopes. Slopes range from 0 to 10 percent. Lakeland soils are classified as thermic, coated Typic Quartzipsamments.

Lakeland soils are on the same general landscape as Blanton, Troup, Fuquay, Wagram, and Lucy soils. The associated soils have a Bt horizon.

Typical pedon of Lakeland sand, 0 to 6 percent slopes:

Ap - 0 to 3 inches (0 to 8 centimeters); yellowish brown (10YR 5/4) sand; single grained; loose; few fine and medium roots; very strongly acid; abrupt wavy boundary.

C1 - 3 to 50 inches (8 to 127 centimeters); very pale brown (10YR 7/4) sand; single grained; loose; few fine and medium roots; most sand grains coated; strongly acid; gradual wavy boundary.

C2 - 50 to 60 inches (127 to 152 centimeters); light yellowish brown (10YR 6/4) sand; single grained; loose; about 15 percent clean sand grains; strongly acid; gradual wavy boundary.

C3 - 60 to 80 inches (152 to 203 centimeters); very pale brown (10YR 7/4) sand; loose; about 20 percent clean sand grains; few coarse sand grains and small pebbles; moderately acid.

The thickness of the sandy layers is more than 80 inches. The soils are moderately acid to very strongly acid throughout.

**Rogers* [4]

3.2 Selection and sampling of the purged water monitoring well

A monitoring well was selected that

(1) contained ground water with relatively high concentrations of TCLP inorganics but with concentrations below trigger levels,

(2) was capable of providing the quantities of water required by the lab and field studies, and

(3) was easily accessible to sampling personnel.

Table 3. Orangeburg Series

The Orangeburg series consists of well drained, moderately permeable soils that formed in loamy marine sediment on the Coastal Plain. These soils are on broad ridgetops; moderately long, smooth side slopes; and gently rolling breaks below gentle side slopes and nearly level ridgetops. Slopes range from 0 to 10 percent. Orangeburg soils are classified as fine-loamy, siliceous, thermic Typic Paleudults.

Orangeburg soils are on the same general landscape as Ailey, Dothan, Fuquay, Lucy, Norfolk, Vaucluse, and Wagram soils. Ailey, Fuquay, Lucy, and Wagram soils are in the arenic subgroup. Norfolk soils have a yellowish brown subsoil. Dothan and Fuquay soils have over 5 percent nodules of plinthite in the subsoil. Vaucluse and Ailey soils have brittle and cemented layers in the subsoil. They are on the complex slope breaks.

Typical pedon of Orangeburg loamy sand, 2 to 6 percent slopes:

Ap - 0 to 6 inches (0 to 15 centimeters); dark yellowish brown (10YR 4/4 loamy sand; weak medium granular structure; very friable; common fine roots; very strongly acid; abrupt wavy boundary.

Bt1 - 6 to 10 inches (15 to 25 centimeters); yellowish red (5YR 4/6) sandy loam; weak medium subangular blocky structure; very friable; few faint clay films on faces of peds; few fine, medium, and large roots; very strongly acid; gradual wavy boundary.

Bt2 - 10 to 17 inches (25 to 43 centimeters); yellowish red (5YR 5/6) sandy loam; weak medium subangular blocky structure; friable; few faint clay films on faces of peds; common fine and few medium and large roots; very strongly acid; gradual wavy boundary.

Bt3 - 17 to 33 inches (43 to 84 centimeters); yellowish red (5 YR 4/6) sandy loam; weak medium and coarse subangular blocky structure; friable; common faint clay films on faces of peds; common fine and few medium roots; few fine holes; very strongly acid; diffuse smooth boundary.

Bt4 - 33 to 56 inches (84 to 142 centimeters); red (2.5YR 4/8) sandy clay loam; weak coarse subangular blocky structure; friable; common faint clay films on faces of peds; few medium roots; few fine holes; very strongly acid; gradual wavy boundary.

Bt5 - 56 to 62 inches (142 to 157 centimeters); red (2.5 YR 4/8) sandy clay loam; few fine prominent strong brown (7.5 YR 5/6) mottles; weak coarse subangular blocky structure; friable; common faint clay films on faces of peds; few small pebbles of quartz; very strongly acid.

Table 4. Lakeland Soil (Sand) Data Summary (PWMP)*

Parameter and Regulatory Level (mg/L)	Lakeland Soil 0-8 in. Initial Conc.	Final Conc. ($4L H_2O$)	Final Conc. ($8L H_2O$)	Lakeland Soil 30-36 in. Initial Conc.	Final Conc. ($4L H_2O$)	Final Conc. ($8L H_2O$)
TCLP Inorganics (mg/L)						
As (5.0)	<0.005	<1.0	<1.0	<0.005	<1.0	<1.0
Ba (100.0)	0.248	<1.0	<1.0	0.204	<1.0	<1.0
Cd (1.0)	<0.005	<0.5	<0.5	<0.005	<0.5	<0.5
Cr (5.0)	<0.008	<1.0	<1.0	<0.008	<1.0	<1.0
Pb (5.0)	<0.030	<1.0	<1.0	<0.030	<1.0	<1.0
Hg (0.2)	<0.002	<0.2	<0.2	<0.002	<0.2	<0.2
Se (1.0)	<0.005	<0.5	<0.5	<0.005	<0.5	<0.5
Ag (5.0)	<0.005	<1.0	<1.0	<0.005	<1.0	<1.0
Total Inorganics (mg/kg)						
As (total)	<0.603	69	107	<0.432	54	123
Ba (total)	11.1	1600	2500	14.6	1300	2900
Cd (total)	<0.301	15	23	<0.0216	12	24
Cr (total)	2.55	79	117	3.52	64	135
Pb (total)	3.22	298	94	<1.29	51	1005
Hg (total)	<0.100	1.4	4.0	<0.100	1.4	1.6
Se (total)	<0.301	13	21	<0.216	11	24
Ag (total)	<0.301	72	109	<0.216	57	126
Other Metals (total mg/kg)						
Aluminun	2900	9700	1200	3700	8800	114,000
Beryllium	<1.0	<1.0	<1.0	<1.0	<1.0	<2.1
Calcium	26	420	332	38	190	270
Copper	1.4	12	17	1.3	9.5	19
Iron	1700	3700	4200	2200	3600	5400
Lithium	1.23	7.3	9.2	1.9	6.4	11
Magnesium	82	580	510	151	450	680
Manganese	30	190	180	2.2	110	210
Nickel	1.1	37	56	79	31	64
Potassium	48	78	55	1.5	570	122
Sodium	<110	700	1000	5.1	73	860
Zinc	5.6	89	132			147
Other Analyses						
pH	4.9	6.55	6.45	5.0	6.53	6.53
CEC	0.3	NA	NA	0.3	NA	NA
Total CEC	0.9	NA	NA	0.7	NA	NA
% org. matter	2.0	NA	NA	<0.1	NA	NA
% sand (1)	98	NA	NA	92	NA	NA
% silt (1)	0	NA	NA	6	NA	NA
% clay (1)	2	NA	NA	2	NA	NA
Dry Bulk Density	1.33	NA	NA	1.54	NA	NA
Particle Density	2.53	NA	NA	2.60	NA	NA
Gravimetric moisture %	2.21	32.9	32.9	2.76	31.4	42.1

Notes:
*PWMP - Purged Water Management Program
(1) - USDA soil classification system
N/A - not analyzed or not applicable
TCLP - Toxicity Characteristic Leaching Procedure (40 CFR Part 261.24)
Soil concentrations are reported on a dry-weight basis

Table 5. Orangeburg Soil (Clay) Data Summary (PWMP)*

	Orangeburg Soil 0-8 in.			Orangeburg Soil 30-36 in.		
Parameter and Regulatory Level (mg/L)	Initial Conc.	Final Conc. (4L H_2O)	Final Conc. (8L H_2O)	Initial Conc.	Final Conc. (4L H_2O)	Final Conc. (8L H_2O)
TCLP Inorganics (mg/L)						
As (5.0)	<0.005	<1.0	<1.0	<0.100	<1.0	<1.0
Ba (100.0)	0.250	<1.0	<1.0	0.289	<1.0	<1.0
Cd (1.0)	<0.005	<0.5	<0.5	<0.005	<0.5	<0.5
Cr (5.0)	<0.008	<1.0	<1.0	<0.008	<1.0	<1.0
Pb (5.0)	<0.030	<1.0	<1.0	<0.030	<1.0	<1.0
Hg (0.2)	<0.002	<0.2	<0.2	<0.002	<0.2	<0.2
Se (1.0)	<0.005	<0.5	<0.5	<0.005	<0.5	<0.5
Ag (5.0)	<0.005	<1.0	<1.0	<0.005	<1.0	<1.0
Total Inorganics (mg/kg)						
As (total)	1.22	60	123	4.03	133	406
Ba (total)	3.51	1400	2900	32.6	2900	10,000
Cd (total)	<0.355	13	23	<0.355	25	68
Cr (total)	4.43	72	131	39.2	213	559
Pb (total)	<2.13	63	105	<21.3	129	339
Hg (total)	<0.100	1.4	2.6	<0.100	2.4	6.6
Se (total)	<0.355	12	23	<3.55	17	84
Ag (total)	<0.355	63	120		125	413
Other Metals (total mg/kg)						
Aluminun	3900	9600	15,000	38,000	68,000	175,000
Beryllium	<1.0	<1.0	2.1	<1.0	<1.0	10
Calcium	68	290	330	220	420	700
Copper	0.96	10	20	9.1	29	91
Iron	2000	3600	5100	25,000	57,000	98,000
Lithium	1.5	6.8	11	11	19	70
Magnesium	87	440	620	380	500	1800
Manganese	89	170	260	39	350	900
Nickel	1.8	33	60	9.8	76	245
Potassium	51	59	92	240	230	750
Sodium	14	660	660	16	3400	6400
Zinc	3.7	78	141	15	167	545
Other Analyses						
pH	5.0	6.54	6.51	5.3	6.44	6.47
CEC	0.7	NA	NA	0.3	NA	NA
Total CEC	0.9	NA	NA	0.7	NA	NA
% org. matter	1.4	NA	NA	<0.1	NA	NA
% sand (1)	90	NA	NA	92	NA	NA
% silt (1)	8	NA	NA	6	NA	NA
% clay (1)	2	NA	NA	2	NA	NA
Dry Bulk Density	1.57	NA	NA	1.54	NA	NA
Particle Density	2.63	NA	NA	2.60	NA	NA
Gravimetric moisture %	2.84	33.5	33.4	13.3	73.7	85.7

Notes:
*PWMP - Purged Water Management Program
(1) - USDA soil classification system
N/A - not analyzed or not applicable
TCLP - Toxicity Characteristic Leaching Procedure (40 CFR Part 261.24)
Soil concentrations are reported on a dry-weight basis

After a review of a SRS monitoring well inventory, monitoring well DCB-1A was selected. Chemical characteristics of this water are provided in Table 1. Note that the inorganic TCLP constituents, in spite of their elevated levels, are orders of magnitude lower than the trigger levels.

In addition to the relatively low pH and high metals concentrations, DCB-1A ground water has elevated levels of sulfate (Table 1). Sulfate concentrations ranged from 540 to 8300 mg/L with an arithmetic average of 4700 mg/L. The effect of elevated sulfate on the formation of insoluble precipitates will be further discussed in the Results and Discussion section.

Approximately 340 liters (90 gallons) of DCB-1A well water was collected for the laboratory study. DCB-1A water was slightly turbid with a yellow-orange cast to it, even after purging the well with the dedicated submersible pump. The historical pH of the purged water from DCB-1A is approximately 2.5. Water was collected and transported in polyethylene containers.

After the water was transported to the treatability laboratory, the concentrations of TCLP inorganics were increased to the trigger levels (Table 1) by adding the appropriate inorganic TCLP constituent in a water soluble form (i.e., silver nitrate to raise the dissolved silver concentration). The entire batch of spiked water was mixed at the same time in a 50 liter polyethylene carboy. This water is referred to as spiked DCB-1A water. Addition of the inorganic stock solutions caused the formation of white, cloudy precipitates (further discussed in the Results and Discussion section). Prior to sampling or use of the water, the 50 liter carboy was thoroughly mixed to resuspend any precipitated materials.

3.3 Laboratory study

Batch studies were conducted for each of the two study soils to determine the amount of TCLP inorganics that they were capable of retaining (e.g, adsorption). CEC data from a previous study [2] were used to design the batch experiments. According to that study, CEC for the Orangeburg soil averaged 6 meq/100 grams while the Lakeland was 3 meq/100 grams [2]. We estimated that 300 grams of Lakeland soil would require about 2.5 liters of purged water (with TCLP inorganics concentrations at trigger levels) to equal maximum adsorption capacity. The Orangeburg soil, because of its greater CEC, was estimated to require twice as much purged water (5 liters) per 300 grams of soil to achieve maximum inorganics adsorption. Thus, we decided to use 4 and 8 liters of spiked DCB-1A water per 300 grams of the study soil.

Selection of an equilibrium time is difficult for a complex system such as soil, inorganics, and water. Adsorption is generally regarded as a fast reaction. Solute (e.g., metals) adsorption processes are often initially rapid, while further reduction in solute concentration continues at a decreasing rate, asymptotically approaching a constant concentration.

Based on these considerations, the Environmental Protection Agency (EPA) has suggested an operational definition of equilibrium for soil/water adsorption studies [5]. This operational definition of equilibrium time is the minimum amount of time needed to establish a rate of change of the solute concentration in solution equal to or less than 5% per 24-hour interval. Studies conducted and reviewed by the EPA have shown that 24 to 48 hours are often adequate to meet this operational definition of equilibrium. The American Society for Testing and Materials (ASTM) Method 4319 specifies a minimum contact time of 3 days and a maximum of 14 days or longer [6]. Soils and spiked purged water in this study, therefore, were continuously mixed for approximately two weeks.

A straight factorial experimental design was employed: two soils, two horizons, and two soil/water mixture ratios (i.e., 8 combinations). Three hundred grams of surface and subsurface Lakeland and Orangeburg soils were each added to four and eight liters of spiked DCB-1A ground water. Thus, there were eight combinations of soil and water. For example, the Lakeland experimental series was:

- Lakeland 0-8 inch soil with 4 liters of spiked water
- Lakeland 0-8 inch soil with 8 liters of spiked water
- Lakeland 30-36 inch soil with 4 liters of spiked water
- Lakeland 30-36 inch soil with 8 liters of spiked water.

Soil was air dried prior to use (i.e., not oven dried). Soil and water were mixed in 4 liter polyethylene, wide mouth jars. For the 8 liter volumes, 150 grams of soil was placed with 4 liters of spiked water in a polyethylene jar. After mixing, the pH of the resultant soil/water mixture was adjusted to 6.5 with 4 N NaOH. The soil/water mixtures were placed on a rotary extractor at 30 rpm and the pH periodically adjusted to 6.5 with either 4 N NaOH or 2 N HCl.

A pH of 6.5, although higher than the soil pH (approximately 5.0 for both soil types), was selected as a worst case scenario. Using the geochemical equilibrium model GEOCHEM [7], the TCLP metals (Ba, Cd, Cr, Hg, Pb, Ag) are predicted to be less soluble at a pH of 6.5 than at a pH of 5.0. We reasoned that at a lower solubility (higher pH), there was a greater chance that the TCLP metals would be retained by the soil and thus approach the regulatory level when subjected to the TCLP test. Arsenic (metalloid) and selenium (nonmetal) do not necessarily behave the same as the TCLP metals. For example, pentavalent arsenic adsorption has been shown to be greater at lower pH values [5]. Adsorption of trivalent arsenic, used in these experiments, was not pH dependent from pH 4 to 10 [5].

At the completion of the mixing, the containers were allowed to stand undisturbed for at least 48 hours. The supernatant was decanted from the soil, filtered (0.45 micron), and submitted for total TCLP inorganics analysis to determine the dissolved concentrations of TCLP inorganics remaining in the supernatant.

Soil was vacuum filtered through 15 cm diameter filter paper (Whatman 41) to remove as much of the water as possible. Samples of the soil were submitted for total and TCLP inorganics analysis. This step collected both adsorbed and precipitated TCLP inorganics, similar to the application of purged water to the soil surface at a monitoring well.

3.4 Field study

A field study is being conducted to better predict the effect of long-term application of purged water on SRS soils. Test plots were not disturbed (e.g., graded, vegetation removed) prior to application of purged water. Basically, the field study involves the stepped-up application rate of unspiked DCB-1A ground water to the study plots. Many monitoring wells at SRS are sampled every 3 months or less frequently. Thus, application of purged water at those wells occurs at most 4 times per year. The field study involves application of purged water on a weekly basis, or 13 times more frequently than the most heavily sampled monitoring wells.

DCB-1A water is uniformly applied to the test plots by pouring water into high density polyethylene trays (46 x 66 x 23 cm deep or 18 x 26 x 9 inches deep) that are arranged in a group over the study plots. Each group of trays is 8 trays long (429 cm or 169 inches) by 2 trays wide (137 cm or 54 inches). Application area for the field plots is therefore 5.9 square meters (63 square feet). The trays are supported by a PVC pipe rack such that the bottom of the trays are approximately 10 cm (4 inches) above the ground surface. Each tray has 12 evenly spaced 3/32 inch (2.4 mm) diameter holes in the bottom to allow the purged water to slowly flow out onto the study plot. For each application of purged water, 9.5 liters (2.5 gallons) of freshly collected DCB-1A water is poured into each plastic tray (described above). This results in an application rate of 2.5 cm (1 inch) of purged water per study plot per week. Time to empty a tray is about 15 minutes. Runoff of water on the ground surface is minimal. If the ground appears to be hydrophobic (i.e., dry), potable water is first applied as a fine mist to wet the soil and hence reduce runoff from under the trays.

Surface and shallow subsurface samples will be collected after approximately 3, 6, and 12 months of weekly application. Soil samples will be analyzed for total TCLP inorganics and TCLP inorganics after the TCLP extraction.

4 RESULTS AND DISCUSSION

4.1 Comparison of spiked water to targeted trigger levels

DCB-1A ground-water data are summarized in Table 1. Spiked water was analyzed to ensure that the TCLP inorganics were added such that their concentrations were at the trigger levels. The inorganics concentrations are close to the targeted trigger levels for arsenic, cadmium, chromium, selenium, and silver (Table 1). There is reasonably good agreement for lead and mercury.

The barium data, however, suggest that little or no barium was added to the DCB-1A water during spiking. Barium in the spiked water was analyzed by two laboratories to confirm these data. However, examination of barium data for soils after reaction with the spiked water (Tables 4 and 5) indicates that the barium was added to the spiked water. A mass balance for barium was done for each of the 8 combinations of study soil as follows:

Initial Ba in Soil (unreacted)	+	Initial Ba in Water (spiked)	=	Final Ba in Soil (reacted)	+	Final Ba in Water (supernatant)

The unknown term was assumed to be the initial barium in the spiked water (second term). Since there are data for the other 3 terms, it was possible to back-calculate the theoretical concentration of barium in the spiked water. The arithmetic average concentration was calculated as 143 mg Ba/L for all eight combinations of study soil, or 105 mg Ba/L if the datum for the Orangeburg 30-36 inch, eight liter combination was excluded. These back-calculated values agree closely with the targeted TCLP trigger level of 100 mg Ba/L.

The GEOCHEM model [7] was run to determine the speciation of barium at pH 2 and 6.5 using the constituents of spiked DCB-1A water as input. GEOCHEM is a thermodynamic model that predicts the speciation of metals in soil solutions and other natural water systems based on data inputs and model assumptions. At both pH values and with an average sulfate concentration of 4700 mg/L in DCB-1A water, barium was predicted to be a solid (barium sulfate). This may account for the laboratory errors in analyzing total barium. Prior to analysis, an aliquot of spiked DCB-1A water was sampled by thoroughly mixing the spiked DCB-1A water in a 50 liter preparation carboy. Concentrated nitric acid was added to this sample to give a final pH of less than 2. However, barium sulfate is a solid even at pH 2. The two laboratories then filtered the sample or took an aliquot of the sample supernatant prior to analysis. In either case, barium concentration would have been significantly underestimated since most of the barium remained in the sample container as a solid. Soil samples, however, were subjected to a thorough digestion with concentrated acid and heat prior to analysis. This would explain why the barium was detected in the soil samples but not in the spiked water sample. Thus, it is evident that the correct concentration of barium was added to the spiked DCB-1A water prior to reacting the water with the study soils.

Tables 4 and 5 present the data for initial (unreacted) and final (reacted) soil. Data are provided for both four and eight liters of spiked DCB-1A water that was reacted with 300 grams of study soil (Lakeland or Orangeburg). Following is a discussion of the soils data.

4.2 Comparison of total TCLP inorganics with TCLP data

As seen in Tables 4 and 5, all inorganic TCLP constituents in unreacted and reacted soils were below detection limits when the TCLP was performed on the soil. In addition, these detection limits are considerably less than the TCLP trigger levels. Total inorganics analyses on the TCLP inorganic constituents, however, revealed that the reacted soils retained significant amounts of TCLP inorganics.

Two major mechanisms are involved in the retention of metals by soils: adsorption and precipitation [8]. Inorganics in the study soils were adsorbed to the soil or were not soluble enough (e.g., precipitated) such that the extraction procedure (TCLP) removed detectable quantities.

Table 6. Estimated Number of Applications of DCB-1A Purged Water to Field Plots to Equal Inorganic Concentrations of Reacted Laboratory Soils

TCLP Inorganic	Avg. Conc in DCB-1A Water (mg/L)	Total TCLP Inorganics in Lakeland 0-8 in. Soil (mg/kg)	Estimated Number of Applications to Equal Lakeland 8L-Reacted Soil Concentration*	Total TCLP Inorganics in Orangeburg 0-6 in. Soil (mg/kg)	Estimated Number of Applications to Equal Orangeburg 8L-Reacted Soil Concentration*
Arsenic	0.07	107	6800	123	7800
Barium	0.018	2500	619900	2900	719100
Cadmium	0.033	23	3100	23	3100
Chromium	0.14	117	3700	131	4200
Lead	0.08	94	5200	105	5900
Mercury	0.0003	4	59500	2.6	38700
Selenium	0.008	21	11700	23	12800
Silver	0.012	109	40500	120	44600

*Data are from the Lakeland and Orangeburg 8-Liter reactors (Tables 4 and 5).

Assumptions

- application area at each plot = 5.9 m^2 (36 ft^2)
- depth of water/application = 2.54 cm (1 in.)
- volume of water/application = 151 liters (40 gal)
- soil depth that retains all TCLP inorganics = 7.62 cm (3 in.)
- soil bulk density = 1.5 g/cm^3
- soil mass = 674 kg/plot

Equations

(Eq 1)

$$\frac{\text{DCB-1A water concentration} \times \left(\frac{\text{volume}}{\text{application} \times \text{number of applications}}\right)}{\text{soil mass}} = \text{soil concentration}$$

(Eq 2)

$$\text{number of applications} = \frac{\text{soil concentration} \times \text{soil mass}}{\text{water concentration} \times \left(\frac{\text{volume}}{\text{application}}\right)}$$

The mass of TCLP constituents from spiked DCB-1A water that were retained by the soils are orders of magnitude greater than the mass received by an experimental plot over many applications. Table 6 presents estimates for the number of applications that would have to be done for an experimental field plot to retain an equal concentration of inorganics as retained in the laboratory experiments. It is assumed that all of the TCLP inorganics are retained by the upper 3 inches of soil. From Table 6, thousands of applications of purged DCB-1A water would be necessary for the field plots to retain TCLP inorganics at the same concentration as the laboratory experiments. Recall that all TCLP constituents in the reacted laboratory soils were below detection limit (below regulatory levels) when subjected to the TCLP extraction and analysis (Tables 4 and 5). Furthermore, the reacted soil was kept at a pH of 6.5 while the field plots have a natural pH of around 5. Most of the TCLP inorganic constituents are most likely more mobile at the lower pH found in the soil (pH 5). Thus, in the field plots, metals would be expected to migrate downward and not accumulate at the surface to the extent that the laboratory data indicate.

For these reasons, Lakeland and Orangeburg field plots are predicted not to exceed TCLP regulatory levels after repeated application of inorganics-contaminated ground water. Furthermore, soils adjacent to SRS monitoring wells that receive purged water will most likely not accumulate inorganics such that they become a hazardous waste according to the RCRA TCLP definition.

4.3 Comparison of total cation exchange capacities (CECs)

Initial data indicated that the CECs were 3 and 6 meq/100 grams for the Lakeland and Orangeburg soils, respectively [2]. However, CEC data from this study were two to three times lower (Tables 4 and 5). The highest CEC and clay content was in the subsurface (30-36 inches) Orangeburg soil with a total CEC of 3.5 meq/100 grams. The total CEC for both the Lakeland and Orangeburg surface soils (0-8 inches) was 0.9 meq/100 grams. The Lakeland subsurface soil had a total CEC of 0.7 meq/100 grams.

4.4 Comparison of initial, 4-liter, and 8-liter data

The intent of the laboratory experiment was to react the study soils with a large mass of TCLP constituents and then determine if the soil would still pass the TCLP test. As discussed, we did not intend to perform adsorption isotherms and determine equilibrium adsorption coefficients (Kd).

Final concentrations of inorganics in the surface Lakeland soil were similar between the 8 and 4 liter combinations. Results for the subsurface Lakeland and the surface Orangeburg soils indicate that the concentration of inorganics in the 8 liter reactors was about two times greater than those in the 4 liter reactors. The soil able to retain the greatest mass of TCLP inorganics was the Orangeburg subsurface soil. The 8 liter reactor results for total inorganics was generally 3 to 4 times higher than the totals for the 4 liter reactor. The mechanism of inorganics retention was not determined.

4.5 Comparison of total inorganics between the surface soils

The concentration of inorganics retained by the Lakeland and the Orangeburg soils is similar (Tables 4 and 5). This is expected since both surface soils have similar chemical and physical properties (e.g., CEC of 0.9 and primarily composed of sand).

4.6 Comparison of total inorganics between the subsurface soils

The Orangeburg subsurface soil retained from two to five times more inorganics than the Lakeland subsurface soil, with the exception of aluminum and iron. The Orangeburg subsurface soil retained from 10 to 20 times more aluminum and iron than the Lakeland subsurface soil, possibly due to the higher CEC of the subsurface Orangeburg soil (CEC for the Lakeland and Orangeburg subsurface soils are 0.7 and 3.5 meq/100 grams, respectively, or a factor of five difference).

CONCLUSIONS

The following conclusions were made:

(1) The study soils are capable of retaining (adsorption and precipitation) a relatively large amount (up to several thousand parts per million) of inorganic constituents, including the TCLP inorganics.

(2) In spite of the soils' ability to retain relatively large amounts of inorganics, the soils did not exceed TCLP regulatory levels when subjected to the Toxicity Characteristics Leaching Procedure.

(3) Lakeland and Orangeburg surface soils both retained approximately the same amount of inorganics.

(4) The subsurface Lakeland soil retained only slightly more inorganics than the surface Lakeland and Orangeburg soil.

(5) The subsurface Orangeburg soil retained the greatest amount of inorganic constituents, the increased retention presumably due to adsorption since it had the highest CEC.

(6) Lakeland and Orangeburg field plots will not exceed TCLP regulatory levels after repeated application of inorganics-contaminated ground water.

(7) Soils adjacent to SRS monitoring wells that receive purged water will not accumulate inorganics such that they become a hazardous waste according to the RCRA TCLP definition.

(8) Field experiments are currently being conducted to confirm laboratory studies.

Acknowledgments: This paper was prepared in connection with work done under Contract No. DE-AC09-88SR18035 with the Department of Energy. The authors thank Mr. Virgil Rogers (WSRC) for help with soil selection and classification, Ms. Teresa Jordan (WSRC) and Mr. Sean Asquith (Sirrine) for assistance with field work, and Dr. David Hargett for review.

REFERENCES

1. Westinghouse Savannah River Company. "Purged Water Management Plan," Prepared for Westinghouse Savannah River Company by Sirrine Environmental Consultants, (April 1990).

2. Westinghouse Savannah River Company. "Geochemical and Physical Properties of Soils and Shallow Sediments at the Savannah River Plant (U)," WSRC-RP-90-1031. (August, 1990).

3. Rhoades, J.D., "Cation Exchange Capacity," in *Methods of Soil Analysis, Part 2: Chemical and Microbiological Properties, 2nd Ed.* A.L. Page, Ed. American Society of Agronomy and Soil Science Society of America, (Madison, WI 1982).

4. Rogers, V.A., "Soil Survey of Savannah River Plant Area, Parts of Aiken, Barnwell, and Allendale Counties, South Carolina," USDA, (June, 1990).

5. United State Environmental Protection Agency (USEPA). "Batch-Type Adsorption Procedures for Estimating Soil Attenuation of Chemicals," Draft Technical Resource Document for Public Comment. USEPA Office of Research and Development, Cincinnati, Ohio. EPA/530-SW-87-006 (NTIS No. PB87-146155 1987).

6. American Society for Testing and Materials. "1991 Annual Book of ASTM Standards, Method D 4319-83," Standard Test Method for Distribution Ratios by the Short-Term Batch Method, Vol. 4. 8:619-624 (1991).

7. Sposito, G. and S.V. Mattigod, "GEOCHEM: A Computer Program for the Calculation of Chemical Equilibria in Soil Solutions and Other Natural Water Systems," Dept. of Soil and Environmental Sciences, University of California, (Riverside, 1980).

8. Evans, L.J., "Chemistry of Metal Retention by Soils," *Environ. Sci. Technol.* 23:1046-1056 (1989).

A

acid-labile 259
acidity 244, 247, 257
activity 2, 96, 108, 113, 130, 133, 138, 139, 147, 154, 161, 168, 191, 194, 196, 296, 297
adiabatic 125
adsorption 1, 18, 21, 41, 47, 50, 53, 59, 66, 67, 70, 72, 76, 84, 88, 89, 91, 92, 93, 102, 106, 113, 146, 192, 209-211, 214, 217, 218, 219, 222, 223, 230, 237, 244, 261, 283, 287, 290, 293, 298-300, 303, 304, 313, 315, 318, 321, 324, 334, 340-342, 344, 348, 349
aggregation 230, 235
Ah-horizon 229, 230, 233
Aiken, South Carolina 331, 332
alkali silicate 18
aluminosilicate 37
aluminum 36, 37, 65, 67, 69, 70, 80, 81, 88, 96, 114, 118, 119, 154, 229, 230, 232, 280, 294, 300, 321, 324, 335, 348
americium 94
amine 37
amino 2, 185
aminocarboxylic acid 258
ammonium 20, 193
amorphous 69, 92, 104, 183, 189, 191, 280, 281, 291, 295, 296, 302-304, 314, 321
andesite 278
annite 119
Apache County, Arizona 240
Aquaspirillum 195
aquifer 20, 59, 72-76, 78, 79, 88, 92-97, 99, 100-102, 104, 294
Argonne National Laboratory 104, 272
arsenic 7, 8, 15, 18, 20, 230, 244, 250, 280, 286, 321, 331, 332, 335, 342, 343
arsenate 214
ASTM 341
ASV 316, 318
Australia 94
Azotobacter vinlandii 188

B

Bacillus licheniformis 185, 187
Bacillus subtilis 185, 189, 197
bacteria-clay 183, 198
Baratron gauge 242
barium 331, 332, 335, 341, 343, 344
Baruch Forest Science Institute 101
beryllium 335
beech-oak 229
beidellite 118, 163
bentonite 163, 261
BES 211
biodegradability 258
biofilms 196
biotite 96, 144
birnessite 37
Blackfoot River 276, 277
Bloch decay 242
Bo Hai Bay 313, 316
Boltzmann's constant 98
boron 5, 20
Brownian diffusion 97
brucitic 118
BtCvck-horizon 229
Butte Mines, Inc. 278

C

cadmium 2, 4-8, 10-12, 15, 19-21, 26, 39, 76, 214, 259, 262, 280, 283, 292, 316, 318, 321, 324, 327, 328, 331, 332, 335, 341, 343
Caijiawan 320, 321, 324, 325, 328
calcite 184, 189, 193-195, 285-287, 292, 293
calcium 5, 11, 19, 37, 194, 195, 199, 229, 262, 280, 283, 285-289, 292, 335
Calvin-Benson cycle 194
Cambrian Meagher limestone 278
Cambrian Wolsey shale 278
Canada 8
Cape Cod aquifer 100

carbon 10, 20, 96, 101, 102, 211, 229
carbonate 11, 12, 18, 36, 94, 162, 189, 192-196, 211, 229, 233, 277, 280, 285, 290-293, 316, 318, 324
carboxyl 37, 104, 183, 185, 187
Caribou National Forest 276, 277
cation 10, 15, 19, 21, 66, 90, 92, 113, 119, 146, 150, 153, 154, 160, 168, 183, 187, 188, 227, 238-240, 244, 246, 247, 250, 262, 292, 316, 324, 335
CBD 163
CEC 3, 10, 15, 227, 230, 335, 340, 347-349
celadonite 118
cement-kiln 257
cerium 94
cesium 37, 94, 239, 327
Chalk River Nuclear Laboratory 94
Chalk River, Ontario 78
Chelex resin 316
chert 277, 279, 290
Cheto montmorillonite 240
chloride 5, 7, 21, 60, 240, 259, 283, 288, 292, 315
chloro-alkali 313
Chromic Luvisol 229
chromium 3, 7, 10, 11, 88, 119, 146, 154, 168, 188, 199, 259, 261, 280, 292, 293, 331, 332, 335, 341, 343
  chromate 259
citrate 162, 259
citric acid 146
Clark's Fork 276, 278
coagulation 22
cobalt 3, 78, 79, 83, 94, 99, 146, 243, 244, 280
concentration-jump 46
Cooke City, Montana 276, 278
copper 2, 3, 6-8, 10, 11, 15, 19, 39, 76, 78, 79, 119, 187, 199, 229, 240, 243, 244, 246, 259, 262, 276, 278, 280, 283, 294, 298, 299-304, 306, 316, 318, 320, 321, 324, 325, 327, 328, 335
CPS 103
Crown Butte 278
crystalline 95, 183, 189, 195, 280, 281, 291, 318
Custer National Forest 278
cyanobacteria 183, 192, 193, 196
Cyperus esculentus 13

D

dacite 278
Daisy and Fisher Creek 276, 278, 279, 281, 283, 294, 295, 297-303, 305, 306
Daisy Creek 278, 279, 297, 299
Daisy Pass 278
Dawu River 320, 321, 324
DEC 240
Degussa aluminum oxide C 80
Denmark 7
deuteron 239
Dexing Copper Mine 320, 327
dextran 100
dielectric 154, 238
diffractometery 320
dithiocarbamic acid 280
DLM 283, 304
DLVO theory 98
Dystric Cambisol 229, 230

E

EDS 189, 192
electric field pulse 46
electron microscopy 185, 187, 189, 192, 193
electrophoretic 103, 104
electropositivity 149
electrostatic 21, 37, 50, 67, 97, 113, 139, 149, 150, 187, 209, 249, 250
Elovich equation 41
EPA 7, 15, 256, 280, 308, 324, 341
Escherichia coli 185, 197
ESR spectroscopy 163, 164
ethylenediaminetetraacetic acid (EDTA) 70, 72, 74, 75, 78, 79-

81, 199, 229, 230, 233, 255, 258, 259, 261-263, 265, 266, 268, 269, 271, 272, 280, 281, 287, 290, 291, 293
Eulerian 97
Eutric Cambisol 229

F

Fayetteville Green Lake 183, 192, 193, 195, 196
Fayetteville, New York 192
Federal Republic of Germany 4
Fermi level 144-146
ferrihydrite 192
ferripyrophyllite 118
ferruginous 163
Fick's equation 43
FIESTA 207
first-order 41, 46, 50
Fisher Creek 277-279, 297, 299
Fisher Mountain 278, 306
flocculation 22, 183, 196, 324
flotation 256, 257
fluorescence 65, 81, 320
fly ash 3, 5, 17, 18, 255, 257
Forest Service 275
Freundlich 35
FTIR 104, 237, 239-244, 248, 250
fulvic acid 37, 146, 149, 197, 199, 315

G

Gallatin National Forest 276, 306
gallic acid 147
Galvanic potential 143
gamma-glutamyl 187
GC-MS 314
GEOCHEM 341, 344
Georgetown, South Carolina 101
Gibbs free energy 126, 128, 129-131, 135, 137, 139, 145, 161
gibbsite 37
Glatt River 75
Glattfelden 75, 78, 79
glauconite 118
Glengary Mine 278, 279, 297
Glutamic acid 147
glutamine 147
goethite 37, 39, 47, 50, 54, 56, 209, 210, 219
gold 276, 278
Gouy-Chapman diffuse layer model 89
Green Lake 193
Green Lakes State Park 192
Gurney electron 139
gypsum 5, 18, 184, 194, 195, 285, 286, 292

H

Hamburg 8
Hamburg harbor 12, 13
Hamburg/Georgswerder 6
Happel model (1958) 98
hectorite 118, 119
Helmholtz free energy 126
hematite 67, 69, 70, 80, 111
Henderson Mountain 278
Henry's Law 132
heterogeneity 39, 40
heterogeneous 36, 41, 42, 44, 140, 141
heterovalent 119
hexacyano 149
histidine 147
Hobcaw Field 101
homogeneous 42, 44, 141
homogenization 229
homoionic 239-241
homovalent 119
HQS 65, 67, 69, 80, 81
HSAB 149, 150, 155
humic acid 11, 37, 313-315
humin 37
hydration 97, 113, 149, 153, 238, 239, 242, 244, 247-251
hydrologic 207, 276, 277, 292, 293, 303, 305, 318, 324, 332
hydrolysis 37, 90, 149, 262
hydrophobic interaction 97
hydrous 283, 294-298, 300, 301, 304, 305, 306, 318
hydrous ferric oxide 89, 90, 92
hydrous oxide 21, 36, 88, 91, 96, 209, 258, 316, 328

hydroxide 21, 37, 39, 69, 87, 88, 94-96, 102, 104, 114, 189, 196, 321, 324
hydroxyl 37, 50, 66, 74, 88, 90, 141, 149, 161, 187, 194
hysteresis 244

I

Ichi and Maruyama River basin 5
ICP 320
ICPES 280, 281
Idaho 276
Igeo-class 327
illite 118
imino 2
interlamellar 238, 244, 247-250
interstitial 238, 315
IR 239, 241, 242, 244, 249, 250
iron 2, 10-13, 19-21, 23, 36, 37, 50, 53, 67, 69, 70, 72, 78-81, 87, 88, 91, 94, 96, 101, 102, 104, 114, 118-121, 146, 147, 149, 155, 156, 160, 161, 163, 164, 166, 168, 189, 191-193, 209, 210, 229, 231, 258, 259, 262, 280, 281, 283, 291, 294, 295-302, 304-306, 316, 321, 324, 328, 335, 348
  ferrous 13, 90-92, 96, 102, 163
isotherm 41, 209, 222, 223, 242-244, 250, 318, 334, 348
isotope 239
isotropic 100, 142
Itai-itai disease 5

J

Japan Sea 5
Ji Yun River 311, 313, 315
Jiangxi 320
Jintsu River 5

K

kaolinite 36, 37, 118, 197, 261
Kaolinite-Serpentine group 117
kinetics 19, 36, 37, 39-41, 45, 46, 55, 56, 60, 63, 67, 70, 74, 97, 99, 105, 113, 140, 219
Klebsiella aerogenes 188
Klopman quantum mechanical perturbation theory 151
KOE 162

L

Lakeland 331, 335, 340, 341, 344, 347-349
Langmuir expression 63
Langmuir, Temkin 35
lead 2, 3,6-8, 10-12, 14, 19, 21, 24, 39, 44, 88, 91, 135, 154, 183, 189, 209- 211, 214, 217-219, 221, 223, 259, 261-266, 268, 269, 271, 272, 280, 294, 316, 318, 321, 327, 328, 331, 332, 335, 341, 343
Lekkerkerk 6
lepidocrocite 70, 80, 81
Leptothrix 191
Leptothrix discophora 191
Levenberg-Marquardt nonlinear least square optimization technique 211
Levich's (1962) solution 98
Lewis acids and bases 149-151, 153, 154
LFER 90, 91
lithium 118, 119, 335
lichen 335
lime 3, 4, 18, 21, 24, 257
limestone 17, 306
Lindsay-Sadiq "pe + pH" scale 139
Lo An River 320, 321, 324, 327, 328
Los Alamos National Laboratory 94
Love Canal 6
Lulu Pass 278
lysozyme 199

M

macropore 40, 100, 101
magnesium 5, 20, 36, 118, 119, 185, 262, 280, 283, 285, 288, 289, 292, 335
magnetite 189

MALs 15
manganese 2, 10, 11, 19, 21, 37, 229, 258, 259, 280, 281, 287, 291, 294, 300, 303, 316, 324, 328, 335
manganese oxide 192
manuronic acid 188
Martin Marietta Energy Systems 105
MAS 242, 247
Maxey Flats 78, 79
Maxey Flats, Kentucky 78
Maybe Canyon 276-280, 283, 292-294
McLaren and Glengary Mines 276, 278, 279, 294, 297, 298
Meagher limestone 278
mercury 2, 6-8, 10, 18, 60, 82, 154, 210, 311, 313-316, 331, 332, 335, 341, 343
metallo-ions 184, 188
metalloenzymes 2
micro-erosion 95, 96
microbial 2, 18, 20, 140, 141, 166, 189, 192, 193, 195-197, 256, 305, 306
microemulsion 94
micropore 40
microscopic 36, 196, 249, 250
microscopy 185, 187
microsphere 101
mineralization 2, 184, 189, 192, 194, 195, 197, 287
mineralogical 292
mineralogical properties 36
mineralogy 293
MINTEQA2 277, 283, 285, 287, 290, 292, 294, 296-301, 304, 324
Misono dual parameter scale 151
molybdenum 5, 10, 280, 283, 289, 292
molybdate 39, 47
Monocotyledonous 13
Montana State Lands Board 306
montmorillonite 39, 94, 118, 164, 197, 238, 240, 241, 243, 244, 246, 249, 250
monzonite 278
Mossbauer spectra 162
mudstone 277, 279, 290, 294
muscovite 119

N

N-acetylglucosaminyl 185
National Priority List of 1986 7
natural organic matter (NOM) 91, 93, 95, 96, 101, 102, 104
Natural Sciences and Engineering Council of Canada 200
neptunium 94
neutralization 183, 187, 321
Nevada Test Site 94
New Kensington, Pennsylvania 210
nickel 3, 7, 10, 11, 119, 259, 280, 283, 285, 287, 289-293, 335
Nievenheim 8
nitrate 7, 15, 20, 75, 80, 81, 199, 211, 263, 340, 344
nitrilotriacetic acid (NTA) 255, 258, 263, 265, 266, 268, 271, 272
nitrogen 211, 242
NMR 238, 239, 242, 247, 248, 249, 250
nodule 229, 232
non-adiabatic 125
non-detrital 262
non-nucleophilic 211
nonferrous 311, 316
nontronite 118, 160, 163, 164
Noranda Minerals Corp. 278

O

Oak Ridge 78, 79
Oak Ridge National Laboratory 105
Oak Ridge, Tennessee 78
octahedral 36, 37, 114, 117-119, 155, 160, 164, 165
octahedron 114, 248
oligosaccharide 185
Orangeburg 331, 335, 340, 341, 343, 344, 347-349
organochlorine compounds 18

outer-sphere 37, 47, 50, 53, 54
oxidation 17, 19, 22, 59, 61, 88, 101, 102, 104, 114, 119, 121, 136, 138, 141, 146, 147, 150, 154-156, 163, 168, 256, 257, 287, 297, 305, 306, 321
oxide 18, 35, 37, 39, 47, 65, 66, 69, 70, 72, 78, 81, 87, 88, 96, 101, 104, 140, 210, 216, 259, 280, 283, 287, 291, 295-298, 300-306

P

paramagnetic 165, 237
Park shale 278
PCS 103, 104
pedon 125
peptidoglycan 185, 199
phenathroline 146
phenol 147
Phlegar valve corer 327
phlogopite 119
phosphatic shale 277, 279, 290, 294
phosphorus 242, 277
  phosphate 4, 5, 12, 18, 93, 94, 96, 185, 189, 196, 276, 287, 294
phosphoryl 183, 185
photocatalytic 140
photon correlation spectroscopy 103
photoreduction 305, 306
photosynthetic 184, 192, 194
phyllosilicate 37, 114, 117-119, 140, 149, 155, 156
physical chemical 88, 156, 167
physical-chemical 237
physico-chemical 313, 315, 321, 324
physicochemical 19, 22, 97, 196, 200
physisorb 238, 239
Pilgrim limestone 278
planktonic bacteria 187
planktonic microorganisms 196
plutonium 78, 79, 94
polarization 247
polymer 185, 187, 188, 191, 192, 199
polysaccharide 185, 187
porosimetry 218
porosity 98, 99, 216, 217, 218
porous media 46, 97, 98, 100, 101, 113, 114, 139
porphyry 278
potassium 11, 20, 33, 37, 86, 156, 163, 243, 244, 280, 335
Poyang Lake 320, 328
pressure-jump 36, 46, 47, 50, 53, 55
prokaryotes 193
protolytic equilibria 54
platinum 241
PWMP 331, 332
pyrite-chalcopyrite 278
pyrogallol 147
pyrophyllite 118
pyrrolidine 280

Q

QCC 248, 249
quadrupolar 242, 247, 249
quadrupole 248, 249
quartz 99, 189, 315
quartz-latite 278

R

radioactive 59, 78, 257
radionuclide 59, 75, 78, 93, 94, 96, 105, 331, 332
Raoult's Law 132
rate law 40, 45, 56, 63, 141
RCRA 332, 334, 347, 349
REDEQL-2 318
REDOX 257
resorcinol 147
Rhine River 5, 8
rhyolite 278
Rotterdam harbor 13
ruthenium 94

S

SAED 189
salt-free clay 240
saponite 238

sauconite 119
Savannah River Site 331, 332
SAz-1 240, 242-244, 246, 247, 249, 250
scanning electron microscopy (SEM) 103, 104
Scotch Bonnet Mountains 278
selected area electron diffraction 189
selenium 5, 280, 331, 332, 335, 342, 343
 selenate 39, 47, 50
 selenite 39, 47, 50, 53, 54
Seto Inland Sea 5
siliceous colloid 94
silicon 37, 88, 114, 118, 119, 159
 silicate 37, 83, 94, 196
silt 12, 271, 272
silver 24, 121, 153, 199, 278, 331, 332, 335, 340, 341, 343
smectite 36, 37, 117-119, 156, 160, 164, 197, 238, 239, 244
Snake River, Colorado 294, 295, 303, 306
snowmelt 277, 278, 283, 292, 306
snowpack 279, 283, 292
Soda Springs, Idaho 276, 277
sodium 5, 15, 20, 21, 155, 163, 165, 239, 240, 242-244, 246-250, 261, 263, 280, 283, 292, 335
Source Clays Repository 240
Spartina alterniflora 13
spectroscopy 65, 81, 189, 242, 249, 250
spectrum 95, 104, 163, 242-244, 247, 248-250
Sphaerotilus 191
SRS 331, 332, 334, 335, 340, 342, 347, 349
SSB 247
steric force 97
sticking coefficient 98, 99
Stillwater 276, 278
Stock clay 240
stoichiometry 47, 119, 134, 136, 138, 162, 209, 296
strontium 280, 283, 292
sulfhydryl 2
sulfur 13, 21, 26, 32, 33, 147, 189, 193, 204, 280, 283, 289, 294, 300, 303, 315, 321, 324
 sulfate 18, 19, 37, 47, 93, 161, 189, 285, 292, 340, 344
 sulfide 12, 18, 19, 23, 140, 161, 163, 165, 278, 281, 291, 294, 306, 311, 315, 316, 320, 324, 328
sulphoxylate 166
Superfund 7
surfactant 261
Synechococcus 183, 193
syringic acid 147

T

talc-pyrophyllite 117
tannic acid 147
tartaric acid 147
TCLP 331, 332, 334-336, 340-349
teichoic acid 185
teichuronic acid 185
temperature-jump 46
Tertiary/Pleistocene 20
tetrahedral 36, 37, 114, 117-119, 155, 160, 163, 164, 248
thallium 8
The Netherlands 6, 7, 33, 324
thermodynamic 73, 74, 113, 114, 121, 125, 128, 129, 136, 137, 139, 140, 143, 155, 344
Thiobacillus ferrooxidans 13
Thiobacillus thiooxidans 12
thiosulfate 165
Tianjin 313
Titrisol, Merck 80
TLM 53
TOC 316
toxicity 2, 11, 30, 184, 258, 331, 332, 347
turbidity 95, 102, 104

U

U.S. Army Waterways Experiment Station 13

U.S. Department of Agriculture 275
ultrafiltration 297, 316
ultrasonic 316
unapodize 241
univalent 262
uranium 94, 96, 189
uronic acid 187

V

vadose zone 114, 168, 334
vanillic acid 147
vanillin 147
vermiculite 37, 39, 118, 249
Vicksburg, MS 13
volkonskoite 119

W

Walktax meter 279
water-smectite 239
West Germany 6, 20
Wolsey shale 278
WSRC 331, 332, 349

X

Xiang River 311, 316, 318, 319
Xiawan 316, 318
XRF 316

Y

Yangtze River 316
Yellowstone National Park 192
Yellowstone Rivers 276, 278

Z

zinc 2, 5-8, 10, 11, 14, 19-21, 24, 39, 76, 119, 229, 262, 280, 283, 285, 287, 289-293, 316, 321, 324, 325, 327, 328, 335
ZnSe 241
Zoogloea ramigera 187, 202
Zr 44, 94